INDUSTRIAL PROTEOMICS

INDUSTRIAL PROTEOMICS

Applications for Biotechnology and Pharmaceuticals

EDITED BY
Daniel Figeys, Ph.D.
Department of Biochemistry, Microbiology, and Immunology
University of Ottawa
Ottawa, Ontario
Canada

A JOHN WILEY & SONS, INC., PUBLICATION

This book is printed on acid-free paper. 

Published by John Wiley & Sons, Inc., Hoboken, New Jersey.
Published simultaneously in Canada.

Library of Congress Cataloging-in-Publication Data:
Industrial proteomics : applications for biotechnology and pharmaceuticals / edited by Daniel Figeys.
p. ; cm.
Includes bibliographical references and index.
ISBN 0-471-45714-0 (pbk. : alk. paper)
1. Proteins—Biotechnology. 2. Proteomics.
[DNLM: 1. Proteomics. 2. Biotechnology. 3. Drug Industry. 4. Technology, Pharmaceutical. QU 58.5 I42 2005] I. Figeys, Daniel, 1968–
TP248.65.P76I535 2005
660.6′3—dc22

2004013223

Printed in the United States of America

10 9 8 7 6 5 4 3 2 1

CONTENTS

PREFACE

In his book, *What Remains to Be Discovered*, John Maddox (former editor of *Nature*) describes the importance of knowing more about the diversity of biomolecules and concludes: "In all of nature, diversity disguises simplicity." The development of genomic technologies, and in particular their ability to rapidly and accurately obtain biological information, was the first step toward deciphering human biology. However, nature did a good job at using diversity to disguise simplicity and therefore genomic tools are far from sufficient. Understanding the diverse biological processes at the molecular level in human and other species poses technical and scientific challenges that are beyond the reach of the human genome sequence alone. It will take an international effort to study the diverse biomolecules present in cells and to rapidly zoom in on specific pathways and biomolecules.

The proteome was introduced as the complement of the genome at the protein level. However, the resemblance ends there. The proteome is in constant change throughout the production, modification, translocation, and degradation of the proteins that are part of complex processes that are modulated by internal and external pressures. Although it may sound chaotic, it is actually a well-organized and efficient ensemble of processes that keep proteins under constant regulation.

Proteins were long studied one at a time. Classical biochemical and molecular approaches have led to our current understanding of cellular processes and provided the majority of drug targets. Proteomics has been defined as the study of the proteome, often seen as requiring high-throughput approaches. Although our ability to probe the proteome is limited compared to genomic approaches, great progress has been achieved over the last 10 years.

Many biotechnology companies and most pharmaceutical companies rapidly adopted proteomics as a way of further enhancing genomics capabilities. Proteomics can impact target discovery, drug discovery, and development all the way to clinical applications. Initial efforts were focused on target discovery using two-dimensional (2D) gel electrophoresis coupled to mass spectrometry. OGS and Large-Scale Biology were pioneers in the industrial application of 2D gel based proteomics. More companies entered the proteomics arena as it rapidly expanded from a 2D gel focus to expression proteomics, functional proteomics, and structural proteomics capable of combining different tools from molecular biology and phage display to mass spectrometry.

Although several books on proteomics are now available, we saw the need for a focused treatment of industrial applications of proteomics. Although the technology might be similar, the scale and direction of proteomics in industry are very different than in academia. Proteomics in industry is generally focused on application in target discovery and pharmaceutical pipelines, thereby requiring proteomic processes that are robust, well characterized, under quality control (QC) and producing statistically significant results. It also requires the capacity to handle significant amounts of samples. For example, a simple clinical proteomic study might require an analysis by expression proteomics from a minimum of 36 to hundreds of complex samples.

This book contains 11 chapters, each of which addresses specific aspects of industrial application of proteomics. In the first chapter I cover the basics of mass spectrometry (MS)-based proteomics. Functional proteomics is covered in Chapters 2 and 3. In Chapter 2, I discuss the MS-based approaches of mapping protein interactions. In Chapter 3, Annan and Zappacosta discuss the protein posttranslational modifications, particularly, protein phosphorylations. Structural proteomics is covered in Chapters 4 and 5. In Chapter 4, Tarin, Jennings, and McRee from Syrrx and ActiveSight cover the use of high-throughput crystallography and in silico methods for structure-based drug design. In Chapter 5, Hamuro, Weber, and Griffin from ExSAR Corporation describe the use of hydrogen/deuterium exchange mass spectrometry for high-throughput protein structure studies.

The first applications of proteomics were in target discovery. In Chapter 6, Hale, Ou, Shiyanov, Knierman, and Ludwig from Eli Lilly discuss the utilization of proteomics technologies for the identification as well as the validation of protein targets.

The latest application of proteomics has been for the discovery of disease or drug-related biomarkers. In Chapter 7, Massé and Gibbs from MDS—Pharma Services provide an overview of biomarker discovery and validation while Rose from GeneProt, in Chapter 8, details plasma biomarker discovery using proteomics.

Proteomics can also be approached from the small-molecule world (i.e., drugs), particularly to find proteins that interact with drugs. In Chapter 9, Doberstein, Hammond, and Hubert from Xencor present chemical genomics/chemical proteomics and discuss the different approaches.

In Chapter 10, Ahrens, Jespersen, and Schandorff, from MDS—Denmark, present a protein friendly bioinformatic approach and important factors to consider when developing such an approach. In Chapter 11, Wilson and Nock from Promab address the promising field of protein arrays, by introducing the different approaches and discussing the challenges and success.

I believe that the future of proteomics will be in the development of systems biology in academia and in preclinical and clinical applications in pharmaceutical research. This book is not just for people in industry. The notion that proteomics should be carried out in a systematic manner using well-characterized quality controlled (QCed) processes to produce statistically meaningful results will also be key to valid academic research and discovery. The industrial "experience" in proteomics will see widespread applications in systems biology and drug discovery.

CONTRIBUTORS

Christian Ahrens MDS Inc.—Denmark, Staermosegaardsvej 6, Odense M, DK-5230, Denmark

Roland S. Annan Proteomics and Biological Mass Spectrometry, Department of Computational, Analytical, and Structural Sciences, GlaxoSmithKline, King of Prussia, Pennsylvania

Steve Doberstein Five Prime Therapeutics, Inc., 951 Gateway Boulevard, South San Francisco, CA 94080

Daniel Figeys Department of Biochemistry, Microbiology, and Immunology, University of Ottawa, Ottawa, Canada

Bernard F. Gibbs Applied R + D Department, Early Clinical Research and Bioanalysis, MDS Pharma Services, Montreal, Quebec, Canada

Patrick R. Griffin ExSAR Corporation, 11 Deer Park Drive, Suite 103 Monmouth Junction, New Jersey, 08852

John E. Hale Enabling Biology Department, Lilly Research Labs, Indianapolis, Indiana

Philip W. Hammond Xencor, 111 W. Lemon Ave., Monrovia, California, 91016

Yoshitomo Hamuro ExSAR Corporation, 11 Deer Park Drive, Suite 103, Monmouth Junction, New Jersey, 08852

René S. Hubert Xencor, 111 W. Lemon Ave., Monrovia, California, 91016

Andy J. Jennings Syrrx, Inc., 10410 Science Center Dr. San Diego, California, 92121

Hans Jespersen MDS Inc.—Denmark, Staermosegaardsvej 6, Odense M, DK-5230, Denmark

Michael D. Knierman Enabling Biology Department, Lilly Research Labs, Indianapolis, Indiana

James R. Ludwig Enabling Biology Department, Lilly Research Labs, Indianapolis, Indiana

Robert Massé Applied R + D Department, Early Clinical Research and Bioanalysis, MDS Pharma Services, Montreal, Quebec, Canada

Duncan E. McRee ActiveSight, 4045 Sorrento Valley Blvd., San Diego, California, 92121

Steffen Nock Promab Biotechnologies, Inc., San Leandro, California, 94577

Weijia Ou Enabling Biology Department, Lilly Research Labs, Indianapolis, Indiana

Keith Rose GeneProt, Inc., 2 rue Pré-de-la-Fontaine, CH-1217 Meyrin/Geneva, Switzerland

Soeren Schandorff MDS Inc.—Denmark, Staermosegaardsvej 6, Odense M, DK-5230 Denmark

Pavel Shiyanov Enabling Biology Department, Lilly Research Labs, Indianapolis, Indiana

Leslie W. Tarr ActiveSight, 4045 Sorrento Valley Blvd., San Diego, California, 92121

Patricia C. Weber ExSAR Corporation, 11 Deer Park Drive, Suite 103, Monmouth Junction, New Jersey, 08852

David S. Wilson Promab Biotechnologies, Inc., San Leandro, California, 94577

Francesca Zappacosta Proteomics and Biological Mass Spectrometry, Department of Computational, Analytical, and Structural Sciences, GlaxoSmithKline, King of Prussia, Pennsylvania

1

PROTEOMICS: THE BASIC OVERVIEW

Daniel Figeys

Department of Biochemistry, Microbiology, and Immunology, University of Ottawa, Ottawa, Canada

Industrial Proteomics: Applications for Biotechnology and Pharmaceuticals, edited by Daniel Figeys
ISBN 0-471-45714-0

List of Abbreviations

2D	Two dimensional
β-gal	β-galactosidase
bp	base pair
CZE	capillary zone electrophoresis
ESI	electrospray ionization
EST	expressed sequence tag
FTMS	Fourier transform mass spectrometer
HPLC	high-performance liquid chromatography
i.d.	inner diameter
ICAT	isotope-coded affinity tag
IPG	immobilized pH gradient
MALDI	matrix-assisted laser desorption ionization
MHC	major histocompatibility complex
Micro-ESI	microelectrospray ionization
MS/MS	tandem mass spectrum
MS	mass spectr(um/ometer)
MW	molecular weight
Nano-ESI	nanoelectrospray ionization
o.d.	outer diameter
ORF	open reading frame
ppm	parts per million
psi	pound per square inch
PVP-40	polyvinylpyrrolidone
RF	radio frequency
SAGE	serial analysis of gene expression
SNP	single-nucleotide polymorphism
SPE	solid-phase extraction
TOF	time of flight

INTRODUCTION

The astonishing pace of scientific discovery in the twentieth century has had direct implications on our daily lives. The increase in quality of life and in life expectancy

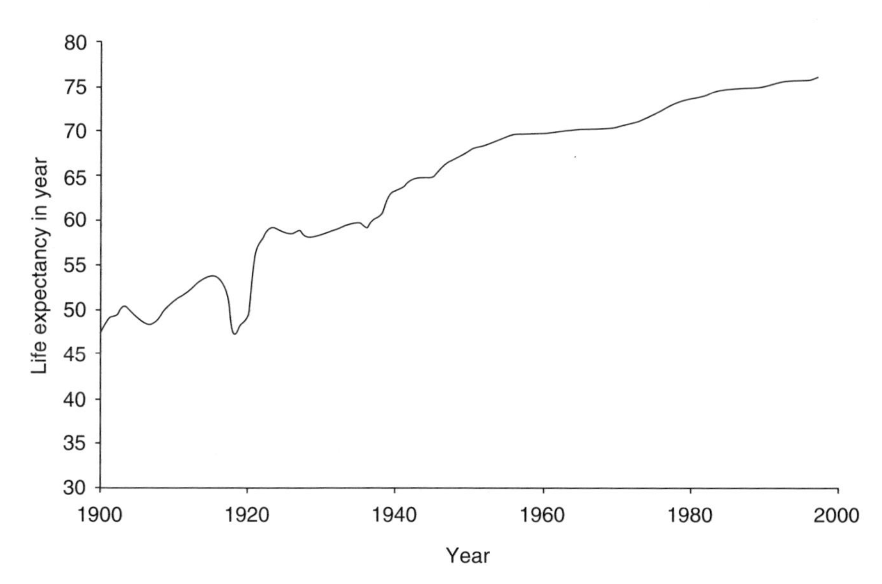

Figure 1.1. Life expectancy in the United States from 1900 to 1999. The data were averaged according to gender and race and smoothed for a 3-year period. Data were obtained from the *National Vital Statistics Report*, Vol. 47, No. 28, December 13, 1999. US Department of Health and Human Services.

seen in the industrialized countries correlate with the discoveries made in medicine and biological sciences (Fig. 1.1). At the onset of the twentieth century, medicine and biology were descriptive sciences aimed at understanding macroscopic phenomena. By the end of the century, these sciences had accessed the micro and nanoscopic worlds with an increased understanding of the implications involved in the biological processes of our lives. The dawn of the new millennium is ushering in a new paradigm for understanding system biology and complex multigenic diseases, as well as the accumulation of massive amounts of genomic information.

In the last quarter of the twentieth century, a few visionaries realized that all the parts were available to create a deoxyribonucleic acid (DNA) sequencing engine that would become capable of sequencing the human genome. Numerous sequencing projects were started with the promise of creating technology to accelerate the pace of sequencing. It took longer than expected to fill this promise, but by the end of the twentieth century new technologies were available to accelerate the pace of sequencing (Fig. 1.2).

In particular, instrumentation was developed to sequence large genomes in reasonable time periods (Dovichi, 1997). This instrumentation provided the framework for developing comprehensive approaches to heighten our understanding of diseases and biology. Furthermore, it was quickly realized that the complexity of biological processes could not be resolved only with genomic sequencing. It became apparent that the diseases remaining to be cured were increasingly more complex, were often trig-

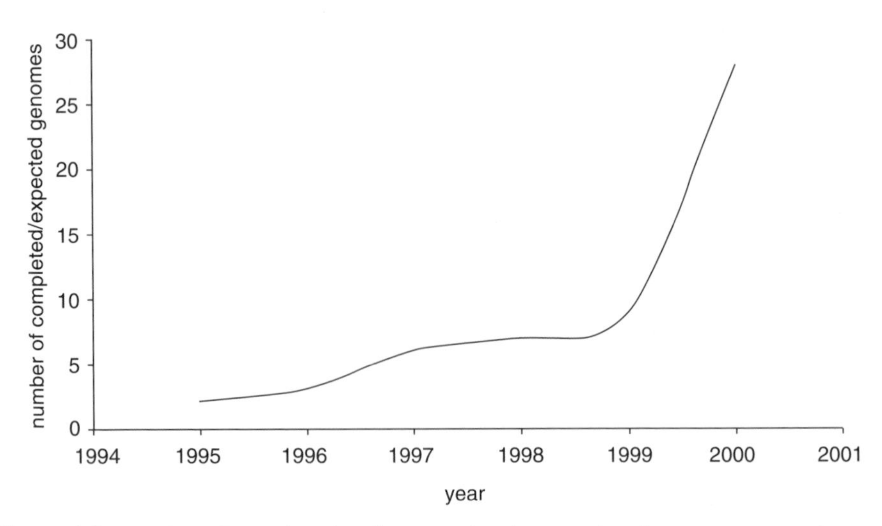

Figure 1.2. Number of completed and expected to be completed genome sequencing projects from 1995 to 2004. The intensification of sequencing efforts is visible from 1999 forward.

gered by a series of genes, and were dependant on the genetic makeup of individuals. Genomic sequencing can rapidly provide the list of the parts of the genome, but it cannot provide the instruction on how these parts fit together. This insufficiency created a disproportion between the amount of information available and the tools available for large-scale studies of molecules involved in biological processes.

Conventional biochemical approaches were relied on to access the functions and interactions of translated material. The key to capturing the value of the human genome is to find approaches that can rapidly link the genomic information to drug discovery and diagnostics. The development, however, of high-throughput tools for assessing the expression levels and functions of transcripted and translated genetic materials has globalized the study of cellular processes that may provide the link between genomic information and drug discovery and diagnostics.

Deoxyribonucleic Acid to Ribonucleic Acid

A series of tools became available for the rapid and quantitative analysis of expression levels in ribonucleic acid (RNA). For example, the emergence of the DNA/RNA array technology and the serial analysis of gene expression (SAGE) have facilitated the acquisition of quantitative expression profiles for complete sets or subsets of RNA (Desprez et al., 1998; Marshall and Hodgson, 1998; Ruan et al., 1998; Service, 1998; Velculescu et al., 1995; Madden et al., 1997; Matsumura et al., 1999, Neilson et al., 2000; Lal et al., 1999; Stein et al., 2004; Weeraratna et al., 2004). Different RNA expression profiles can be acquired using these techniques, and genes that are "up"- or "down"-regulated when comparing different cells or different cell states can also be detected.

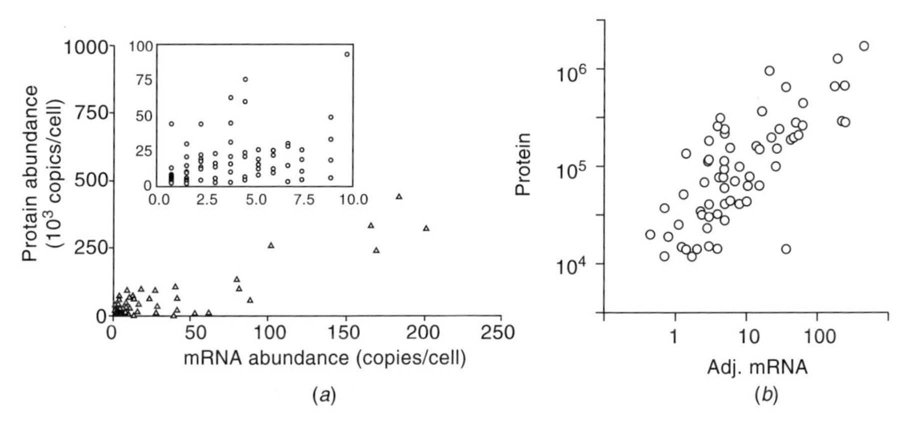

Figure 1.3. Correlation between protein and mRNA levels in *S. Cerevisiea*. (*a*) Correlation between protein and mRNA levels for 106 genes in yeast. The inset shows the low-end portion of the main figure. The Pearson product moment correlation for the entire data set was 0.935 and 0.356 for the inset. Reproduced with permission from Gygi et al. (1999). (*b*) Correlation of protein abundance with adjusted mRNA abundance in yeast. The Pearson product moment is 0.76. Reproduced with permission from Futcher et al. (1999).

Already, with one step away from the genomic sequence, a tremendous amount of information is extracted by studying RNA.

It is clear that these high-throughput DNA/RNA screening technologies provide a rapid and quantitative overview of the genes that are differentially expressed. It turns out, however, that the RNA expression on its own is not sufficient for understanding biological processes and gene functions. Evidence has been collected from different research groups (Gygi et al., 1999b; Gygi and Aebersold, 1999; Anderson and Seilhamer, 1997; Futcher et al., 1999) indicating that the expression of RNA has a poor linear relationship with the changes happening at the protein level (Fig. 1.3) while other experiments indicate a good correlation for higher abundance proteins (Kern et al., 2003). This is due to the numerous regulation mechanisms in place during protein expression and postexpression.

Furthermore, a recent large-scale study of the yeast genome has clearly demonstrated a poor relationship between DNA chip results and protein expressions (Ross-Macdonald et al., 1999). This study, performed by transposon tagging and gene disruption, found 31 meiotic genes, which were detected at the protein level by "in-frame" *lacZ* fusion and assay for β-galactosidase (β-gal) activity. Out of the 31 meotic genes, only 17 had been previously reported to be induced by at least twofold during sporulation. This was achieved by detection based on the DNA microarray representing all the annotated open reading frames (ORFs) in yeast. For the remainder of the meotic genes, the DNA microarray analysis failed to find any significant induction during meiosis. Therefore, not only is the relationship between the expression level of RNA and the expression level of proteins a complex matter, but it is also misleading to rely solely on the RNA expression patterns to predict cellular functions. Clearly, the amount of protein related to a gene can dramatically change without any change in the RNA expression level if this is achieved through downstream regulation mechanisms.

Ribonucleic Acid to Protein

Although serious efforts have been made to develop efficient genomic technologies, the study of proteins cannot be avoided in the quest to understand biological processes. The justification for studying proteins goes even deeper than just the lack of correlation between the expression levels of RNA and proteins. The presence of posttranslational modification, posttranslational truncation of proteins, and protein–ligand interactions are a few examples that illustrate the complexity at a protein level.

The study of proteins has always been done on a relatively small scale, partly because of the lack of methods to unambiguously and easily verify the protein identity. Experiments had to be carried out with great care to ensure that only the protein of interest was isolated. All of this has changed over the last 10 years with the development of technology capable of performing large-scale analyses and identification of proteins (Issaq et al., 2002; Wang and Hanash, 2003). This achievement has opened the door for comprehensive studies of proteins related to a genome (proteome) (Wilkins et al., 1996a).

PROTEOME HANDLING IN CLASSICAL AND FUNCTIONAL PROTEOMICS

Two-dimensional (2D) gel electrophoresis is typically used in profiling proteomic studies, and its most popular implementation is the differential displays of proteins expressed under different conditions. It turns out that conserving the proteome to obtain a truly representative 2D gel pattern is not trivial, and this is probably the most important experimental step in proteomic studies. For example, great care must be taken during the extraction of cells from their environment and during cell lysis to reduce the influence of the sample extraction protocol on the observed state of the proteome. Mistakes are often made while manipulating a proteome, thus seriously affecting the conclusions from the experiments. Therefore, the history of the sample is a prerequisite in order to assess the validity of a sample.

PROTEIN PURIFICATION

The extraction of protein from a cell lysate is a critical step for establishing a stable proteome. It is well known that once the cells are lysed the enzymes that would normally be compartmentalized are brought in contact with other proteins and then rapidly degrade the proteome. Fortunately, sets of well-characterized methods are available to cover the majority of the needs in protein extraction from cells. The harvesting of soluble proteins is simply performed by lysing the cells and collecting the supernatant. Different cell lysis methods are easily accessible (see *www.expasy.ch*). It is best to choose the simplest approach that is directly compatible with the immobilized pH gradient (IPG) isoelectric (i.e., minimum salt and ionic surfactant contents).

The current protein extraction protocols are not universally applicable to all biological samples, and they have limitations in terms of the protein representation in the extracted proteome. First, hydrophobic proteins are not easily extracted and represented on a 2D gel pattern (Wilkins et al., 1998). Figure 1.4 shows the number of proteins visualized on a 2D gel of *Saccharomyces Cerevisiae* versus the gravy hydrophobicity scale. Clearly, a significant portion of the predicted proteins in a proteome is hydropho-

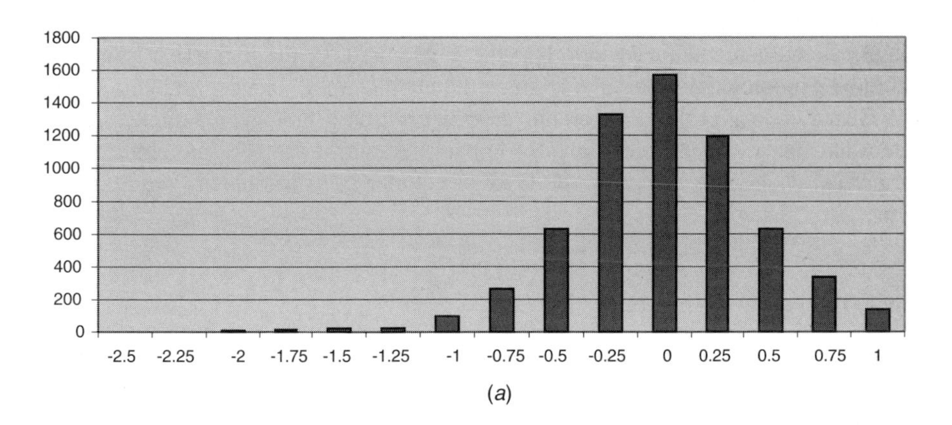

(*a*)

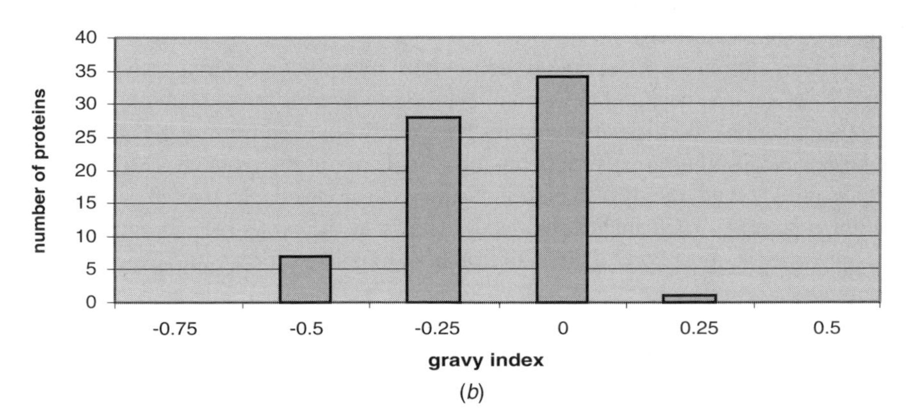

(*b*)

Figure 1.4. Hydrophobicity vs. protein observed. (*a*) The gravy index was calculated for all the known yeast ORF. (*b*) The gravy index was calculated from some of the observed proteins on a 2D gel electropherogram. Data were compiled from Gygi et al. (2000) and Perrot et al. (1999). A negative gravy index indicates hydrophilicity while a positive gravy index indicates hydrophobicity.

bic in nature; however, the compilation of the proteins that have been identified by 2D gel electrophoresis and mass spectrometry indicate a serious lack of hydrophobic proteins. This means that 2D gel electrophoresis fails to display the hydrophobic portion of the proteome. In recent years, significant efforts from the group of Rabilloud (Adessi et al., 1997; Blisnick et al., 1998; Chevallet et al., 1998; Goldberg et al., 1996; Rabillout, 1998; Rabillout et al., 1999; Santoni et al., 1999; Tastet et al., 2003; Luche et al., 2003) provided improved protocols and chemicals for the recovery of hydrophobic proteins. The extraction of hydrophobic proteins, however, is still a considerable challenge in proteomics. Comprehensive reviews have been published on the extraction and solubilization of proteins from biological samples for 2D gel electrophoresis (Dunn and Corbett, 1996; Rabillout, 1996, 2002; Ramagli, 1999).

Current protein extraction protocols often fail to provide a unique proteome. This is particularly a problem when dealing with tissues that are composed of many different cells at different stages and each having its own proteome. Classical protein extraction techniques would lyse the whole tissue or tissue fraction generating a scrambled proteome composed of all the original proteomes. Recent proteomic studies have clearly indicated that the cellular diversity within a biological sample, and the protein localization within cells greatly impact the conclusion of a proteomic study. The present challenges related to the level of cellular mass complexity are presented below.

Cell Culture

Cultured cells have been the most popular source of protein for proteomic studies. They are mainly used because they can provide a wealth of information that is often difficult to obtain from tissues and primary cultures. Furthermore, cell cultures are easily accessible at large curated collections of cells, such as the American Type Culture Collection (*www.atcc.org*).

Another significant advantage of cell culture over tissue material is the controllability of the environment and the growth conditions, as well as the potential to use different stimulations. Furthermore, cultures are usually obtained from single colonies and are homogeneous, thus greatly reducing the complexity of the proteome. Moreover, the culture growth and the lysis conditions are also controllable, which significantly minimize batch-to-batch fluctuations and artifacts in the proteome. To date, cell cultures provide the most controllable environment to perform proteomic experiments.

Other Source of Samples

Proteomes have also been analyzed from more complex samples, such as clinical samples, plants, and animals (Ostergaard et al., 1997; Wimmer et al., 1996; Lubec et al., 2003; Aebersold and Mann, 2003). The conclusions, however, that can be reached from such samples are often affected by many parameters that are not easily controllable. Typically, large numbers of samples need to be studied for any hope of finding the relevant proteins. Many factors increase the complexity of the analysis. The first factor is the methodology used to extract and store the clinical samples. The failure of clinical samples in proteomics is often traced back to the first step in the study, that is, the extraction and storing of the clinical samples. The proteome can often be seriously changed due to extended storage at room temperature. Even though large collections of clinical samples have been established, they are often of limited value for proteomic studies because they were not stored properly shortly after their extraction. Proteomic studies often require the revision of methodology for clinical and other samples.

Once proper methodologies are in place to limit the effects of sample isolation on the proteome, other factors can still make the study difficult. Tissue samples are heterogeneous in nature (i.e., composed of different cell types, volumes, and cell ratio compositions). This can cause large variations in the protein contents within the same tissues and for tissues of different sources. In reality, tissue samples are composed of many different proteomes. The lysis of these samples generates a scrambled mixture of proteomes. Therefore, multiple experiments need to be performed in order to extract the relevant proteins, or else the scrambling of proteomes will shadow the important proteins, thus making it impossible to interpret the results.

Examples are available that demonstrate the feasibility of extracting valuable information from scrambled proteomes. In a proteomic study of bladder cancer, Ostergaard and co-workers (Ostergaard et al., 1997; Wimmer et al., 1996; Celis et al., 1999) have demonstrated that new disease markers associated with the different cancer stages can be determined. They processed hundreds of well-preserved bladder cancer. By applying a proteomic approach, they found a handful of proteins that were differentially expressed during the different progression stages of the disease. Antibodies were raised against these protein markers (mainly keratinocyte markers) and were then used as a diagnostic tool to ascertain the different cancer stages within cryostat sections of biopsies from bladder cystectomies. Although the study was successful, the identified disease markers were actually high abundance proteins, and their changes in expression would be noticeable in scrambled proteomes. In any case, it would have been significantly more difficult to identify low- to mid-abundance proteins involved in the disease.

Two-dimensional gel-based analysis also introduces a third limiting factor that is related to the sample capacity of the technique (See Table 1.1). Typically, up to 100 μg of proteins are used during 2D gel electrophoresis. It is known that 2D gel electrophoresis can at least separate up to 11,000 proteins. That would represent about 9 ng for every protein, which is in the low subfemtomole level for most proteins displayed by 2D gel electrophoresis. In most cells, however, about 10 percent of the proteins represent 90 percent of the protein mass, while the remaining 90 percent of proteins represent only 10 percent of the protein mass. Assuming that cells express about 10,000 proteins, then the mid- to low-abundance proteins would only be, in the best case scenario, at about 1 ng on the gel. In reality the number would be even less because of the sample losses associated with 2D gels and low-abundance proteins. In a recent study, Gygi et al. (2000) demonstrated that for a typical 2D gel of yeast, protein identification could not be achieved for proteins with a codon bias of less than 0.1, thus indicating the lack of identification of low-abundance proteins. Furthermore, they demonstrated that they could discover low- to mid-abundance proteins on a 1D gel, but

TABLE 1.1. Effect of Different Cell Composition and Protein Concentration on the Level of Protein Observed

	[ ] in Cell *X*	[ ] in Cell *Y*	% of Cell That Are *X*				
			10	30	50	70	90
			Observed protein concentration				
Protein 1	1	1	1	1	1	1	1
Protein 2	2	1	1.1	1.3	1.5	1.7	1.9
Protein 3	5	1	1.4	2.2	3	3.8	4.6
Protein 4	10	1	1.9	3.7	5.5	7.3	9.1
Protein 5	100	1	10.9	30.7	50.5	70.3	90.1

only once the sample load had been significantly increased. The difference in sensitivity of different proteomics approached applied to yeast was recently illustrated (Ghaemmaghami et al., 2003).

The technique of 2D gel separation of proteins isolated from complex tissue samples has been performed for over 30 years, and recently, tools have been developed to allow the identification of these proteins. These tools have now been integrated in high-throughput platforms to allow serious study of proteomes. As proteomics moves from cell culture to primary cultures and tissues, it is important to keep in mind that these samples when used with conventional solubilization approaches produce complex mixtures of proteomes, each corresponding to an individual subpopulation of cells. The resulting scrambled proteome can be very difficult to make any sense of even with large population studies.

Subpopulation of Cells

The fractionation of subpopulation of cells from tissues can reduce the complexity of the proteome. Without cellular fractions it is very likely that the information generated will be meaningless. This is clearly illustrated in Table 1.1. The different cell compositions drastically affect the observed protein levels. Interestingly, the proteins that are not significantly different in concentration are not scrambled, while the ones that are significantly different are seriously scrambled.

Thus, it is important to know the cellular composition of the tissue being handled and if necessary to use fractionation of subpopulations of cells. Laser capture microdissection is increasingly being used in proteomics studies (Banks et al., 1999; Sirivatanauksorn et al., 1999; Emmert-Buck et al., 2000; Ornstein et al., 2000). As well, immunomagnetic techniques using immobilized antibodies specific to certain cell types have been investigated to separate cellular populations for proteomic studies (Clarke et al., 1994; Gomm et al., 1995). For example, Page et al. (1999) utilized the technique of immunomagnetic beads to separate cells based on immobilized antibodies that bind to known cell surface markers and followed by a magnetic pull down of the beads.

TABLE 1.2. Cell Culture and Estimated Protein Levels[a]

Cell Culture	Copy/Cell	Number of Mole	Nanogram Expressed for a Protein of 70 kDa	Total Mass of Protein[b]
10^7	10	0.16 fmol	0.01	6.5 μg
	100	1.6 fmol	0.1	
	1,000	16 fmol	1	
	10,000	160 fmol	10	
10^8	10	1.6 fmol	0.1	65 μg
	100	16 fmol	1	
	1,000	160 fmol	10	
	10,000	1.6 pmol	100	
10^9	10	16 fmol	1	650 μg
	100	160 fmol	10	
	1,000	1.6 pmol	100	
	10,000	16 pmol	1,000	

[a] Gray area indicates level of proteins that can be detected on a silver-stained 2D gel.
[b] Assumption 1: 90% of the protein mass is representing 10% of proteins highly expressed (>1000 copy/cell). About 10,000 proteins are expressed.
Assumption 2: 60% of the protein mass is representing 10% of proteins. About 10,000 proteins are expressed.

Subcellular Components

The complexity of specific proteome can also be reduced to the subcellular level. The enrichment of subcellular components, such as the organelles and plasma membrane, reduces the complexity of the proteome while increasing the likelihood of observing lower abundance proteins (Howell et al., 1989; Jung et al., 2000; Brunet et al., 2003). We have come to realize that in many instances it would not be possible to distinguish compartment proteins using conventional whole-cell protein displays by 2D gel electrophoresis due to dynamic range limitation.

Free-flow electrophoresis and density gradients are the classical approaches to subcellular fractionations. An interesting example of the utilization of density gradient separation in proteomics was provided by Fialka et al. (1997). They isolated subcellular compartments from murine mammary epithelial cells (EpH4) by continuous sucrose gradient centrifugation. This was then followed by high-resolution 2D gel electrophoresis on the proteins recovered from the sucrose gradient centrifugation. They were able to obtain 2D gel electropherograms of the late endosomes, early endosomes, and the majority of the rough endoplasmic reticulum. This study did not utilize mass spectrometry to systematically identify the isolated proteins.

Garin et al. (2001) utilized latex beads that are incubated with mouse macrophage-like cell line generating phagosomes that contain these latex beads. The phagosome can then be easily purified using a simple flotation approach. The proteins contained in the phagosome were then separated by 2D gel electrophoresis and the isolated proteins analyzed by mass spectrometry. The end result was the discovery of over 140 proteins associated with the latex bead containing phagosomes.

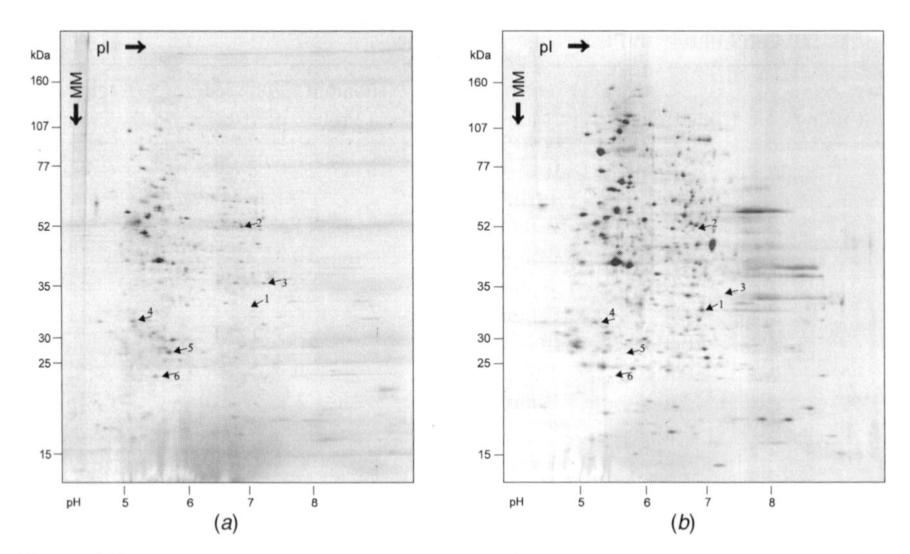

Figure 1.5. Subcellular localization: Magnetic purification. Tetanus-toxin-coated microbeads were incubated with U937 cells for 15 min at 37°C with chased 2 h. Then, the compartments that contained the magnetic beads were magnetically isolated. Two-dimensional gel electropherograms were obtained for the magnetic fraction (*a*) and for the whole-cell lysate. Reproduced with permission from Perrin-Cocon et al. (1999).

An alternative approach was also presented for the recovery of specific compartments based on immunomagnetic purification (Perrin-Cocon et al., 1999; Sarto et al., 2002; Himeda et al., 2004). In their specific application, Perrin-Cocon et al. (1999) isolated intracellular compartments containing endocytosed antigens. The magnetic beads were first covalently attached to a tetanus toxin. The beads were then incubated with U937 cells, allowing the antigen to bind to its receptor. The internalization of the microbeads by the cell was achieved by means of pinocytosis or receptor-mediated endocytosis. The compartments containing the beads were then isolated by the application of a magnetic field. After extensive washes, they were lysed, and their protein content was displayed by 2D gel electrophoresis. Figure 1.5 shows the 2D gel electropherogram obtained for the magnetic purified fraction and for the postnuclear supernatant. About 20 different proteins appeared to be enriched in the isolated compartments.

PROTEIN SEPARATION BY TWO-DIMENSIONAL GEL ELECTROPHORESIS

After carefully assessing the best method to extract the proteome from its medium, the second challenge is to extract information from the proteome, which generally starts by separating the proteins contained in the proteome. Some proteomes are relatively small, containing a few hundred proteins or less, while others contain thousands of proteins. One-dimensional (1D) separation techniques do not have the resolving power to separate complex mixtures. The combination, however, of orthogonal separation techniques can provide the required resolving power. Furthermore, it is important that the quantitative aspect of the proteome be conserved through the separation technique. About 30 years ago, a technique called 2D gel electrophoresis was introduced, and this technique satisfied the resolving power requirement while conserving the quantitative aspect of the proteome. Two-dimensional gel electrophoresis has been the method of choice for the large-scale purification of proteins in proteomic studies. The 2D gel electrophoresis method can potentially separate several thousand proteins in a single experiment (Gorg et al., 1988; Klose and Kobalz, 1995). Although the predictions for the number of genes in some genomes are high, it is generally believed that the number of genes expressed is, on average, between 5000 and 15,000 per cell type. However, these genes can lead to many forms of proteins, greatly increasing the complexity of the proteome. Protein separation can now be achieved as low as 0.1 isoelectric point (pI) unit and 1 kDa in molecular weight (MW).

Principles of Two-Dimensional Gel Electrophoretic Separation

Most people are not aware that 2D gel electrophoretic separation can be performed in different modes. In proteomics, the overwhelmingly popular implementation of 2D gel electrophoresis is the separation of the proteins according to their pI in the first dimension, followed by their separation according to the MW in the second dimension. Clearly, the combination of pI and MW separation offers a truly orthogonal separation technique, which is reflected by the unsurpassed resolving power of 2D gel electrophoresis.

Although 2D gel electrophoresis is a powerful separation technique, it was also initially very tedious and irreproducible. In fact, it was not clear in the late 1970s and early 1980s if the technique would become widely applied and if it would survive. Fortunately, the problems related with reproducibility were greatly reduced in the mid-1980s when the immobilized pH gradient was introduced, and it became commercially available (IPG strip) (Righetti and Gianazza, 1987; Righetti and Bossi, 1997a, 1997b; Righetti et al., 1983; Gorg, 1993; Fichmann, 1999; Matsui et al., 1999b; Sanchez et al., 1999). In the mid-1980s it was realized that the pH gradient could be immobilized and stabilized by copolymerizing different acrylamide monomers carrying ampholyte properties with acrylamide and low levels of cross-linkers. Although the mixture of ampholytes can be relatively complex, a simple system and a computer algorithm were used to, respectively, cast the IPG strip and to accurately predict the pH profile across the strip. IPG strips with various pH gradients and integrated instruments to perform isoelectric focusing became commercially available. The tediousness, however, involved in the technique is still present today, and, although the technique is definitely applied in proteomic groups, it has not been fully implemented in daily biological experiments.

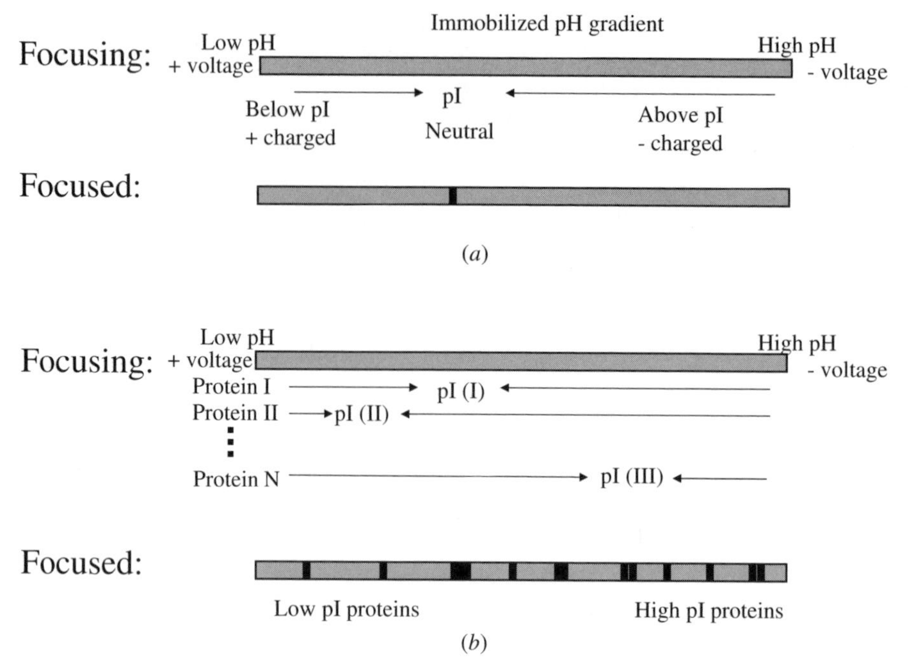

Figure 1.6. Isoelectric focusing on an immobilized pH gradient (IPG). (*a*) Focusing of a single protein. (*b*) Focusing of a mixture of proteins.

The commercial IPG strips are typically available in a dried form, sandwiched between two polymer cover sheets. The first step in using these strips is to remove one of the polymer covers and to inspect the strip for uniformity of the IPG. The peeling of sections of the strips has caused a higher rate of IPG strip rejection in recent years. Once a suitable IPG strip is found, it is first reswelled using a mixture of the protein extract with a loading buffer (Sanchez et al., 1999). This ensures that the protein mixture is uniformly distributed across the whole IPG and limits precipitation due to the excess of proteins in specific areas of the gel. The addition of a small electric field across the strip improves the transfer of the proteins to the IPG strip. Once the strip is reswelled, then an increasingly larger electric field is gradually applied across the IPG (Righetti and Gianazza, 1987) (Fig. 1.6). The proteins that are positively charged (i.e., in the area of the strip with pH below their pI) move toward the cathode and encounter an increasing pH until reaching their pI, at which point they will be neutral. The proteins that are negatively charged (i.e., in the area of the strip with pH above their pI) will move toward the anode and encounter a decreasing pH until reaching their pI. Every protein is concentrated and constantly focused at their respective pI. In theory, the higher the electric field and the longer the focusing is performed, the better the separation. Practically, the power supply, the presence of salts in the sample, and electroosmotic pumping will limit the performance of the separation. We have found that a separation involving 50,000 to 75,000 Vt $\times$ hour is sufficient for the types of samples we have handled. These numbers were typically achieved after 12 to 20 h of focusing. It is important to remember that these numbers need to be adjusted for different proteomes and to keep the time frame of the experiment reasonable.

The second dimension for the 2D separation is typically prepared during the focusing of the first dimension; however, it can also be precast and stored in a fridge with an appropriate buffer for 2 to 3 weeks with no separation problems. The second dimension is generally formed by pouring and polymerizing an acrylamide solution in between two glass plates, spaced using 1- to 1.5-mm spacers. After proper equilibration of the IPG strip performed by in-gel reduction and alkylation of the proteins, the strip is applied to the second dimension. Different systems are available for the second dimension. Some systems provide gels that are slightly larger than a sheet of paper or even larger. The strip can also be cut in different parts and applied to minigels. Regardless of the size of the gel, it is necessary that a tight contact between the strip and the gel be maintained. An electric field is then applied across the gel and the proteins migrate into the second dimension where they are separated according to their MW. Again, systems are commercially available to run the second dimension, and precast, larger gels were recently introduced.

Detection of Proteins Separated by Two-Dimensional Gel Electrophoresis

The ability to detect proteins separated by 2D gel electrophoresis is crucial to its application in proteomics. Over the years different methods have been developed to visualize proteins separated by gel electrophoresis (Rabilloud, 2000) (Fig. 1.7). Although different chemistries are used, the methodology involves either labeling the proteins prior to the separation, labeling the proteins after the first dimension (Jackson et al., 1988; Urwin and Jackson, 1991), or labeling the proteins after the second dimension. Labeling the protein before and after 2D gel electrophoresis has been the dominant approach to the visualization of proteins.

Prelabeling of Proteins. Prelabeling of proteins prior to 2D gel electrophoresis is often performed by adding radioisotopically labeled amino acids to the growth medium of cells to provide in vivo labeling of proteins (O'Farrell, 1975). Typically, ^{35}S-methionine is incorporated in the culture medium for the radiolabeling of proteins. The protein mixture is then separated by 2D gel electrophoresis and visualized using a film or a phosphor imager screen. The radioactivity level used in the in vivo labeling approach cannot be applied for the study of human and animals.

Another common approach to the prelabeling of proteins is the derivatization of the proteins using neutral covalently attached fluorescent dyes. These dyes are covalently attached to the proteins prior to 2D gel electrophoresis (Urwin and Jackson, 1993). Because they are not charged, these dyes have the advantage of minimally disturbing the isoelectric properties of the proteins. The sensitivity of the approach, however, suffers because of poor absorption and inadequate quantum yields. Novel fluorescent dyes provide nanogram levels of sensitivity for proteins separated by 2D

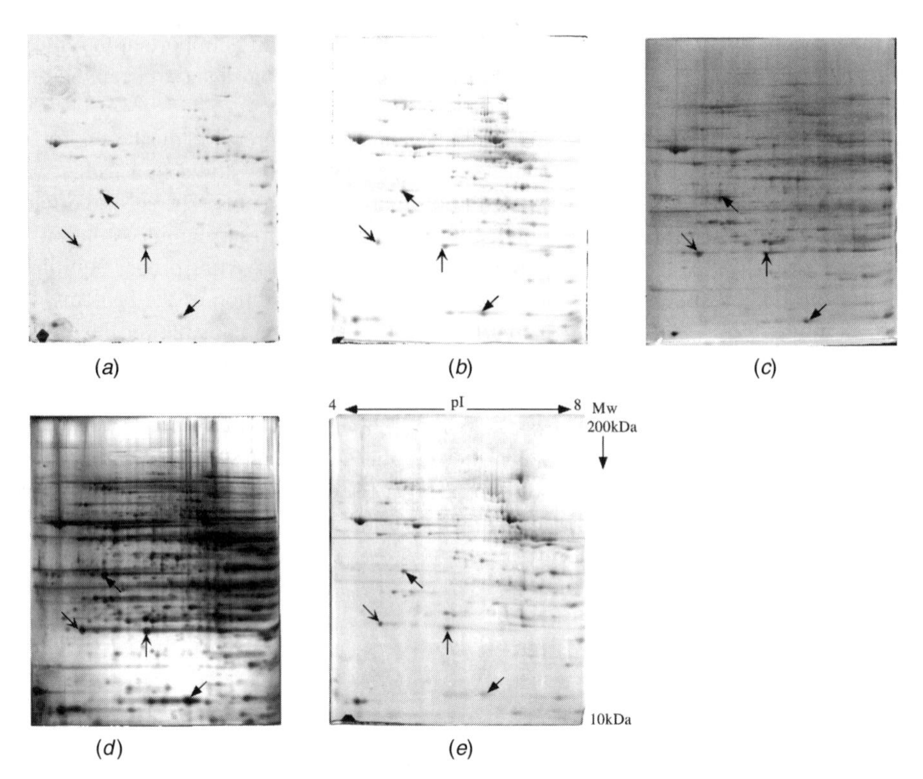

Figure 1.7. Staining approaches for 2D gel electropherogram. Comparison of sensitivities for different staining methods. A series of 2D gel electropherograms of 400 μg of mitochondrial proteins were stained with (*a*) Brilliant Blue G in acid alcohol medium, (*b*) colloidal Brilliant Blue G, (*c*) imidazole zinc, (*d*) silver, and (*e*) Sypro Orange (detected with a fluorescence laser scanner). Homologous spots are marked with arrows. Reproduced with permission from Rabilloud (2000).

gel electrophoresis (Unlu et al., 1997; Tonge et al., 2001; Yan et al., 2002). One disadvantage of this approach is the requirement for a fluorescence detection system and the requirement for an automated spot picker for any further processing of the individual protein spots.

Postlabeling of Proteins. The postseparation detection of proteins has been by far the most preferred route for the visualization of 2D gel-separated proteins. In particular, colloidal coomassie staining and silver staining are the methods of choice because of their ease of use and sensitivity. Colloidal coomassie staining typically provides a limit of detection (LOD) at about 25 to 50 ng of protein (Smith, 1994; Matsui, 1999a), while silver staining routinely provides limits of detection of about 5 ng of protein (Rabilloud, 1990, 1999; Rabilloud et al., 1994; Blum et al., 1987; Richert et al., 2004).

Silver staining is the most sensitive, direct visualization tool for proteins separated by 2D gel electrophoresis; however, it is tedious to manually perform. Instruments have been developed to perform automated silver staining and to significantly reduce the tediousness of the approach. Recently, an improved protocol for silver staining has been reported to provide limits of detection down into the subnanogram range. This was achieved by carefully selecting the chemical utilized for silver staining and by performing an extended rinse of the gel to reduce the background. Each spot on the gel can contain one or more proteins and can be manually or automatically excised from the gel for further analysis.

Fluorescence staining has also been reported for postseparation detection of gel-separated proteins. This technique typically provides a better sensitivity and a wider dynamic range than conventional colloidal coomassie and silver staining (Steinberg et al., 1996a, 1996b, 2000; Steinberg, 1997). The fluorescence-based approach, however, requires the access to a specialized fluorescence-based detection system and access to gel cutting robots when postprocessing of the proteins is necessary. Systems are now commercially available to perform the scanning of fluorescently labeled gels and the extraction of spots for gels.

Software to Handle Two-Dimensional Gel Electropherogram

Differential displays by 2D gel electrophoresis can be extremely tedious and frustrating when manually performed. The gel can display up to 10,000 spots, and from one experiment to another the spots can slightly shift on the pI and MW scale, depending on post-translational modification (PTM), sample processing, and sample composition. These shifts can become a serious problem for studies that require differential expression of proteins on an extended number of samples. To approach this problem, a few software packages have been developed for the alignment of multiple spots. The first challenge is to accurately detect and localize the positions of thousands of spots on a gel. Poor spot detection will affect the number of spots detected and their quantitation. The second challenge is to align multiple gels. This is typically achieved by selecting marker spots that are common to all the gels, and they are then used to realign the gels and the position of the spots. Obviously, when more deviations are present from gel to gel, then the number of markers required is higher. The issues of quantitation of multiple spots and the display of massive amounts of information also need to be addressed.

Commercial software packages are now available to solve some of these problems. In particular, the Melanie package (Wilkins et al., 1996b) from the Swiss Institute of Bioinformatics (*http://www.expasy.ch/*), the Phoretix 2D software from Phoretix (*http://www.nonlinear.com/*), and Gellab II from Scanalytics (*http://www.scanalytics.com/*) can be used to analyze 2D gel patterns.

PROTEIN PROCESSING

The information provided by 1D and 2D gel electrophoresis is far from being adequate to identify proteins. The observed MW on a gel can be easily off by a few thousand

Daltons. Furthermore, the posttranslation modifications of proteins and the truncation of proteins can drastically affect the observed MW and pI. Therefore, the information contained in gels is not conclusive for the identification of a protein, and more information needs to be extracted for unambiguous identification. How this extra information could be extracted from a protein stock in a gel spot was another challenge to overcome.

Blot Digestion of Proteins

Obviously, the first approach to the problem would be to remove the protein from the gel. A technique called electrotransfer, or electroblotting, was specifically designed for this purpose (Patterson et al., 1996; Aebersold et al., 1986, 1987). Once proteins had been separated by 1 or 2D gel electrophoresis, they could be readily transferred to a binding membrane, such as nitrocellulose, by simply sandwiching the gel with a receiving membrane and applying an electric field perpendicular to the plane of the gel. The proteins contained in the gel migrate toward the anode and encounter the binding membrane to which they become attached. The pattern of proteins present on the blot can then be visualized using a staining method such as coomassie or silver staining. In this manner, the separation profile on a gel can be transferred to a blot membrane with fidelity and be available for chemical/enzymatic treatment.

Electroblotting was initially used for protein identification using Edman degradation. The works of Aebersold pioneered this approach (Aebersold et al., 1986, 1987). Edman degradation will be introduced below. Basically, the electroblotting approach linked the gap between Edman degradation and the mainstream approaches for protein separation, that is, gel electrophoresis.

In the early to mid-1990s mass spectrometry (MS) became an attractive approach for the identification of proteins using peptides derived from the protein. The MS techniques will be described below. Electroblotting became the method of choice for processing proteins for MS analysis. The method was modified to allow the enzymatic digestion of the protein present on the blot. Once the spots of interest on a blot were excised, they were treated with a blocking agent such as polyvinglpyrrolidone-40 (PVP-40). The PVP-40 coats the membrane and allows the enzyme solution to freely access the blotted protein without binding to the membrane. Interestingly, the action mechanism of the enzyme is a combination of two reactions: a solid-phase reaction while the enzyme digests the attached proteins and a solution reaction for the further digestion of the released fragments. The end result is a solution of peptides that has been derived from the blotted protein. This peptide solution is then directly compatible with MS analysis.

The utilization of blots for protein identification was the technique of choice in the mid-1990s; however, it had some obvious limitations. Its foremost limitation was the tediousness and time required to perform the experiment. First, the 2D gel of the lysate of interest had to be produced and then electrotransferred to a nitrocellulose followed by staining. It was then followed by the excision of the stained spots and the blocking of the spots using PVP-40, which involves a large number of repetitive washes. Finally, the blotted proteins were digested with trypsin. Altogether, the process could take about 3 to 4 days depending on the number of spots. To curb the tediousness, fluidic stations could be used to perform the majority of the postblot processing of the proteins.

The second limitation was the inconsistent quantitative transfer of proteins from the gel to the blot. This was apparent when comparing stained 2D gel with a similar stained blot. The rate of transfer of some proteins was high while the rate for other proteins was almost nil. The last important limitation was the carryover of PVP-40 during the analysis when not enough rinse steps were included after the blocking of the membrane.

In-gel Digestion of Proteins

The application of in-gel digestion of proteins for Edman sequencing had been developed in the early 1990s (Rosenfeld et al., 1992). Its application to the field of protein analysis by mass spectrometry was first introduced by Wilm et al. (1996). It was an instant success, and the method rapidly replaced the electroblotting approach. Its main advantage was the significant reduction in the number of processes required after gel electrophoresis to obtain digest suitable for analysis. As in the blot approach, a lysate is separated by gel electrophoresis. The difference now is that the gel is fixed and stained to highlight the proteins, the spots of interest are excised from the gels, properly rinsed, and then the enzyme solution is added to the gel pieces. The trick to introduce the enzyme into the gel is to shrink the gel pieces and let them swell in an enzyme solution. After digestion, only a few extraction steps are needed to obtain the peptides. Obviously, the reduction in labor and the reduction in processes offered by the in-gel digestion method made it the method of choice for the generation of peptides from proteins separated by gel electrophoresis.

PROTEIN IDENTIFICATION

Edman Degradation

In the early 1960s a technique called *the protein sequenator* was presented by Edman and Begg (1967) for the N-terminal sequencing of proteins. This chemical degradation technique allowed the extraction of individual amino acids in a cycle-dependant manner from the N-termini of proteins. The retention time observed by high-performance liquid chromatography (HPLC) was then used to identify the individual amino acid sequences. By repeating the chemical degradation cycle, it was possible to obtain the amino acid sequence at the N-terminus of typically up to 20 amino acids. Using this approach, it is possible to painfully de novo sequence a protein or to sequence a sufficient length of the protein to be able to clone the gene. The availability of protein and DNA sequence

TABLE 1.3. Performances of Different Mass Spectrometers for Protein Identification

	Sensitivity in MS/MS	Resolution	Cost	ID Based on MS	ID Based on M/MS
MALDI-TOF	Low	5,000–10,000	Medium	Yes	Sometimes by PSD
MALDI-Pulsar	High	10,000–15,000	Medium	Yes	Yes
ESI-Triple Quadrupole	Low	1,000	Medium	No	Yes
Esi-Ion Trap	Medium	1,000–3,000 higher in zoom scan	Low	No	Yes
ESI-FTMS	High	>50,000	High	Yes	Yes but slow
ESI-Pulsar and Qtof	High	10,000–15,000	Medium	No	Yes

databases has now facilitated the work and allowed protein identification using limited N-terminal amino acid sequencing.

Protein Identification by Mass Spectrometry

Mass spectrometry has been used in its different forms for the analysis of proteins as long as the Edman degradation technique. By today's standard, mass spectrometry was tedious, slow, and required large amounts of samples. Over the years, however, the idea of using mass spectrometers to perform protein identification has evolved with the improvement in instrumentation and in the changes occurring in genomic databases. The introduction of effective matrix-assisted laser desorption ionization and time-of-flight (MALDI-TOF) mass spectrometry and electrospray ionization (ESI) tandem mass spectrometry has revolutionized the field. Manufacturers of these instruments have "beefed-up" their efforts to produce more instruments for proteomic purposes. Therefore, when assessing previous reports, it is important to keep in perspective the tremendous amount of change that has occurred in recent years (Table 1.3).

MALDI-TOF Mass Speetrometry

MALDI-TOF: An Evolving Instrument. Simultaneous development in the field of mass spectrometry has allowed rapid and accurate mass measurements of analytes by a technique called *matrix-assisted laser desorption ionization* (MALDI) and *time-of-flight* (TOF) mass spectrometry. This technique allows the transfer of peptides from a solid state to the gas phase, while the TOF mass spectrometer rapidly separates peptides according to their *m/z* ratio (Fig. 1.8).

Proteins isolated from 2D gel electrophoresis are digested, desalted, and then spotted on a MALDI plate for co-crystallization with a saturated matrix solution. Alternatively, protein digests can be separated by HPLC and eluting peptides deposited on a MALDI plate. The target plate is then introduced in the vacuum chamber of the mass

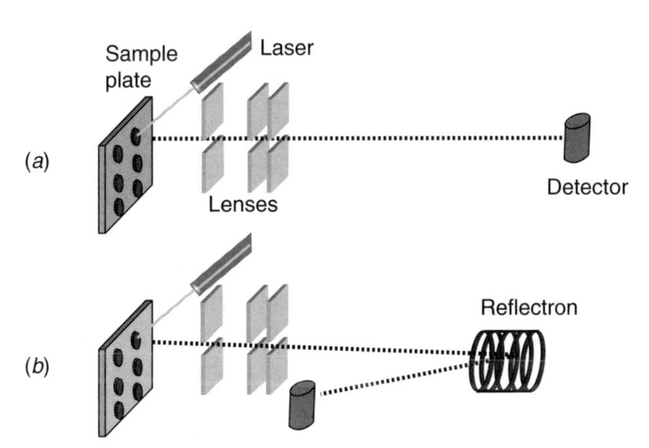

Figure 1.8. Schematic of a MALDI-TOF mass spectrometer. (*a*) Schematic of the linear mode. In the linear mode a portion of the sample is transferred to the gas phase by a pulsed laser. Once in the gas phase, the charge molecules are accelerated for a short distance and enter the field-free region of the tube, where they separate according to their *m/z* ratio. (*b*) Schematic of the reflectron mode. The mechanisms are the same as in the linear mode, except that a reflectron is placed to refocus the ions toward the detector leading to a higher resolution.

spectrometer. An automated translation stage positions individual spots on the axis of the mass spectrometer as well as at the focal point of a laser. A light pulse from a laser beam (1- to 10-ns pulse) of a wavelength tuned to the absorbance of the matrix is focused onto a limited area of the spot. The rapid transfer of energy ejects an ionization plume of material from the plate surface and brings along the peptides into the gas phase. The MALDI plate is biased to high voltages of +20 to +30 kV, with respect to a grounded orifice. This induces the positively charged peptides to accelerate toward the orifice of the flight tube. The peptides are affected by the electric field during the time that they are between the plate and the orifice. They reach the orifice with a velocity proportional to $(z/m)^{1/2}$. Once they pass the orifice, the peptides have all the same kinetic energy but not the same velocity because of their different mass. Thus, molecules that have an identical charge will therefore have a velocity proportional to their masses. Generally, in MALDI the resulting peptide ions are singly charged. Once in the field-free region of the flight tube, the peptides fly through the tube only according to their initial velocity at the orifice. The vacuum in the flight tube is such that the likelihood of a collision with another molecule while in the flight tube is low. Because the peptides are usually singly charged, they traverse the flight tube according to their mass and hit the detector at different time intervals. The mass analyzer, triggered by the laser pulse, records the signal detected versus the time of flight, which can be readily transformed into *m/z* ratios if the mass spectrometer is properly calibrated.

In principle, a longer tube infers a better separation of the different masses. Collisions with gas molecules in the path of the analytes would effectively destroy the separation. Therefore, the longer the path, the larger the vacuum requirement is imposed to maintain a path free of possible collisions. This is why you will see these mass spectrometers equipped with sufficient pumping to achieve 10^{-9} to 10^{-10} torr.

A second factor that affects the resolution of the separation is the initial kinetic energy distribution due to the burst of analytes off the MALDI plate. The technique of *delay extraction* of the ion and the reflectron were added to the MALDI-TOF to improve the resolution, based on a more uniform kinetic energy. The delay extraction acts on the kinetic energy prior to the separation, while the reflectron acts on the kinetic energy distribution during the time of flight. The delay extraction is simply achieved by introducing a time delay prior to the high-voltage biasing of the MALDI plate. This allows the ions ejected from the plate to kinetically cool down providing improved resolution.

The reflectron allows the kinetic refocusing of the ions, which also results in better resolution. It is normal to obtain a peptide mass accuracy down to 10 to 50 ppm and to obtain a 5000 to 15,000 resolution. The reflectron is a piece of hardware that is placed in the path of the ions. Therefore, the design of a MALDI-TOF equipped with a reflectron is very different than the conventional MALDI-TOF instrument. Furthermore, different designs have been fabricated; however, the principle remains the same. The reflectron is a set of ring lenses that are stacked together. Increasingly higher potentials are applied to the rings. In the time of flight each peptide is represented by a pocket of ions that travel with a small range of kinetic energy (i.e., speed). The ions that are at the front edge of the pocket have a higher kinetic energy, while the ions at the trailing edge of the pocket have a lower kinetic energy. The fast moving ions penetrate the reflectron first and start to feel the repulsing electric field imposed by the reflectron. Because of their high kinetic energy, they travel further into the reflectron before reaching a point of zero kinetic energy. Then they are reaccelerated by the repulsing field toward the entrance of the reflectron. Meanwhile, the other ions also enter the reflectron in order of their decreasing kinetic energy. Those ions will travel less of a distance into the reflectron before their directions are reversed. As a result, the slow moving ions are now at the front of the traveling pocket, while the fast moving ions are behind. All the ions then travel back in the field-free region of the TOF tube. While traveling, the pockets of ions become focused by the higher speed trailing ions catching up to the slower leading ions. Obviously, the resolution that is achieved is very dependant on the positioning of the detector at the focal point of the reflectron. The pocket of ions will not have time to focus if the detector is positioned too close to the reflectron. Also, the pocket of ions will start defocusing if the detector is positioned too far from the reflectron.

Different Approaches to Sample Preparation for MALDI. Sample preparation for MALDI is often an art. Numerous protocols have been reported for the spotting of the samples. Here we review the most utilized approaches.

CLEANUP PRIOR TO SPOTTING. The utilization of reverse-phase material, either in a column format or loosely packed into a pipet tip, has proven to be a useful approach for the desalting of samples and their separation from polymers prior to MALDI-TOF analysis. The properties, however, of the reverse-phase material and the packing need to be carefully assessed. For example, Ziptips are pipet tips that have a small plug of extraction material at their distal end. The plug of material is obtained by embedding dispersed reverse-phase material in a polymeric support. To maintain easy flow of liquid through the tip, the reverse-phase beads are well dispersed into the embedding polymer resulting in a low back-pressure. The tips have proven useful for the cleanup of high to mid levels of peptide mixtures. The dispersion, however, of the reverse beads causes the extraction efficiency to be diffusion limited, that is, the linear flow is too high compared with the distance that the peptides have to diffuse to reach the packing material. Therefore, the extraction efficiency drops significantly for lower concentrations of peptide mixtures, and it can therefore be a challenge to analyze lower amounts (<100 fmol) of peptide mixtures.

Spotting Approaches. Over the years different sample preparation methods have been developed for the analysis of minute amounts of peptides. The most popular method has been the *dried droplet method* (Karas and Hillenkamp, 1988). In this method the peptide mixture of interest is dried down and resuspended in an acidified water:organic (acetonitrile or methanol) solution saturated with α-cyano-4-hydroxycinnamic acid. One microliter from this mixture is then deposited onto a MALDI plate and allowed to dry. Although this is a rapid way of preparing samples, it does not provide for the most sensitive analysis of peptides.

A second approach termed the *two-layer method* was recently introduced for the analysis of peptides (Dai et al., 1999). First, a solution of α-cyano-4-hydroxycinnamic acid is pipetted on a MALDI plate and allowed to dry to form a microcrystal layer. Then a solution with the analytes and the matrix is pipetted on top of the first layer and allowed to dry. Although this approach is more tedious, it offers a better representation of the peptides, a cleaner spectra, and it can be automated.

Cleanup Post Spotting. A cleanup method can be directly applied on the MALDI plate either with the dried droplet method or the two layer method. The sample cleanup method consists of performing a cleanup procedure right after the spotting. Briefly, a drop of water is added to individual spots on the MALDI plate for a set time period. The water droplet is then blown away, and the spot is allowed to dry again before being placed in the MALDI-TOF mass spectrometer. This rinse step allows the extraction of salts and other hydrophilic molecules from the dried crystal. This often leads to a decrease in background, therefore an increase in signal/noise ratio. The timing of the rinse is important to avoid unnecessary loss of hydrophilic peptides.

On-Plate Concentration. As mentioned previously, mass spectrometers are concentration-dependant devices. MALDI is more complicated because the ionization process and the involvement of the matrix are not clearly understood. It appears that an increase in the analyte concentration on the surface of the MALDI plate might improve the signal observed; this, however, would only work well if the sample can be sufficiently cleaned up. The amount of sample is limited, so the only way to increase the concentration is to limit the size of the MALDI spot. MALDI plates were developed with spotting sections of small diameters that are hydrophilic, while the rest of the plate is hydrophobic. When a drop of peptide solution is deposited onto the spot, the hydrophobic surroundings and the surface tension force the droplet to take a shape that limits its contact with the surface. As the droplet dries, it focuses into the small hydrophilic patch. The end result is the concentration of a large volume into a small area on the MALDI plate. In itself, this is not sufficient to provide significant improvement because the contaminants, usually salts, are similarly concentrated. The addition of an on-plate cleanup (drop of water method) helps to remove salts while retaining peptides in the crystal. The drop of water is then blown away.

Protein Identification by MALDI-TOF: A Moving Target

USING ACCURATE PROTEIN MASS MEASUREMENT. It was believed that protein identification could be achieved based on the accurate measurement of the protein MW. But, it was quickly realized that the growing size of databases and the accuracy of the mass measurement limited the unambiguous identification of proteins based on their MW. The information carried by the protein MW was insufficient to identify the protein. It was then realized that unambiguous identification of a protein could be readily achieved by accurately measuring the masses of the peptides contained in a protelytic digestion of the protein. Clearly, in itself, the mass of an individual peptide derived from a protein is inadequate to identify the protein, although, the masses of a large set of peptides derived from the same protein is often sufficient to identify the protein. MALDI-TOF MS is the method of choice for the measurement of peptide masses; however, MALDI-TOF MS experiments must be carefully designed to preserve the mass accuracy. In MALDI-TOF, the ions are accelerated for a short distance, and then they freely fly in the TOF tube, sometimes up to a few meters. It is generally assumed that the acceleration space is constant; however, any errors on the acceleration space are propagated in the time of flight and thus affect the accuracy of the measured masses. In reality, the plate fidelity, the positioning, and tilt of the plate holder can slightly change the acceleration space and, therefore, can affect the mass accuracy. Approaches have been developed to reduce this phenomenon. This is normally achieved by adding an internal standard to the sample or correcting the masses by scanning an external calibrant near the sample spot on the MALDI plate.

PROTEIN IDENTIFICATION BY MALDI-TOF AND ACCURATE PEPTIDE MASSES. The identification of proteins by MALDI-TOF mass spectrometry is generally achieved by measuring the m/z ratio of the peptides predominantly of charge +1. The combination of accurate peptide mass measurement with the availability of protein sequence databases forms the basis of protein identification by MALDI-TOF. Figure 1.9 describes the principle behind the identification of proteins by MALDI-TOF. This is often called peptide mass fingerprinting. The measured masses present in tryptic digests are tabulated, and the known contaminants are deleted from the list. The reduced mass list is then used to search protein databases.

Different software packages have been developed for the identification of proteins based on the accurate measurement of peptide masses. The simplest method for scoring

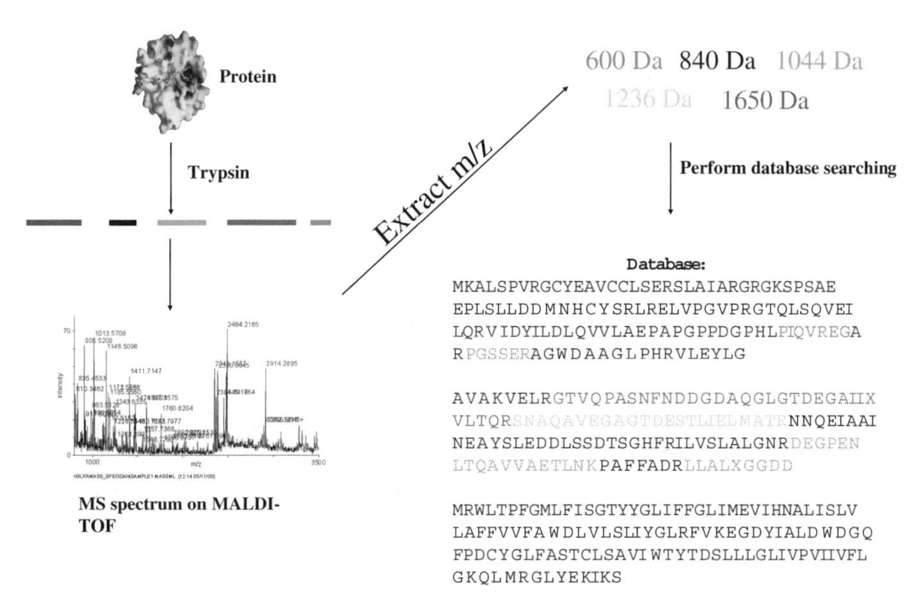

Figure 1.9. Scheme employed for the identification of proteins based on MALDI-TOF and peptide mass fingerprinting. A protein of interest is digested with trypsin and the resulting peptide mixture is analyzed by MALDI-TOF mass spectrometry. The measured *m/z* ratios are then used to search protein/DNA databases leading to the identification of the protein. The matching proteins are ranked according to the number of observed *m/z* ratios that match to their predicted tryptic peptide patterns.

is to add the number of peptide masses that match with the predicted masses for each entry in a protein database. The database entries are then ranked according to the number of hits. This forms the basis behind software such as PepSea and MS-Fit (*http://prospector.ucsf.edu*). Typically, these software packages work well for quality experimental data.

Mascot originated from the software called MOWSE (Pappin et al., 1993). MOWSE uses more information to make its decision on the score by taking into account the protein size and the relative abundance of peptides in the databases. Mascot further incorporates probability scoring for the probability that the match between the data and the entry in the database will be a random event. This score is calculated for every entry in the database. The identification is then established by ordering the proteins with a decreasing probability of being a random match. Mascot can be freely accessed over the web (*http://www.matrixscience.com/*) and is also available for commercial purposes through MatrixScience.

ProFound (Zhang and Chait, 2000) uses a different approach based on Bayesian theory to rank the protein sequences in a database by their probability of occurrence. It is an expert system (i.e., it simulates what an expert in the field would do) that uses detailed information about each protein sequence and empirical information about the distribution of proteolytic peptides that are included in the scoring scheme. ProFound is also free over the Internet (*http://prowl.rockefeller.edu/*).

Generally, all these software packages perform well when good-quality spectra are available. The ones that provide more advanced scoring schemes perform better when less information is available or when the quality of the MS spectra is reduced. Regardless of the software, the identification of the protein depends on the number of peptides observed, the accuracy of the measurement, and the size of the genome of the particular species. For smaller genomes, such as yeast and *Escherichia coli* (*E. coli*), protein identification using the MALDI-TOF mass spectrometer is generally successful. For larger genomes the rate of success drops significantly using MALDI-TOF.

POSTSOURCE DECAY. MALDI-TOF mass spectrometry also offers the possibility of recording the fragmentation patterns obtained from a peptide. This is achieved using a technique called *postsource decay* (PSD). PSD is achieved by increasing the laser power beyond the value needed to generate ions. The precursor ions are transferred from the MALDI plate to the gas phase. The excess energy induces the precursor ions (peptides) to fragment along their backbone. Generally, these ions are not seen in conventional MALDI-TOF analysis because of their lower kinetic energy. Fortunately, on the MALDI-TOF equipped with a reflectron, the lower kinetic energy of peptide fragments can be compensated by changing the settings on the reflectron. PSD is typically achieved by acquiring spectra for specific mass ranges with different settings on the reflectron. All the spectra are then stitched together to make a full PSD spectrum.

Although PSD seems to be a rapid way of obtaining fragmentation patterns of peptides, it seriously suffers in terms of sensitivity. Furthermore, the fragmentation patterns are often difficult to discern and are of poor quality. Therefore, PSD has not been the method of choice for the generation of peptide fragmentation patterns.

Hybrid Instruments

Conventional MALDI-TOF mass spectrometers can also provide fragmentation patterns related to the amino acid sequence of a peptide. This is also done using postsource decay; however, the mass selection of the peptide to be fragmented, the sensitivity of the approach, and the quality of the MS/MS spectra make it difficult to utilize postsource decay for the routine and rapid generation of MS/MS spectra. Recently, a set of novel MS/MS-capable mass spectrometers have been developed based on MALDI ionization. These instruments combine the MALDI ionization technique with the fragmentation of ions by collision-induced dissociation (Shevchenko et al., 2000; Loboda et al., 2000). For example, the recently introduced MALDI-Pulsar from Sciex is illustrated in Figure 1.10. This instrument, as its name alludes to, includes a MALDI ionization interface followed by a set of quadrupoles and a collision cell. The first quadrupole is used as an ion guide. It is followed by a second quadrupole, which is either used as an ion guide, a precursor scan, or for a parent ion selection. The first quadrupole is followed by a collision cell that can be used to fragment ions by collision-induced dissociation. A pulsing grid set at a 45° angle is positioned after the collision cell and deflects the ions upward into a TOF mass analyzer equipped with a

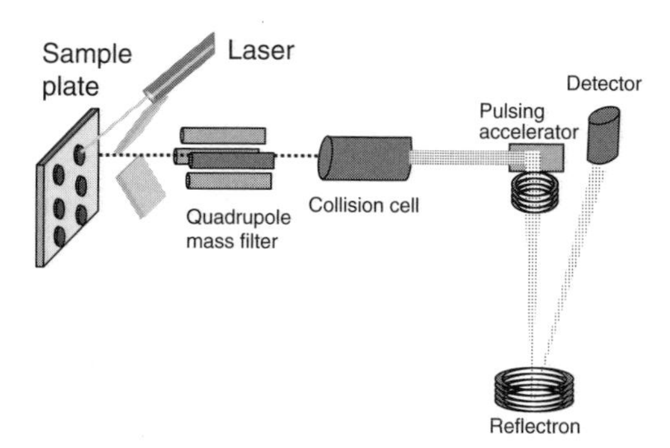

Figure 1.10. Schematic of a MALDI-Pulsar mass spectrometer. A portion of the sample is transferred from the MALDI plate to the gas phase by a pulsed laser. Once in the gas phase, the charged molecules go through a quadrupole, which can be used as a mass filter, lenses, or a mass analyzer. They then enter a collision cell followed by a pulsing plate, which pulse part of the beam of ions into the perpendicular TOF. They enter the field-free region of the TOF in which they separate according to their *m/z* ratio. A reflectron is employed to refocus the ions and to provide better resolution. This instrument is capable of providing MS spectra and good quality MS/MS spectra generated by collision-induced dissociation.

reflectron. Although a pulsing laser is still used to generate ions off the MALDI plate, it is not used for the timing and correction of the masses of the observed ions.

In this design, the ionization of the sample is decoupled from the acceleration of the ions into the TOF tubes. Furthermore, significant collisional cooling is present in the first quadrupole, reducing the distribution of the kinetic energy of the ions. Therefore, the geometry of the plates does not affect the mass accuracy of the instrument. This is a significant difference over conventional MALDI-TOF in which great care must be taken regarding the plate geometry. This means that internal or close external calibrants are not necessary to maintain the mass accuracy of the MALDI-Pulsar system. It also means that totally different plate designs can be constructed.

The Pulsar also allows peptide selection and efficient fragmentation by collision-induced dissociation in the collision cell. Collision-induced dissociation is more reliable than postsource decay for generating fragmentation patterns. MALDI ionization predominantly provides peptides of charge 1+. The fragmentation of 1+ ions generates lower quality fragmentation patterns than what is usually obtained for 2+ and 3+ ions. Therefore, this hybrid approach offers significant advantages over the conventional MALDI-TOF in terms of its efficient fragmentation patterns. It also offers advantages over the ESI-MS/MS for the rapid screening of samples.

Electrospray Ionization Mass Spectrometry. Electrospray ionization is also a very popular approach to introduce protein and peptide mixtures to mass spectrometers. Typically, ESI is used in conjunction with a triple quadrupole, an ion trap, or a hybrid quadrupole-TOF mass spectrometer. ESI is mainly popular because it provides a direct interface between the mass spectrometer and the atmospheric pressure. It can also be readily coupled to separation techniques such as HPLC, liquid chromatography (LC), and capillary zone electrophoresis (CZE), or it can be used for the continuous infusion of samples. Until the introduction of the MALDI-Pulsar, the ESI-based mass spectrometers were the only viable approach for the generation of MS/MS spectra. Furthermore, ESI produces more multiply charged ions that provide richer MS/MS spectra.

PRINCIPLE. Electrospray ionization (Fenn, 1990) allows analytes to transfer from the liquid phase to the gas phase at atmospheric pressure. Generally, the ionization process is achieved by applying an electric field between the tip of a small tube and the entrance of a mass spectrometer. The electric field forces the charged liquid at the end of the tip to form a cone, called *Taylor cone*, that minimizes the charge/surface ratio. Droplets form at the end of the cone and move toward the entrance of the mass spectrometer. Different theories have been put forward to explain what happens after the droplets have been formed and how the transfer of the analytes to the gas phase is achieved. The most popular explanation is that the liberated droplets go through a repetitive process of solvent evaporation, whereby they fragment into smaller droplets to reduce the charge density on them. This leads to a large number of droplets of shrinking sizes until the solvent has disappeared; thus the analytes are left in the gas phase. Furthermore, as the droplets shrink, the pH in the droplets decreases and facilitates the protonation of the analytes.

MASS SPECTROMETER AND INFORMATION. The most important advantage of ESI is that it can be readily coupled online with separation techniques. This has been illustrated over the years by the numerous development of methodologies based on separation techniques coupled to electrospray mass spectrometers. The variety of applications, separation techniques, and mass spectrometers utilized with ESI can be confusing to those new to the field. Therefore, in the next few sections we will introduce the different mass spectrometers with their respective limitations, the different separation techniques that are specifically used for proteomics, and the generation of tandem mass spectra.

MASS ANALYZERS UTILIZED WITH ELECTROSPRAY IONIZATION IN PROTEOMICS. There has been a flurry of novel commercial mass spectrometers over the last 5 years. This is in part related to the constant demand for newer and better mass spectrometers imposed by the biotechnology/pharmaceutical sectors. Even though the number of mass spectrometers is increasing, and the confusion related to their application in proteomics is building even faster, it is still possible to make a logical assessment of their functions, advantages, and limitations.

Triple Quadrupole Mass Spectrometer. The first mass analyzer that was utilized for proteomic studies is the triple quadrupole mass spectrometer. A quadrupole mass analyzer consists of four rods placed at equidistance as if they were placed on the surface of a cylinder. The Mathieu equation was derived to describe the motion of a charged molecule in an electric field. It is also applicable to describe the motion of charge molecules in the triple quadrupole mass spectrometer. Electric fields can be constant [direct current (DC)] or variable [alternating current (AC); different amplitude and frequency]. The Mathieu equation takes into account the combination of DC and AC fields. The combinations that are possible are such that the quadrupole can be used for three functions. The combination of a DC and radio frequency (RF) potential transforms the quadrupole into a mass filter to transmit a specific *m/z* ratio. The application of a RF-only mode will set the quadrupole as an ion guide, while the application of a DC-only mode transforms the quadrupole into a lens element. For every mass, there is a region of stability for the DC, RF amplitude, and frequency. Therefore, the resolution of the instrument and the ion transmission (sensitivity) are intimately linked. Increasing the resolution decreases the number of ions transmitted. Changing the shape of the quadrupole rods affects the electric field and can also improve the resolution. The mass filter mode and the RF-only mode are the two most common applications of the quadrupole. Furthermore, most triple quadrupoles have been fabricated using cylindrical rods for simplicity. More recently, designs for hyperbolic rods have allowed improved resolution, and they have been introduced in the TSQ-quantum (Thermo-Finnigan):

$$d^2x/dt^2 + [\alpha + \beta f(t)]x = 0 \tag{1.1}$$

where α and β are constant and the $f(t)$ is the sinusoidal function of time. This equation is called the Mathieu equation. In a quadrupole mass analyzer only two axis of direction are being affected by the field.

The triple quadrupole mass spectrometer consists of three subsequent sets of quadrupoles, positioned one after the other in a linear fashion toward the detector (Fig. 1.11). The functions of the first and third quadrupoles are either as mass filters or as ion guides. The second quadrupole is typically run in the RF-only mode (i.e., ion guide) and modified to allow the introduction of gas. It is used for collision-induced dissociation of charged molecules. This combination of quadrupole (q_0), collision cell (q_1), and quadrupole (q_2) allows a multiplicity of experiments to be performed.

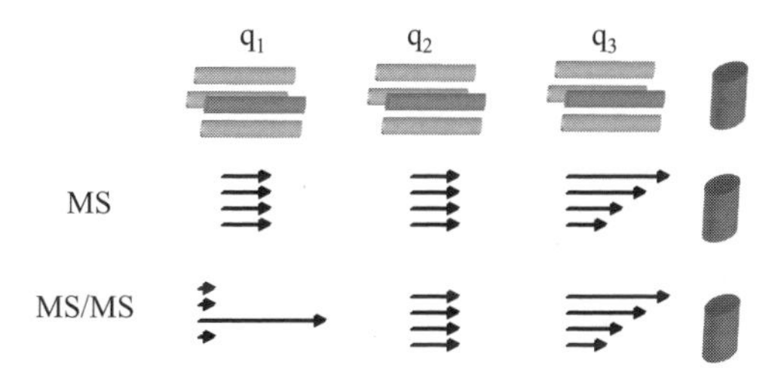

Figure 1.11. Triple quadrupole mass spectrometer for the analysis of peptides. A set of two experiments is performed on the triple quadrupole mass spectrometer for peptide analysis. In the first pass an MS spectrum is acquired to detect the *m/z* ratio of the analytes. Then, a peptide corresponding to a specific *m/z* ratio is selected in a second experiment, isolated in q_1 and fragmented in the collision cell. The resulting fragments are separated in q_3 leading to an MS/MS spectrum.

Although numerous experiments can be concocted with a triple quadrupole mass spectrometer, only two serial experiments are performed for peptide identification. The first experiment consists of running q_0 and q_1 in RF-only mode while scanning the *m/z* on q_2 (DC and RF). In this fashion, q_0 and q_1 let the positively charged analytes pass through while q_2 continuously scans the *m/z*, generating an MS spectra (signal versus *m/z*) for the positively charged ions present in a proteolytic digest.

The second set of experiments, called MS/MS, consists of selecting and fragmenting an ion followed by the separation of the generated daughter fragments. This is achieved using q_0 as a mass filter to allow only a narrow *m/z* window around the selected ion to pass. A small amount of neutral gas is then added to q_1, while the potentials, applied to lenses, are changed to provide increased kinetic energy to the ions entering q_1. The ions collide with the small gas molecules and fragment by collision-induced dissociation. The generated ions are then separated by q_2. The end result is an MS/MS spectrum that contains the selected ion fragmentation patterns.

The combination of both experiments (MS and MS/MS) allows peptides to be detected, selected, and fragmented. Typically, the triple quadrupole mass spectrometer is either manually or automatically cycled through MS and MS/MS acquisition.

The application of the triple quadrupole mass spectrometer in proteomics was supplanted by the introduction of more sensitive instrumentation such as ion trap mass spectrometers and hybrid mass spectrometers. The advantage of the triple quadrupole mass spectrometer with collision-induced dissociation is the efficient conversion of precursor ions into product ions. The disadvantages of the triple quadrupole mass spectrometer are its low resolution, the mass discrimination (i.e., peak height dependency on mass), the narrow window, and the strong mass dependency on collision energy for efficient fragmentation of peptides. Generally, the information generated in MS mode is too low of a resolution to be useful for database searches. Furthermore, the mass accuracy is generally too low to allow adequate de novo sequencing of proteins when the identification by database searching fails.

Ion Trap Mass Spectrometer. The second analyzer introduced for proteomic studies was the ion trap mass spectrometer. The ion trap mass spectrometer utilizes a combination of electrodes for the accumulation of ions in a space defined by the shape of the electric field present in the trap. Similar to the triple quadrupole mass spectrometer, the ion trap mass spectrometer follows the Mathieu equation. Because of the shape of the electrodes and the definition of the electric fields, the resolution obtained on the trap can be higher than on conventional quadrupole instruments. An example of a complete ion trap mass spectrometer is described in Figure 1.12. It consists of a heated stainless steel capillary followed by a skimmer, an ion guide, by the ion trap, and a detector. All of these parts are maintained under vacuum. The ion guides have a dual function: to guide the ions toward the ion trap and to gate the ions into the trap. The number of ions present in the trap is critical and needs to be controlled. Otherwise, a phenomenon called *space charging* can occur that causes the performance of the instrument to be distorted. The ion optic is used to control the injection of ions into the trap. The reversal of the polarity applied to the optics deviates the ions from their paths and prevents the ions from entering the trap. In reality, a scan from an ion trap consists

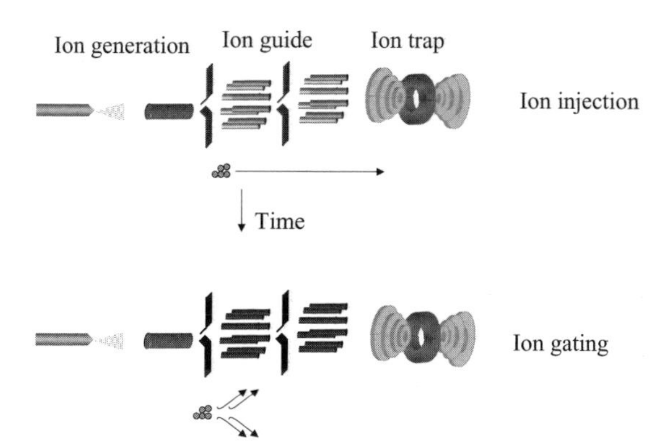

Figure 1.12. Schematic of the ion trap mass spectrometer. The schematic of the trap illustrates how the ions are guided to the trap and how the gating approach is used to sufficiently fill the trap prior to MS and MS/MS experiments.

of at least two experiments. In the first experiment a short burst of ion is injected into the trap and then ejected to the detector. The signal observed by the detector is used to automatically calculate the maximum allowable injection time at which space charging can still be avoided. Then the ion optics are open to inject the ions into the trap for the calculated injection time. The ions injected into the trap are not stationary but have an orbit that resembles a boomerang shape. Depending on the *m/z* ratio of the ions, the orbit will occupy more or less space in the trap. Once enough ions have been accumulated, they can be sequentially ejected from the trap by changing the RF amplitude applied to the ring electrode. This creates instabilities in the orbits of specific ions, pushing them further away from the center of the trap, and finally drawn to the ejection point toward the detector. The lower *m/z* ions are ejected at a lower RF amplitude, while the higher *m/z* ions require higher RF amplitudes. Typically, the RF amplitude is calibrated using analytes of well-defined masses.

Ion ejection and ion fragmentation can be achieved by *resonance excitation*. This approach consists of creating a resonance effect in the orbit of the ions to increase its axial kinetic energy and to increase the axial dimension of the orbit. Resonance excitation is achieved by applying a small DC voltage to the endcap electrodes at the same time as applying an RF to the ring electrode. This creates a cumulative increase in the kinetic energy of the selected ions, thus leading to an unstable orbit and ejection. Furthermore, the increased kinetic energy can be funneled to fragment the selected ions by introducing a small amount of inert gas into the trap.

The operation of the ion trap mass spectrometer in MS mode consists of a rapid scan for automatic gain control, followed by the timed injection of ions, and finally the scanning of the RF amplitude to successively eject the ions from the trap. The end result is an MS spectrum with the RF amplitude on the bottom axis, which then can be converted into *m/z* by calibration.

The operation of the ion trap mass spectrometer in MS/MS mode consists of a rapid injection/ejection for automatic gain control, a timed injection of ions and an ejection of undesirable ions (except a preselected mass window), followed by a resonance excitation of the selected peptide in the presence of small gas molecules, and, finally, by a scanning of the RF amplitude to successively eject the fragmented ions from the trap. The end result is an MS/MS spectrum. Although the fragmentation of ions is also obtained by collision-induced dissociation, the quality of the MS/MS spectrum is different than the one obtained on a triple quadrupole mass spectrometer. This can be explained by the fact that only the selected ions are sensitive to the RF and to the resonance excitation; therefore, once fragmentation occurs, the resulting fragments are of different *m*/*z* and are invisible to the RF/resonance excitation. This is not the case for triple quadrupoles because all the generated charged fragments are still subjected to the acceleration voltage and can gain sufficient kinetic energy to proceed into further collision-induced dissociation.

The ion trap mass spectrometer has seen wide acceptance in proteomic application mainly due to its level of automation, sensitivity, and cost. The information provided in MS mode is insufficient for protein identification based solely on peptide masses. The information provided by MS/MS spectra is, however, generally sufficient to identify proteins by database searching. De novo sequencing of peptides is also difficult on ion trap mass spectrometers due to its low resolution and its lower mass cutoff in MS/MS mode.

More recently, new linear-type ion trap mass spectrometers have been introduced. These offer a larger trap space allowing wider dynamic range as well as better sensitivity and resolution over conventional traps. Furthermore, some of these instruments can be operated either as a trap or as a triple quadrupole mass spectrometer (QTrap from MDS-Sciex).

Ion Cyclotron Mass Spectrometer. The third analyzer is the *ion cyclotron Fourier transform* mass analyzer (Comisarow and Marshall, 1974). At the basic level this analyzer is also an ion trap mass spectrometer. It typically consists of an ion source, followed by an ion guide, and, finally, by a Fourier transform cell contained in a superconducting magnet (Fig. 1.13). As in the ion trap mass spectrometer, the purpose of the ion guide is to focus the ions and transfer them into the Fourier transform cell. In some designs the ion guide is also utilized to perform the accumulation and selection of ions, for example, using a fragmented quadrupole before pulsing them into the trap. The purpose of the superconducting magnet is to provide a uniform magnetic field in the trapping region of the cell. The cell itself forms a box made of two detector electrodes facing each other, another set of two RF electrodes facing each other, and a third set consisting of trapping plates. In the cyclotron, the ions have to pass through significant fringing fields created by the superconducting magnet before progressing into the trap. The ions that enter the Fourier transform cell are trapped by the magnetic field and the trapping plates. Furthermore, once in the trap, they will be limited only to circular orbits. It turns out that every ion of a specific mass and charge has a different cyclotron frequency, defined by:

$$w = \beta z m^{-1} \tag{1.2}$$

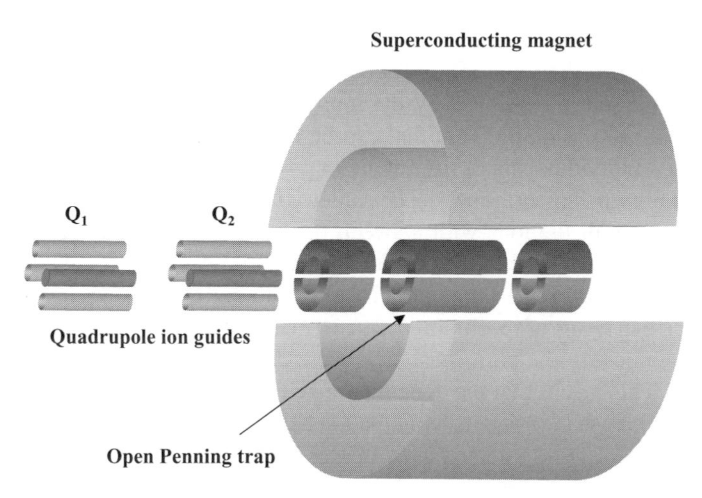

Figure 1.13. Schematic representation of a Fourier transform mass spectrometer. The schematic illustrates the utilization of the quadrupole as ion guide. The trap consists of an open penning trap inserted into the bore of a superconducting magnet. A set of electrodes in the trap is used to bring the ions to their cyclotron frequency. A second set of electrodes is used to measure the frequencies.

where w is the cyclotron frequency in radians per second, β is the magnetic field strength in tesla, z is the charge, and m is the mass of the ion. This equation is only valid once the ions are in orbits, but the ions do not spontaneously spring into the orbits. By applying an oscillating electric field that corresponds to the cyclotron frequency of the ions to the two opposite electrodes of the cell, a cyclotron resonance can be induced. This cyclotron resonance induces the ions to travel into an increasing circular orbit. After a short time, the electric field is turned off, and the ions reach a stable orbit determined by their m/z ratios and the magnetic field. The stabilization and resolution of different circular orbits require a very homogeneous magnetic field created by a superconducting magnet. Typically, the cyclotron resonance is provided by a sine wave signal generator. Ions of different m/z ratios have different resonance frequencies, and they will only be accelerated if the right resonance frequency is applied to the electrodes.

The ions in orbit pass by the electrodes and induce a small, but detectable, current at the electrode, called the *image current*. The end result of the motion is an alternative current in phase with the frequency of the ion orbits. Furthermore, the amplitude of the image current is proportional to the number of ions in orbit. An amplification system is used to transform the small AC signal into a larger and easier signal to process.

Obviously, this would be a relatively useless approach if only one ion could be trapped and detected at a time. It turns out that in the beginning of the ion cyclotron design, the ions were individually detected, but it could take up to 20 min to acquire a full MS scan. The twist around this problem is to careful look at Eq. (1.2). If all the ions could be accelerated into their respective orbits, they would all have a resonance frequency related to their m/z ratio. Therefore, the induced current would be a composite of all the currents induced by the ions at their respective frequencies. Clearly this is a signal processing problem that can be handled by Fourier transform analysis to extract each individual signal from the composite.

Today, Fourier transform mass spectrometry (FTMS) spectra are rapidly acquired by first activating all the ions with a sweep of the RF to accelerate all the ions. The transient image current (due to the multitude of m/z) is accumulated, and the MS spectrum is reconstructed by the Fourier transform. Although the FTMS instrument is the ultimate choice in terms of resolution and sensitivity, it has not seen widespread acceptance in proteomics. In the past, most cells were designed to trap a defined fraction of the ion beam. This means that the dynamic range of the cell was limited. Methods were developed to increase the ion density in the ion beam by trapping ions and by the pulsed injection of ions using fragmented quadrupoles prior to the cyclotron. Again, the challenge is to have enough ions for detection while avoiding space charging. For most real-life samples this can still be an issue forcing the mass accuracy to be above the 1 to 1.5 ppm level. The cost of FTMS instruments and their robustness have also been serious issues. The information, however, provided in MS mode is often enough for protein identification based solely on accurate peptide masses. The information provided in MS/MS mode can be used for database searches or for de novo sequencing of proteins; however, the generation of MS/MS fragments by collision-induced dissociation in the trap is relatively slow in FTMS and precludes its utilization for rapid analysis.

Recently, hybrid instruments that combine external selection and fragmentations of ions have been developed. For example, Thermo-Finnigan developed a hybrid linear ion trap coupled to an FTMS. The advantage of this approach is the rapid fragmentation and analysis of peptides. As the fragmentation is performed in the linear trap, there is no need to wait for long ion cooling period in the FTMS. Other types of hybrid FTMS have been introduced by IonSpec and Bruker.

Hybrid Mass Spectrometers. The last analyzer is the hybrid quadrupole time-of-flight mass spectrometer. The Pulsar mass spectrometer from Sciex and the Qtof mass spectrometer from Micromass are two examples of hybrid instruments. The Pulsar utilized with ESI is the same as the one previously described for the MALDI-Pulsar. It utilizes an ESI source instead of a MALDI source. Please refer to the MALDI section for the description of the Pulsar instrument.

TECHNIQUES TO INTRODUCE PEPTIDE MIXTURES TO ESI-MS/MS. The MALDI-TOF mass spectrometry is definitely a high-throughput platform for the identification of proteins. So why rely on the more cumbersome and lower-throughput ESI approaches? ESI provides multiply charged peptides that can be efficiently fragmented by collision-induced dissociation. The MALDI-TOF provides singly charged peptides that are more difficult to fragment even when collision-induced dissociation is available. Furthermore, ESI can be readily coupled with infusion and concentration/separation techniques. Hence, intense efforts have been focused in recent years on the development of infusion and separation techniques for efficient analysis of protein digests. These techniques differ in sensitivity, the handling of contamination, the quality of the MS/MS spectra, and the throughput. In the following sections we will review the different approaches to sample introduction, their disadvantages, and their advantages.

Continuous-Infusion ESI-MS/MS. Within the last 10 years, ESI has become compatible with the analysis of low-level protein digests. The main challenge was to make the flow and concentration detection of conventional electrospray approaches compatible with the volume and concentration that were obtained with blot, or in-gel digestion of separated proteins. Conventional electrospray requires a few microliters/minute flow rate to be stable, while the final volume of a digest is in the low microliter levels. Therefore, efforts were placed in developing new "lower-flow" versions of the electrospray process. The microelectrospray (100 to 500 nL/min), the nanoelectrospray (1 to 100 nL/min), and the picoelectrospray (<1 nL/min) methods were developed. They all have the reduction of the flow requirement in common by reducing the internal and external diameter of the sprayer. The introduction of these novel electrospray approaches has greatly facilitated the analysis of protein digests by providing longer analytical time windows, better ionization, and better transfers to the mass spectrometer.

In particular, the "low-flow" requirement of the nanoelectrospray became rapidly attractive for the identification of proteins based on the continuous infusion of a sample (Shevchenko et al., 1996a, 1996b, 1997). From as little as 1 µl, an analytical time of 30 min to 2 h could be easily obtained. This would leave plenty of time to acquire good-quality MS spectra, to select large number of ions, and to successively generate their MS/MS spectrum. Furthermore, when the nanoelectrospray technique was introduced, most mass spectrometers did not provide automated control of the different experiments (i.e., data-dependant experiments), thus frequent user interventions were required for the analysis of peptides (Shevchenko et al., 1997). This means that on the mass spectrometers, the operator had to generate and interpret the MS spectrum, select the peptides from the MS spectrum, switch the mass spectrometer to collision-induced dissociation, and then successively generate the MS/MS spectrum of the ions by manually changing the parameters for every ion. Even for the most experienced user, this can easily take 5 to 15 min to perform per peptide.

In the continuous-flow infusion nanoelectrospray, the peptide mixture of interest is inserted into a glass needle of 1 mm in diameter that has been pulled to a closed and tapered end using a pipette puller. Furthermore, the pulled needles are also coated with a layer of conductive material such as gold, to which a high voltage is applied (Fig. 1.14). The needle is installed in front of the mass spectrometer, and a slight pressure is applied. Then, the end of the needle is opened by touching it against the front plate of the MS, producing an open tip of only 1- to 10-µm inner diameter and a slightly bigger outer diameter. Alternatively, a nanospray needle that is tapered down, but not closed, can also be purchased. The peptide mixture is then continuously delivered at the end of the needle by a gentle gas pressure. The peptide solution that reaches the tip of the nanoelectrospray needle is electrosprayed into the mass spectrometer at low flow rates

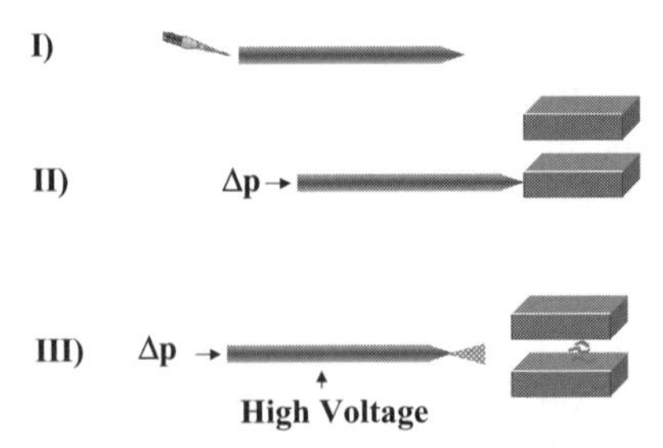

Figure 1.14. Schematic representation of the nanoelectrospray analysis of peptide mixtures. The samples of interest are pipetted into a nanoelectrospray needle. The needle is then mounted in front of the mass spectrometer. The distal end of the needle is open by applying a slight gas pressure at the other end of the column and gently touching the front plate of the mass spectrometer. Once, the open needle is positioned to electrospray into the mass spectrometer.

(few nanoliters/minute). Roughly 30 min to 1.5 h analysis can be performed on 1 μl of sample.

Separation Coupled to ESI-MS/MS. Electrospary ionization can also be coupled with separation techniques. This is often preferable for low levels of samples and for contaminated samples. Furthermore, separation techniques are readily automatable, while the nanoelectrospray approach has evaded automation due to its inherent skill requirement. Separation techniques offer the advantage of concentrating the analytes into shorter and separated analytical windows. While nanoelectrospray presents all the analytes at the same time to the mass spectrometer, separation systems subsequently present the analytes to the mass spectrometer; however, a very narrow analytical window (typically 1 to 30 s) is available for every analyte, during which all mass spectrometric measurements for an analyte must be performed.

Automated Mass Spectrometers. Evidently, the manual triggering of the mass spectrometer, as conventionally done in nanoelectrospray, is not possible with separation techniques. Therefore, the utilization of an automated mass spectrometer is essential for the success of these experiments. A set of standard data-dependent features has emerged. (1) The mass spectrometer should offer the automated selection of precursor ions based on a threshold or a signal-to-noise ratio. (2) The *n*th-most intense ion should be selectable for fragmentation as predefined by the user. (3) A static exclusion list of ions should be predefinable. The static exclusion list allows the exclusion of known contaminants, such as clusters, trypsin autolytic peptides, and other known peptides. (4) The ions for which *m* number of MS/MS spectra have been generated should be added to a dynamic exclusion list until their intensity in MS mode falls below a threshold value or until a time period has elapsed. The dynamic exclusion of ions allows the selection of the next most intense ion for the next round of MS/MS spectra. (5) The dynamic isotope exclusion is also an important option, especially for higher resolution instruments. This feature ensures that the isotope peaks from an ion do not trigger the acquisition of an MS/MS spectrum. (6) The exclusion of charge states, such as the 1+ ions, can also be convenient because it avoids the generation of MS/MS spectra for 1+ ions, which are typically not very informative. (7) The automated selection of fragmentation energy is an important aspect for adequate fragmentation of peptides. Although most mass spectrometers utilize collision-induced dissociation to fragment peptides, the ways in which the kinetic energy is ascribed to the precursor ion are different. In triple quadrupole and hybrid quadrupole TOF mass spectrometers, the fragmentation pattern is only adequate in a limited kinetic energy range. Excess energy increases the internal fragmentation of the precursors and causes poor transmission of ions through the collision cell, thus resulting in a poor-quality MS/MS spectrum. Therefore, the automated adjustment of the collision energy is an important feature on these mass spectrometers. The fragmentation pattern is less sensitive to the kinetic energy on the ion trap and cyclotron mass spectrometers, as long as sufficient energy is provided. Therefore, the automated adjustment of the collision energy is less important in ion trap and cyclotron mass spectrometers.

High-Performance Liquid Chromatography–MS/MS. High-performance liquid chromatography has been the technique of choice for the separation of analytes online with ESI mass spectrometers (LaCourse, 2000). In reverse-phase mode, HPLC concentrates and separates analytes according to their hydrophobicity. Different approaches have been developed to introduce peptides into a mass spectrometer by HPLC. They all have in common peptide solutions that are loaded and separated on an HPLC column made of C18-like material. They differ, however, in terms of their flow rates, their sample paths, robustness, and sensitivity.

Conventional HPLC. Conventional microbore HPLC coupled to ESI-MS is a well-established approach for the identification of protein levels higher than picomoles (Hunt et al., 1981, 1986; Gibson and Biemann, 1984). Microbore HPLC is usually achieved using a 0.3-mm inner diameter (i.d.) HPLC column installed on a low-flow HPLC system. The flow rate generated through these columns is typically in the low microliter per minute range. The integration of an autosampler capable of handling less than 10 μL of sample completes the automation of the procedure. Therefore, sequential introduction, separation, and analysis of protein digests on this automated system can be routinely performed.

Microflow HPLC. High-performance liquid chromatography is a concentrating technique, while mass spectrometry is a concentration-dependent device for constant flow rates. Therefore, improving the concentration of analytes delivered to the mass spectrometer provides better sensitivity. To improve the concentration when a limited amount of analytes is available, only two approaches can be taken. The first approach consists of improving the separation and decreasing the peak width of the eluting analytes. Although this is feasible, practically it does not provide sufficient improvement in sensitivity. The second approach consists of decreasing the size and flow rate of the column. For example, reducing the size of an HPLC column from 1 mm to 100 μm in diameter offers an improvement by a factor of 100 in the concentration of the analytes.

This is a very attractive alternative for low-level samples. It is important to keep in mind that the HPLC theory also predicts that the volume amount of injected analytes should be correspondingly reduced. In reality, there is a tendency to disregard this warning and to inject as much volume as possible at the price of losing the lower affinity analytes.

Although attractive, low-flow HPLC (<200 nL/min) was not directly compatible with conventional electrospray. Earlier attempts to couple low-flow HPLC with conventional electrospray interfaces by using sheath liquid were successful; however, they did not demonstrate the expected improvement in sensitivity. Fortunately, microelectrospray interfaces compatible with the low-flow technique provided by μ-HPLC columns were developed, and, these greatly improved the usefulness of this technique (Chervet et al., 1996; Yates et al., 1996).

Fabrication of Columns. Recently, microcolumns (150 μm i.d. or less) have been commercially available; however, the range of available packing materials and the cost of these columns are often limited. Therefore, it is still common to fabricate in-house columns. Two standard techniques have been developed for the fabrication of microcolumns.

The first technique consists of fabricating a column in a capillary tubing with a constant inner diameter. In this design, small borosilicate beads are dry packed at the end of a capillary tubing of 50 to 150 μm i.d. by tapping the capillary in an aliquot of beads. Typically, about 1 mm of the capillary tubing is filled with beads. The end that contains the beads is rapidly sintered in a Bunsen burner. This results in the formation of a small porous frit at the end of the capillary tubing. It takes some practice to appropriately sinter the beads. The packing material for the column is pressure forced at the other end of the capillary in the form of a slurry. This requires the utilization of a bomb that can be pressurized up to a few 1000 psi. Alternatively, the slurry can be installed in a large volume union and pushed into the column using an HPLC pump. The pressure is operated until the required length of column packing is achieved; then it is slowly reduced to atmospheric pressure. It also helps to place the part of the column being packed in a sonification bath. The end result is a packed capillary column (typically 5- to 10-cm of packing) terminated by a frit. The disadvantage of this approach is that it does not include a microelectrospray needle. Typically, the fritted end of the column is installed in a low dead-volume union connected to a micro-ESI needle. To minimize the band broadening, a short needle that has a smaller inner diameter than the capillary column and a reduced diameter at the tip (5- to 15-μm opening) is utilized to generate the electrospray (see Fig. 1.15).

The column can be directly formed in a microelectrospray needle to avoid the dead-volume issues of the packed capillary approach. This requires a frit of some sort to be fabricated in the tip of a microelectrospray needle. Typically, a needle of 50 to 150 μm i.d., sharply terminated at the tip to 5 to 15 μm i.d., is used for the fabrication of the column. These needles are made from a capillary tubing that is stretched using a laser pipette puller. Then again, the needle can be pulled by placing a capillary tubing in a vertical position, attaching a weight to it, and heating a section of the capillary with a flame. Both of these needle pulling techniques can be tedious to perform, while creating a frit at the end of the needle can be even more challenging. Therefore, the best

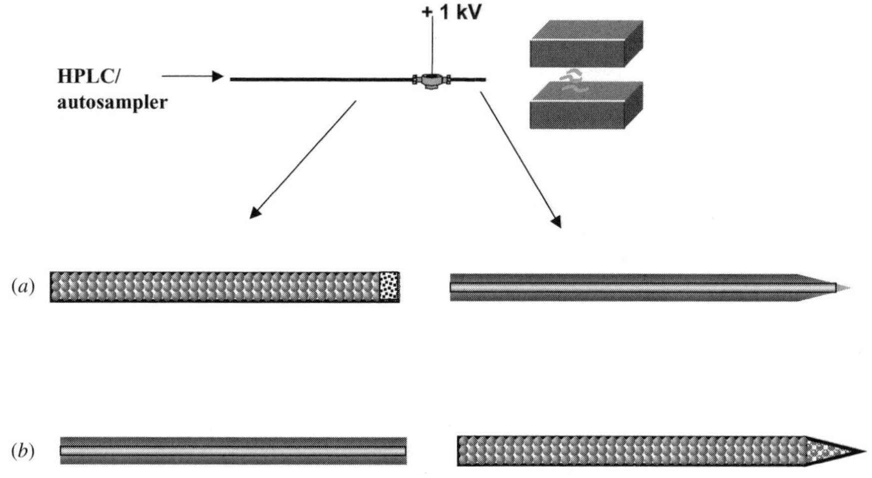

Figure 1.15. Schematic of nanoflow HPLC-ESI-MS/MS. (*a*) In this design an HPLC column is fabricated in a capillary tubing (50 to 150 μm i.d. × 5 cm long) terminated with a sintered glass frit. This capillary column is then connected to a low inner diameter nanoelectrospray tip by a low-dead-volume union. (*b*) In this design a low inner diameter capillary tubing is used to connect an HPLC pump to a low-dead-volume union. An HPLC column is then fabricated in a (50 to 100 μm i.d.) nanospray needle with a tip inner diamater of 5 to 15 μm. The HPLC needle is then installed at the other end of the low-dead-volume union.

alternative is to buy microelectrospray needles (e.g., New Objectives needles) with a frit already fabricated at the tip, and then to pack the column with the reverse-phase material of choice. The packed needle is then installed on a low-dead-volume union also connected to a nanoflow HPLC system (see Fig. 1.15).

More exotic techniques have been developed for the fabrication of columns (Fig. 1.16). In particular, needles without frits have been packed with a reverse-phase material. This was done either by pulling long needles of progressively decreasing inner diameters or by creating an isthmus by differentially pulling the capillary tubing (Martin et al., 2000). These two methods allow the needles to be packed without having to create a frit. Davis and Lee (1998) used a laser-pulled microelectrospray needle (150 μm i.d.) with a 5-μm orifice and introduced into the needle a short piece of fused silica capillary [25 μm i.d. × 150 μm outer diameter (o.d.)]. The piece of capillary was pushed as far as possible in the needle and then backed with a membrane frit. The reverse-phase material was then packed behind that frit. Although these approaches are interesting alternatives for HPLC needle fabrication, they have not yet seen widespread application and thus remain the forte of only a few groups.

More recently, we have seen the appearance of monolithic columns that are fabricated by direct polymerization of the stationary phase inside the capillary tubing. These monolithic columns are obtained by the in situ polymerization of a mixed polymer such as styrene and divinylbenzene. The polymerization occurs in an oven at a controlled temperature. The advantage of the monolithic column is its lower "multiple-path" effect, which provides a good number of theoretical plates and a good resolution for

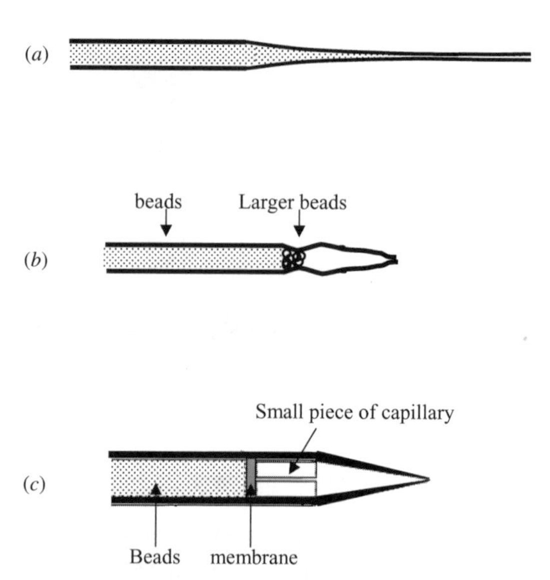

Figure 1.16. Schematics of different packed nanospray needles that have been fabricated. (*a*) Needle with no frit obtained by gradually narrowing down the inner diameter of the needle. (*b*) Needle with an isthmus to retain the reverse-phase beads (Martin et al., 2000). (*c*) Needle with a capillary frit to retain the beads (Davis and Lee, 1998).

short column beds. Comparisons with conventional packed columns have indicated similar, if not better, separations. Other advantages of the monolithic columns are that they are easier to fabricate and to mass produce than the conventional packed HPLC columns. The disadvantage, however, of the monolithic column is that it requires an organic synthesis, which has to be carried out with caution. Therefore, this approach is less amenable for the in-house and widespread fabrication of columns by research laboratories.

High-Performance Liquid Chromatography Setup. The microcolumn system would not be complete without the addition of an HPLC gradient delivery system. The reverse-phase microcolumn typically requires a flow rate of 100 to 200 nL/min. Although conventional HPLCs can deliver such flow rates in isocratic mode, they cannot deliver gradients or provide efficient mixing at that flow rate. The simplest solution to this problem has been to "flow split" the eluent from an HPLC system that pumps at 5 to 50 nL/min down to 100 to 200 nl/min. This is typically achieved by adding a tee after the HPLC system and by having one of the two exits going to waste while the other exit is connected to the microcolumn. The split ratio is obtained by carefully gauging the tubing at both exits. The advantage of this method is that it can be easily adapted to conventional HPLC. Its disadvantage is that the split ratio is affected by the back pressure on the microcolumn. Therefore, as the column ages and the back pressure increases, the split ratio changes. Regardless of the disadvantages, the splitting method is still the preferred approach for the generation of nanoliter per minute gradients. In fact today's commercially available nanoflow HPLCs are based on the splitting principle.

High-performance liquid chromatography systems can be divided into two broad types: low-pressure mixing systems that mix the solvents prior to a single pump and high-pressure mixing systems that mix the solvents in a static or dynamic mixer placed after the pumps, or they can mix the solvents directly into the pump head. For both systems, the important parameters for μ-HPLC-MS/MS are the delay time in the gradient, the prepressurization performance of the pump, and their performance at lower pressures.

The low-pressure mixing systems can have low void volumes following the pump because no mixers are placed after the pump. Therefore, flow splitting can be performed right after the pump, reducing the dead time of the experiment; however, low-pressure mixing systems have the disadvantage of obtaining the solvent ratios by mechanically proportioning the solvents. This happens either by switching valves or by proportional valves, using the vacuum action of the pump refill stroke. These approaches can perform poorly for low and high solvent ratios, generating less reproducible gradients.

The more common HPLC systems are the ones that perform high-pressure mixing of solvents and have the mixer placed after the pump or directly in the pump head. In these systems, the pumps are driven at a few microliters per minute, and the flow splitting required to achieve 100 to 200 nL/min occurs after the mixer and the pressure damper. Generally, the void volume is fairly large from the pumps to the column. For example, a void volume of 100 μL is not unusual and can represent 20 min delay in the gradient, assuming a flow rate of 5 μL/min. Furthermore, the back pressure can be as little as 100 to 1000 psi, which can cause some check valves used in HPLC to fail to operate properly.

Coupling to the Electrospray Mass Spectrometers. In recent years three approaches have emerged for the coupling of μ-HPLC columns to ESI-MS. The first approach consists of establishing an electrical contact directly at the end of the electrospray needle. This is typically achieved by gold coating a part of the tip. This allows the direct application of a high-voltage potential to the tip of the column, which is necessary for the generation of the ESI process. The direct application of the high voltage generates, by far, the most stable electrospray process; however, the thin gold layer applied to the tip will be removed over time, and the electrical connection can be lost.

Alternatively, the high voltage can be indirectly applied to the electrospray needle by using a liquid junction before the needle. This is typically achieved using an ultra-low-dead-volume connector placed between the transfer line and the electrospray needle. This approach has the advantage of being more robust in terms of maintaining the electrical connection; however, it has the disadvantage of generating a less stable, although still adequate, electrospray process.

The last approach is a combination of the previous two approaches, and it consists of using a fully external gold-coated needle coupled to a ultra-low-dead-volume connector to which a high voltage is applied. In this fashion, the electric connection is always maintained while an optimum electrospray process is obtained.

The electrical liquid junction in commercial connections or home-made connections can cause debilitating problems when used improperly. It is important to understand that electrolysis does occur in the liquid junction, generating a small amount of gas. Generally, the linear flow is strong enough to carry the gas forward with no formation of bubbles; however, in improperly fitted connections, the linear flow could be reduced when entering the liquid junction to a level that allows the formation of gas bubbles.

Application to Protein Identification Based on Peptide Mass Spectrometry. Once all the elements of the system are in place, the protein identification experiment can be performed. In a typical experiment, a few microliters of the protein digest is pressure loaded onto the μ-HPLC column. Then, the HPLC system delivers a solvent gradient of increasing hydrophobicity to the column. The peptides are eluted in order of hydrophobicity, transferred to the microelectrospray interface, and then moved to the MS. The eluting analytes successively trigger the MS to select one of the analytes and to generate the MS/MS spectra. The MS/MS spectra generated for the analytes are then used to search protein databases.

This method has established itself over the years as the method of choice for protein identification by mass spectrometry. Although, the continuous infusion by nanoelectrospray has been shown to provide exquisite sensitivity, it still remains a slow and manual technique. HPLC offers the advantages of being rapid and automatable, while still providing comparable limits of detection. The most popular combination has been to couple μ-HPLCs with ion trap mass spectrometers. Figure 1.17 shows the analysis of 50 fmol of bovine serum albumin (BSA) on such a system. The analysis was completed in about 35 min. The insert shows the typical quality of fragmentation patterns that can be obtained at that level of protein. It has been reported that approximately 100 fmol of protein present on the gel can be analyzed with this system.

Two-Dimensional HPLC-MS/MS Techniques. One-dimensional HPLC-MS/MS does not provide a high enough resolution for complex samples. Although longer gradients can be performed, one-dimensional HPLC has an inherent separation limitation. Historically, it has not been a significant problem because the complexity of the samples was always reduced through 1D and 2D gel electrophoresis. The limitations, however, of gel separation have pushed research toward novel two-dimensional separation systems.

Jorgenson's group has been pioneering the multidimensional separation from the analytical point of view (Hooker and Jorgenson, 1997; Lewis et al., 1997; Moseley et al., 1989; Opiteck et al., 1997, 1998). Techniques have been recently developed for online multidimensional chromatographic separation of peptides and their subsequent analysis by mass spectrometry. In particular, multidimensional chromatographic approaches, based on ion exchange chromatography, have been coupled to reverse-phase chromatography online with an ESI mass spectrometer for protein identification (Washburn et al., 2001). This system allows the analysis of complex peptide mixtures, such as the mixtures obtained by proteolytic digestion of whole-cell lysates.

Setup. A bed of reverse-phase material is packed in a capillary tubing followed by a second packing of a bed of ion exchange material. The end result is a column con-

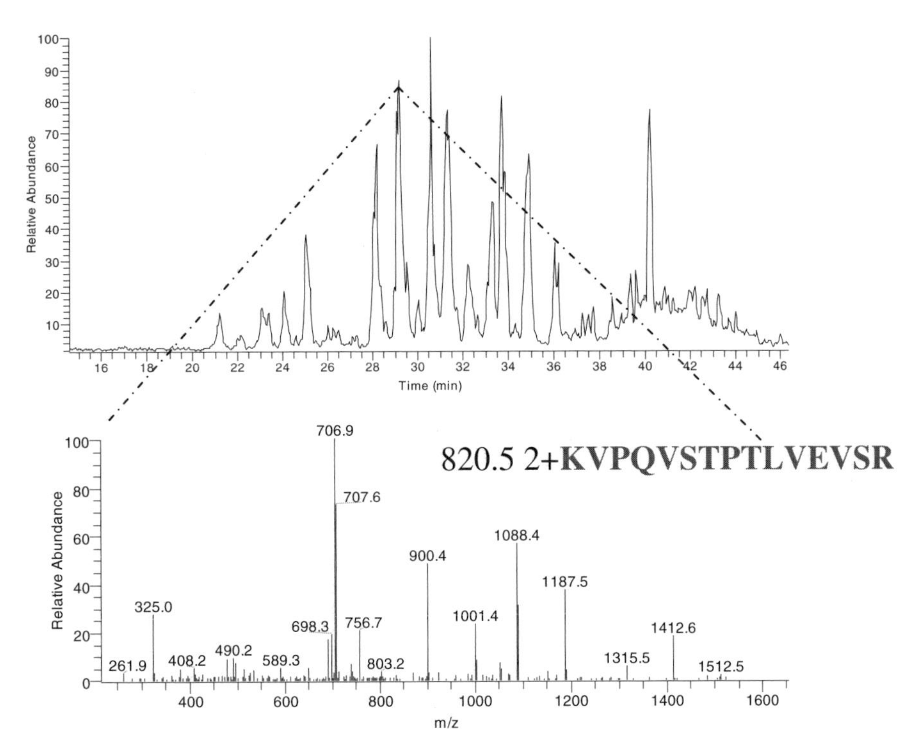

Figure 1.17. Example of the analysis of protein digest by HPLC-ESI-MS/MS. (*a*) Base peak chromatogram for 50 fmol of BSA tryptic digest analyzed by HPLC-ESI-MS/MS on an electrospray LCQ ion trap mass spectrometer using a 75 μm i.d. × 360 μm o.d. × 5 cm long C18 packed needle column. (*b*) The inserts show an MS/MS spectra automatically generated for *m/z* = 820.5 of charge 2+. Sufficient MS/MS spectra were generated to identify the protein by searching protein sequence databases.

sisting of two beds of different materials. This system is then used to address the complexity of large-scale protein digests, such as the ones obtained from the proteolytic digestion of whole-cell lysates.

The peptide mixture is then loaded on the dual-mode column and accumulates on the ion exchange resin. Groups of peptides are then subsequently eluted off the ion exchange resin by a ionic step gradient. In reality, the ion exchange material is used only for the extraction of the peptides, and it does not provide efficient separation; however, during each step of the ionic gradient, the eluted peptides are captured at the head of the reverse-phase bed. Each step in the ionic gradient is followed by a rinse and an organic-phase gradient to perform the reverse-phase separation of the peptides. The peptides are then separated on the reverse-phase column and eluted to the ESI mass spectrometer. The mass spectrometer is used to detect the peptides and perform collision-induced dissociation of selected peptides.

Microfabricated Devices Coupled to Mass Spectrometer. The application of microfabricated devices has dramatically increased in the 1990s, and it has grown from an esoteric technique only applied in a few laboratories to a recognized technique. Its application in the field of proteomics is more recent because of the difficulty to couple microfabricated devices to an external detector such as a mass spectrometer. In this section, we will describe some of the microfabricated systems that have been used for the identification of protein by mass spectrometry. It is, however, important to remember that all these systems are far from being robust and still have some serious hurdles to overcome. Therefore, at this point, use the information provided below as an eye opener to a novel technology that might, one day, be a strong platform for proteomics.

Microfabricated devices are typically made by photolithographic techniques to create various patterns of reservoirs, channels, and reaction chambers on a planar substrate such as glass or polymer. Then a second plate, with access ports for the reservoirs, is bonded to seal the channels. These devices were originally designed to have in situ detection of the analytes. The analysis, however, by mass spectrometry requires the transfer of the analytes to the mass spectrometer. In the case of ESI mass spectrometers the standard microelectrospray interface can be used for the transfer; however, the interface still needs to be connected to the microfabricated device through a transfer capillary. This is still the most challenging aspect of interfacing microfabricated devices to MS analyzers. Various ways of making the connectivity have been published; however, they are mostly incompatible with the mass production of the devices.

Continuous Infusion of Peptides for Protein Identification by Mass Spectrometry. The simplest application of microfabrication is the continuous infusion of peptide digests to a mass spectrometer and the fragmentation of the individual peptides. For example, Figeys et al. (1997) performed protein identification via infusion from a microfabricated device into an MS. In their design, a 12-cm pump capillary was connected to a simple three-reservoir device and to a micro-ESI source. The device contained two independent sample reservoirs and a buffer reservoir. The sample mobilization and their direction were done by controlling the voltage at each of the three reservoirs. In this fashion, one sample can be mobilized toward the MS, while the other sample is retained in its reservoir through an unfavorable electric field. The mass spectrometer is then used to provide MS/MS analysis of the eluting peptides. They demonstrated that reasonable levels of standard peptides (33 fmol/μL) can be analyzed through this system while retaining a high signal-to-noise ratio.

In a subsequent implementation, Figeys et al. (1998b) put into practice a more complex nine-reservoir device with computer-controlled voltage applied to each reservoir (Fig. 1.18). Furthermore, a software was developed to control the chip device and to trigger the mass spectrometer for automated acquisition of the peptide MS/MS. The only user intervention that was required was the manual pipetting of samples into their individual reservoirs on the microfabricated device before triggering the software. The system sequentially mobilized the sample using directed electroosmotic pumping, while the MS software controlled the data acquisition and subsequent database searching. The system was successfully used to identify yeast proteins previously separated by 2D gel electrophoresis.

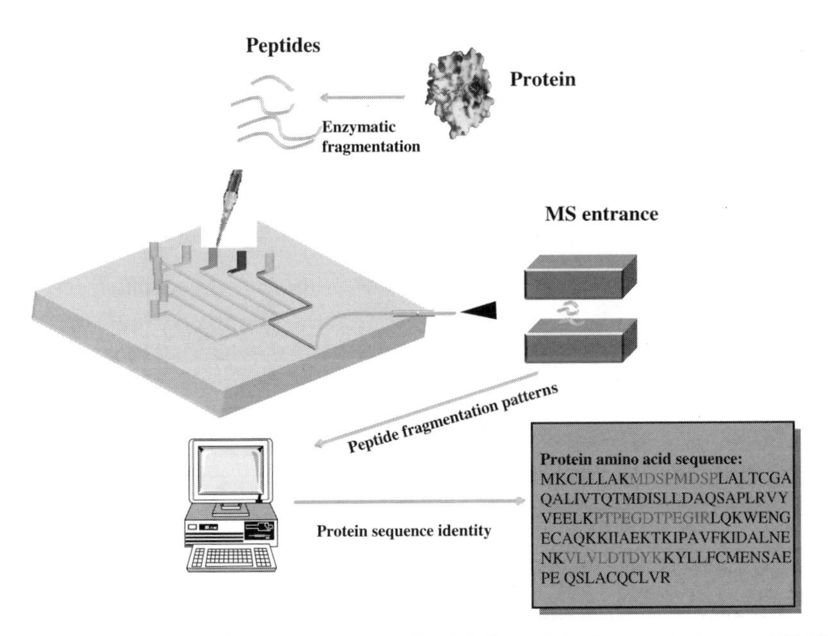

Figure 1.18. Schematic of a nine-position microfabricated device coupled to an ESI-MS/MS mass spectrometer for the automated analysis of protein digests. The protein digests of interest are pipetted into the different reservoirs. Then, the samples are successively mobilized by the application of a directed electric field from their specific reservoir toward the mass spectrometer where the peptides are analyzed by MS and MS/MS. The analysis proceeds until all the samples have been successively analyzed. Reproduced with permission from Figeys et al. (1998b).

Separation of Peptides on Microfabricated Devices Coupled to Mass Spectrometer. Electrophoresis has been the method of choice for the separation of analytes on chips. Furthermore, it is also compatible with the microelectrospray interface. The same limitations, however, that were present in CZE are also present in the microfluidic system. Therefore, discontinuous methods such as sample stacking, isotachophoresis, and solid-phase extraction (SPE) are required to increase the amount of sample injected on the microfabricated device.

Electrophoretic separations with offline sample concentration (Zhang et al., 1999) and online sample concentration (Figeys et al., 1998a; Li et al., 1999; Wang et al., 2000) have been reported. In particular, a microfabricated system for the separation of peptides by electrophoresis was coupled to a mass spectrometer by Li et al. (2000). In this approach, the protein digests of interest are introduced one at a time on the microfabricated device. An electric field is applied to fill an injection cross with the sample. Once the cross has been filled, an electric field is applied from the buffer reservoir to the microelectrospray needle driving the separation of analytes toward a microelectrospray interface. Rapid separation of protein digests has been obtained through this approach; however, the limit of detection for real samples still needs to be improved (Fig. 1.19).

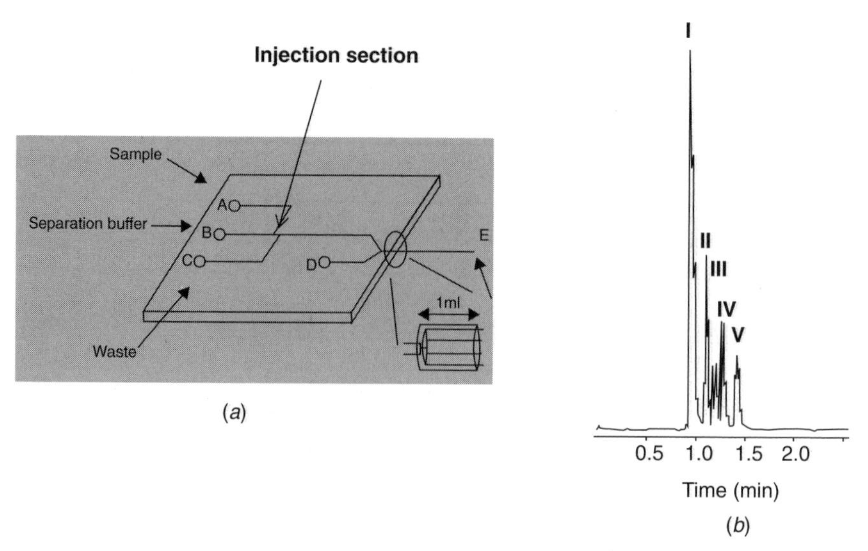

Figure 1.19. Schematic of a four-position device coupled to an ESI-MS/MS mass spectrometer for the separation of peptide mixtures and their analysis by mass spectrometetry. (*a*) Schematic of the device coupled to a triple quadrupole mass spectrometer. A portion of the sample is mobilized from the sample reservoir toward the waste, filling a small injection line. Once the injection section is filled, a potential is applied between reservoir B and the nanospray emitter driving the separation and the electrosrpay process. (*b*) Example of a separation for a peptide mixture in less than 2 min. Separation of leu-enkephalin (I), LHRH (II), somatostatin (III), angiotensin II (IV), and bradykinin (V) all at 10 μg/mL. Reproduced with permission from Li et al. (2000).

The development in the field of microfluidics coupled to mass spectrometry has been tremendous in the late 1990s. There are still, however, some serious challenges to the technology that still need to be addressed before it can become routinely applicable in a robust format.

Bioinformatics for the Identification of Proteins Based on MS/MS

All the ESI-based tandem mass spectrometers are generally used to perform MS and MS/MS spectra. Although the mass accuracy, resolution, sensitivity, and quality of the fragmentation patterns are different, they contain information that can be used to identify a peptide provenance by protein/DNA database searching. The MS/MS spectra contain the fragmentation patterns related to the amino acid sequence of specific peptides. The analysis of MS/MS spectra is more intensive than the interpretation of MS data. The approaches that are used for the interpretation of these spectra can be classified into three subgroups according to the level of user intervention required.

No Interpretation. Clearly one would like to be able to identify proteins without having to do any interpretation of the MS/MS spectra. For high-throughput analysis, this becomes essential. A few algorithms have been developed to search protein/DNA databases with uninterpreted MS/MS spectra. They all have in common the requirement for the partial or full sequence of the protein to be already known and included in a database. All of the software packages provide a list of possible matches between individual MS/MS spectrum and peptide sequences obtained from the database; however, the scoring algorithms used to determine the validity of the matches are very different. These algorithms are explained in more detailed below.

Mascot by Matrix Sciences (*www.matrixscience.com*) can be freely accessed over the web (for noncommercial entities) (Pappin et al., 1993). It was built based on the MOWSE scoring algorithm from Papin. The MOWSE algorithm computes a matrix of probability in which each row represents an interval of 100 Da in peptide mass, and each column represents an interval of 10 kDa in intact protein mass. As each sequence entry is processed, the appropriate matrix elements are increased to accumulate statistics on the size distribution of peptide masses as a function of the protein mass. Therefore, this matrix represents the peptide distribution by protein MW present in the database. The matrix will change as the database increases. From this matrix, a score is ascribed based on the measured masses. The same algorithm is used for peptide MS/MS. In Mascot they have further incorporated a probability based on the probability that a match is a random event. This allows the probabilistic evaluation and ordering of all the potential matches, as well as a probabilistic evaluation of the best match.

ProteinProspector from UCSF (*http://prospector.ucsf.edu/*) is also an example of a web-based MS/MS search engine. The identification of the protein is typically unambiguous, achieved by the number of peptides that matches to the same protein.

Another algorithm that is popular is Sequest (Eng et al., 1994; McCormack et al., 1997; Yates et al., 1995). Sequest is not freely accessible over the web and must be purchased. This algorithm performs a two-pass search of the database. In the first pass, for every MS/MS spectra submitted, Sequest searches protein/DNA databases for the top 500 isobaric peptides using some of the information on immonium ions. In the second pass, the 500 predicted spectra corresponding to the isobaric peptides are generated and are rapidly matched against the measured spectra. This is achieved by multiplying in the frequency domain of the fast-Fourier transformation, the measured spectrum with each individual predicted spectrum for the 500 isobaric peptides, and by pairwise multiplication of the measured and predicted transforms.

The multiplication in the Fourier domain is the same as doing a convolution in the time domain. This means that this approach verifies how well each predicted spectrum matches to the measured spectrum. Correlation parameters that indicate the quality of the match between predicted and measured spectra are then deduced from the results of the transform multiplication. A high cross-correlation indicates a good match between the predicted spectrum and the measured spectrum. More importantly, the cross-correlation value is independent of the size and nature of the database. Therefore, the same peptide sequence found in a small database or a large database will return the same correlation coefficient when matched against the measured spectrum. This is an important aspect of Sequest that makes it unique. All the other approaches provide scoring schemes that are very dependent on the size and nature of the database.

Furthermore, although protein identification has been performed with as little as one peptide using this algorithm, unambiguous identification of the provenance of a protein is often achieved by the multitude of peptides that matches with the same entry in a database. The Sequest software is computing intensive, and for high-throughput demand, it can rapidly paralyze the best dual-CPU server. In reality, the slowness of Sequest is due to the recurrent scans of the selected database to find the top 500 isobaric peptides. The larger the database, the longer it takes to scan the databases. An improved version of the software, called Turbo-Sequest, predigests and orders the databases and has greatly improved the search speed.

Sonars by Proteometrics can search uninterpreted MS/MS spectra of peptides against protein and DNA databases (Field et al., 2002). This software uses Bayesian theory to rank the protein sequences in a database by their probability of occurrence. It also provides a scoring scheme for the differentiation of the match. In our hands, it has proven to be the most rapid system for the analysis of peptide digests. X!Tandem (*http://www.proteome.ca/x-bang/tandem/tandem.html*) (Graig and Beaivis, 2004).

Partial Interpretation. Although the fully automated approaches are favorable for automation, historically the computing power was not available for the rapid identification of proteins.

Another subgroup of database search engines, based on the partial interpretation of the MS/MS spectra, were developed to perform faster searches of databases, while requiring human intervention. The most popular partial interpretation of MS/MS spectra is the "sequence-tag" approach (Wilkins et al., 1996a; Mann and Wilm, 1994) (Fig. 1.20). It consists of reading the mass spacing between specific fragments of an MS/MS spectrum. This allows the generation of a short peptide sequence (tag). The tag is then used to pinpoint the possible peptide sequences from isobaric peptides in databases, while the residual mass information before and after the tag confirms one or a few peptide matches. Every MS/MS spectrum requires the generation of a tag followed by database searching. Unambiguous identification of the protein is established by the multitude of peptides that matches to the same protein.

De novo Sequencing. Finally, the last option is the full interpretation of the MS/MS spectra, often called de novo sequencing (Papayannopoulos, 1995; Shevchenko et al., 1997). Obviously, the other automated and semiautomated approaches are used prior to this approach. Although many genomes have been recently sequenced, the genomic data only represent a fraction of the world's genomic pool, and often no databases are available for specific organisms. The requirements for de novo sequencing are more stringent. The MS/MS spectra must be of good quality in terms of intensity and coverage of the peptide sequence. The MS/MS spectra of peptides contain ladder-type information, which, in principle, indicates their amino acid sequence. Experienced mass spectrometrists can manually extract the peptide sequence from the MS/MS spectra.

In large-scale proteomic studies, the throughput of analysis is a critical factor. Therefore, once enough MS/MS spectra have been generated to unambiguously identify a protein, generating MS/MS spectra on the residual peptides can be a waste of time. In some cases, however, like when dealing with expressed sequence tag (EST) or protein mixtures, it is important to increase the number of MS/MS generated.

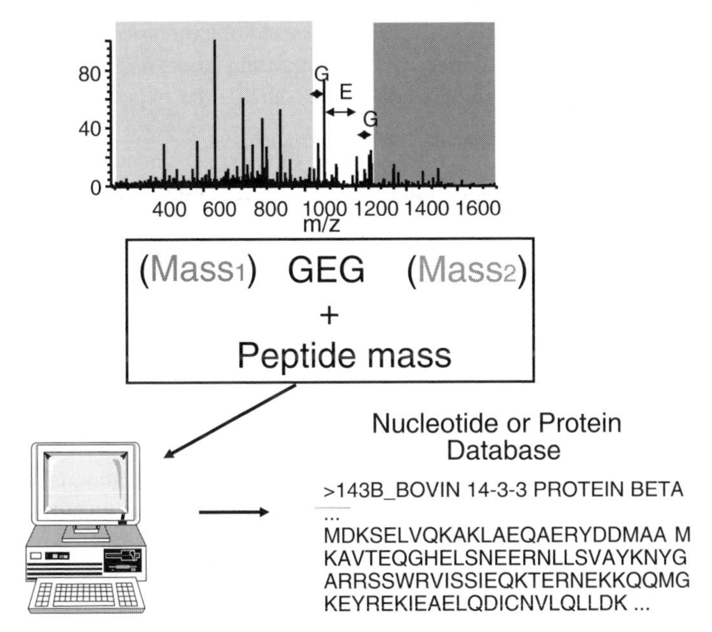

Figure 1.20. Schematic of protein identification by sequence tag. A small stretch of an individual peptide sequence is read from the MS/MS spectrum. The peptide mass, the short amino acid sequence, and the residual masses before and after the sequence are used to search protein/DNA databases.

CURRENT CHALLENGES IN PROTEOMICS

Classical Proteomics

There has been a lot of confusion in the field of proteomics on the issue of the complete proteome coverage and low-abundance proteins. This confusion is, in part, due to a bit of propaganda about the power of the technology, as well as some confusion related to the definition of *proteome* and *proteomics*. The idea of the comprehensive study of the proteome is more an idealism than a reality.

Also, 2D gel-based proteomics is still facing a serious number of challenges. In particular, although 2D gel electrophoresis is a powerful technique, it still lacks the capability for discovering hydrophobic proteins, basic proteins, and low-abundance proteins.

The presence or absence of low-abundance proteins on 2D gels has been controversial. Fortunately, for *S. cerevisiae* and a few other species, indexes can be calculated for all the ORF entries in their corresponding database that reflects the abundance at the protein level. One such index is the codon bias index. The codon bias index is calculated based on a compiled index for codon usage in messenger RNA (mRNA). The codon usage is typically measured for a few ORF that are translated into highly expressed proteins. The codon bias index is thought to be a good indication of the expression levels of proteins. Codons that correspond to low-level transfer RNAs (tRNAs) during translation decrease the potential yield of expression; on the other hand, codons that utilize higher abundance tRNA during translation will increase the potential yield of the protein expression. Figure 1.21*a* shows the expected range of codon bias in yeast calculated using the predicted ORF. Clearly, a good portion of the yeast proteome is expected to fall in the low codon bias (<0.1) and therefore, has a low expression. Figure 1.21*b* shows a compilation of the codon bias index for the proteins that have been identified by 2D gel electrophoresis of *S. cerevisiae* [compiled from Gygi et al. (1999b) and Perrot et al. (1999)]. At a first glance, low-abundance proteins appear to be detected; however, if we only plot the proteins that were identified by mass spectrometry, the result is different. Furthermore, some of the reported proteins with 0.1 codon biases are suspicious because they represent proteins that were present at the picomole level on the yeast 2D gel, and this level is far from being low abundance. Therefore, they probably represent erroneous identifications (Fig. 1.21*c*). The lack of identification of low-abundance proteins has been reported to be caused by a limited load capacity on 2D gel electrophoresis. In order to see low-abundance proteins, the amount of sample that needs to be loaded exceeds the capacity of 2D gel electrophoresis.

Hydrophobic proteins have also been a challenge for proteomics. Hydrophobic proteins have the tendency to precipitate once they reach their pI during the electrofocusing on an IPG, and therefore they do not transfer to the second dimension. Although, efforts have been made to improve the solubilization of hydrophobic proteins (Rabilloud, 1999; Santoni et al., 1999), the display of hydrophobic proteins by 2D gel electrophoresis still remains a challenge.

The efficient gel digestion of low-level proteins can, as well, be a challenge for proteomics. The kinetic of digestion is dependent on the concentration of the substrate (protein to be digested). Therefore, as the concentration of substrate decreases, the efficiency of the digestion also decreases. Furthermore, sample loss is also a critical factor for low-level proteins.

Gel-Free Analysis: An Alternative Differential Display

The limitations of 2D gel electrophoresis can be alleviated with the gel-free analysis of complex protein mixtures. The analysis of complex proteolytic mixtures is now possible due to the improvement of online peptide separation and the improvement of mass spectrometry software. In this approach, the lysate of interest is directly digested to provide a complex mixture of proteins. This offers the advantage of reducing sample lost and contamination, therefore expanding the dynamic range of protein analysis toward lower abundance proteins. Characteristic hydrophilic peptides can also be found from proteins that have a sequence in preponderance of hydrophobic amino acids. The

Figure 1.21. Codon bias and the yeast proteome. (*a*) Distribution of codon bias calculated for all the yeast ORF. (*b*) Compilation of the codon bias for the proteins that have been observed on 2D gel electropherograms of yeast. (*c*) Compilation of the codon bias for the proteins identified from a 2D gel electropherogram of yeast using only mass spectrometry.

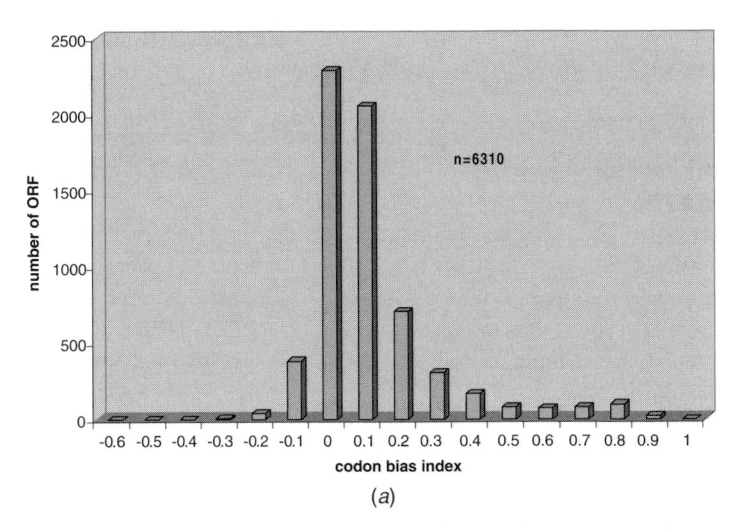
2500
2000
1500
1000
500
0
number of ORF
n=6310
-0.6 -0.5 -0.4 -0.3 -0.2 -0.1 0 0.1 0.2 0.3 0.4 0.5 0.6 0.7 0.8 0.9 1
codon bias index

(*a*)

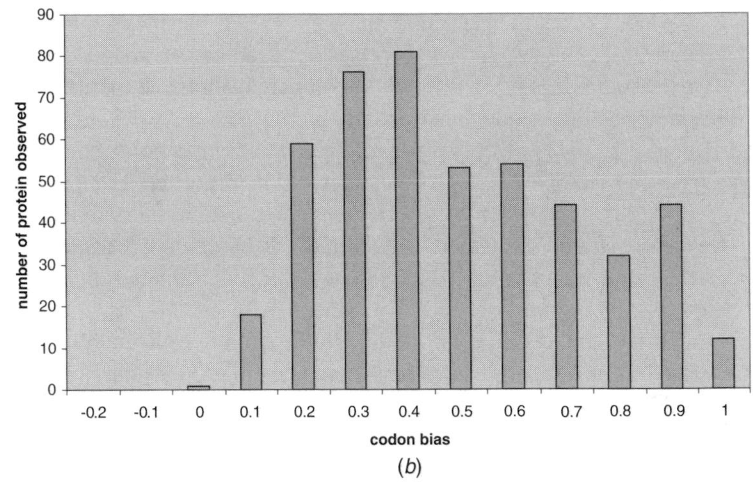
90
80
70
60
50
40
30
20
10
0
number of protein observed
-0.2 -0.1 0 0.1 0.2 0.3 0.4 0.5 0.6 0.7 0.8 0.9 1
codon bias

(*b*)

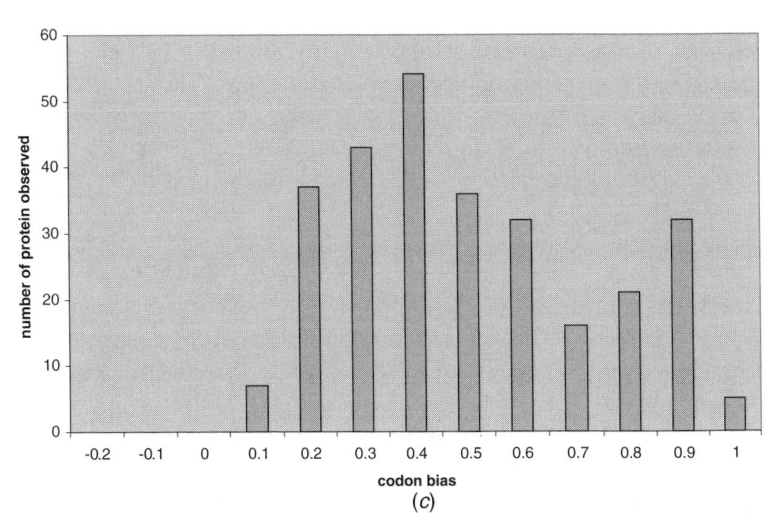
60
50
40
30
20
10
0
number of protein observed
-0.2 -0.1 0 0.1 0.2 0.3 0.4 0.5 0.6 0.7 0.8 0.9 1
codon bias

(*c*)

complex mixture of proteins is then separated online either by one- or two-dimensional HPLC and ESI mass spectrometry.

Quantitation. The relative quantitation of the proteins that are differently expressed between two sets of proteins is an important aspect of proteomics. The gel-based separation of proteins provides direct visualization and relative quantitation of separated proteins based on the staining intensities on the gel. Therefore, choosing the proteins of interest is as simple as comparing two lanes, or two gels, and looking for changes in staining intensity.

The gel-free analysis of proteins is typically a dynamic process, which also requires the relative quantitation of peptides. No staining is performed in gel-free approaches, and the quantitation relies on the mass spectrometric signal. Mass spectrometers have been used for many years to do quantitative studies; however, this is always done by establishing response curves for particular analytes. This is hardly possible for a constantly changing set of analytes, such as peptides obtained from a proteolytic digestion of proteins. Furthermore, it is often difficult to compare even two experiments run one after the other in terms of peptide signal intensity.

The differential analysis of complex protein mixtures in a gel-free approach can be achieved by using isotope tagging of peptides to allow a direct comparison of the changes in peptide levels between two proteomes in a single experiment. Gygi et al. (1999a) have demonstrated a method called ICAT (isotope-coded affinity tag) that allows the differential quantitation of proteomes using cystein-containing peptides. Munchbach et al. (2000) proposed a more general way for N-terminal labeling of peptides for the differential quantitation. Although the method by Munchbach et al. was only demonstrated for gel-separated proteins, it appears that the reaction would just be as valid for gel-free analysis of proteins.

Briefly, the ICAT (Gygi et al., 1999a) method consists of labeling cystein-containing peptides from different samples with a light and heavy form of a reactive chemical. The ICAT reagent consists of a biotin group followed by a linker and is terminated with a cystein reactive group. The only difference between the light and the heavy tag is the presence of hydrogen or deuterium. This results in a similar response at the mass spectrometer while a diagnostic m/z spacing occurs on the MS spectrum. Furthermore, because the heavy/light peptides are quasi co-elute and are identical in sequence, their relative quantities can be obtained by the ratio of their intensity.

Typically, in the ICAT approach, a lysate sample first is reduced and reacted with the light form of the ICAT reagent. A second lysate sample is also reduced and reacted with the heavy form of the ICAT reagent. Both ICAT reagents preferentially react at the sulfhydryl group present on the cysteins. Once both labeling reactions have been performed, the two protein lysates are combined and enzymatically cleaved to generate peptide fragments. Then, the cystein-containing peptides are purified using a monomeric avidin column. The purified cystein-containing peptides are separated on a nanoflow HPLC system, online with an ESI mass spectrometer. Pairs of peptides with identical amino acid sequences that are tagged, respectively, with the light and heavy form of the ICAT are different in mass by 8 mass units. These are easily differentiated on a mass spectrometer and eluted with just a slight delay from each other. Furthermore, the automated generation of MS/MS fragmentation patterns of the peptides identifies the peptide sequence and its protein provenance by database searching. The relative quantitation is extracted from the MS spectrum using the ratio of the signal intensity observed for the light/heavy labeled peptides.

Munchbach et al. (2000) have introduced a more generic approach for the labeling of peptides at their N-termius using 1-(nicotinoyloxy)succinimide esters. They have synthesized a heavier form of the 1-(nicotinoyloxy)succinimide ester and introduced four deuteriums, which provide a 4 mass unit difference with the normal 1-(nicotinoyloxy)succinimide ester. The light/heavy pair of chemicals can be used for the differential labeling of proteins and their relative quantitation. Briefly, both samples are enzymatically digested to generate peptide fragments. The peptide fragments are then succinylated to block the ε-amino group of lysines. Then the first sample is reacted with a light form of 1-(nicotinoyloxy)succinimide ester, and the second sample is labeled with the heavy form of the 1-(nicotinoyloxy)succinimide ester. The two samples are then combined and analyzed by MALDI-TOF mass spectrometry.

It is to be expected that many more pairs of chemicals will become available for the relative quantitation of proteomes. It is important to realize that the expression profile of a proteome covers a few orders of magnitude. All the chemical tagging approaches have a defined dynamic range. Therefore, it is important to keep in mind that for low levels of proteins, these reactions can fail to provide linear responses. Therefore, although both of the labeled heavy and light forms of the peptide might be observed, the calculated relative quantitation for the low-level proteins might be inaccurate. Side reactions can also provide significant challenges.

Fourier Transform Mass Spectrometery for Gel-Free Analysis and Single-Peptide-Based Identification

The FTMS has attracted renewed attention from the proteomic community. The instrumentation is now more robust, it offers fragmentation, it has a larger dynamic range, an improved sensitivity, and it has recently been coupled to MALDI. The FTMS is by far the best mass spectrometer in terms of mass accuracy, resolution, and sensitivity. Peptide masses can be measured down to 1 ppm or less, depending on the sample. We have come to appreciate the advantages that accurate mass measurement offers in terms of protein identification.

Depending on the size of the genome, it is possible to unambiguously identify a protein solely based on the accurate mass measurement of one of its peptides (mass tag). For example, Figure 1.22*a* shows the plot of the number of isobaric peptides versus the peptide masses for a 10 ppm tolerance and a 1 ppm tolerance for the yeast proteome. This was predicted for all of the known yeast open-reading frames. Figure 1.22*b* also shows the percentage of unique peptides versus the peptide masses. It is clear that for small and medium genomes, protein identification based on a single peptide seems to be a possibility. It is also clear that the homology level is low in yeast. We have performed an $N \times N$ blast of the yeast genome and obtained about 20% of the protein in yeast with at least one homolog at an homology level of 80%. We expect this number to be significantly larger for more complex organisms. Furthermore, the idea of identifying a protein solely based on a single peptide does not take into account splicing

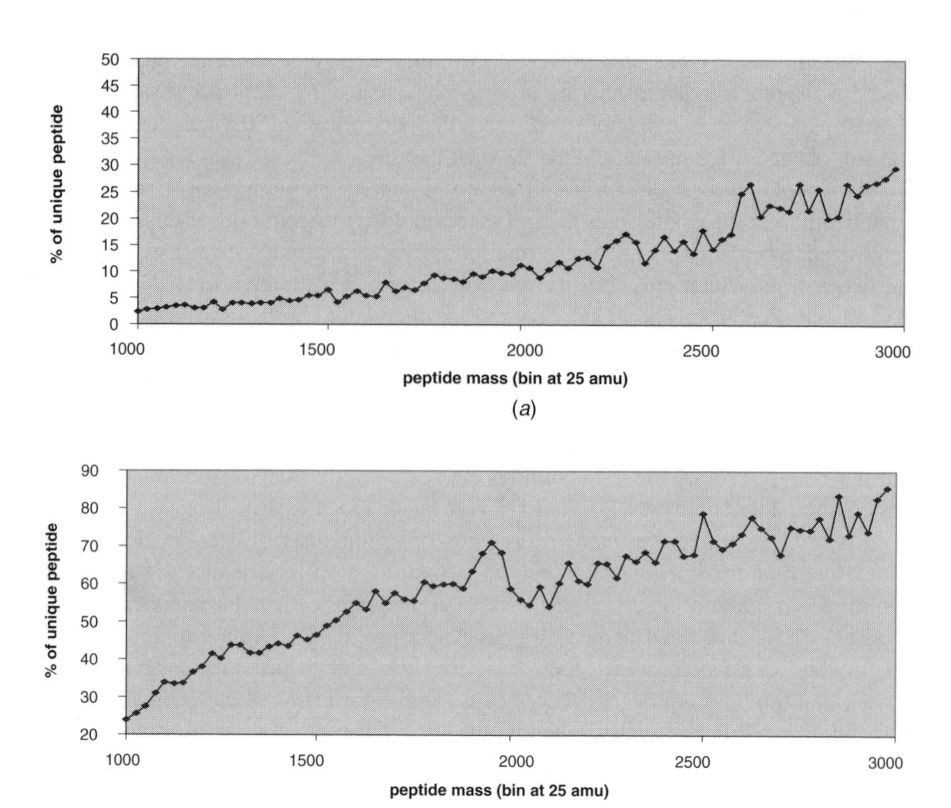

Figure 1.22. Effect of mass accuracy observed on the mass spectrometer on the probability of unique identification.

and point mutation. Also, it is one thing to say that a specific peptide is unique to a protein; however, it is another thing to find it.

One of the possible applications of this approach is the identification of protein contained in complex mixtures in a gel-free manner based on HPLC-ESI-MS/MS. Smith et al. have demonstrated the potential of the protein identification based on a single-peptide analysis by ESI-FTMS. Furthermore, they have also demonstrated that the addition of the ICAT reagent improves identification of peptides by restricting the search to the cystein-containing peptides. This also means that proteins can be identified in complex mixtures, based only on the presence of one of their peptides. It is not clear yet if this approach will be useful for larger genomes, such as the human genome.

REFERENCES

Adessi, C., Miege, C., Albrieux, C., and Rabilloud, T. (1997). Two-dimensional electrophoresis of membrane proteins: A current challenge for immobilized pH gradients. *Electrophoresis* **18**(1):127–135.

Aebersold, R., and Mann, M.(2003). Mass spectrometry-based proteomics. *Nature* **422**:198–207.

Aebersold, R., Leavitt, J., Saavedra, R., Hood, L., and Kent, S. (1987). Internal amino acid sequence analysis of proteins separated by one- or two-dimensional gel electrophoresis after in situ protease digestion on nitrocellulose. *Proc. Natl Acad. Sci. USA* **84**(20):6970–6974.

Aebersold, R., Teplow, D., Hood, L., and Kent, S. (1986). Electroblotting onto activated glass. High efficiency preparation of proteins from analytical sodium dodecyl sulfate-polyacrylamide gels for direct sequence analysis. *J. Biol. Chem.* **261**(9):4229–4238.

Anderson, L., and Seilhamer, J. (1997). A comparison of selected mRNA and protein abundances in human liver. *Electrophoresis* **18**:533–537.

Banks, R., Dunn, M., Forbes, M., Stanley, A., Pappin, D., Naven, T., Gough, M., Harnden, P., and Selby, P. (1999). The potential use of laser capture microdissection to selectively obtain distinct populations of cells for proteomic analysis—preliminary findings. *Electrophoresis* **20**(4–5):689–700.

Bartel, P.L., Roecklein, J.A., SenGupta, D.J., and Fields, S. (1996). A protein linkage map of *Escherichia coli* bacteriophage T7. *Nat. Genet.* **12**:72–77.

Blanchard, A. (1998). Synthetic DNA arrays. *Genet. Eng. (N Y)* **20**:111–23.

Blisnick, T., Morales-Betoulle, M.E., Vuillard, L., Rabilloud, T., and Braun Breton, C. (1998). Non-detergent sulphobetaines enhance the recovery of membrane and/or cytoskeleton-associated proteins and active proteases from erythrocytes infected by plasmodium falciparum. *Eur. J. Biochem.* **252**(3):537–541.

Blum, H., Beier, H., and Gross, H.J. (1987). Improved silver staining of plant proteins, RNA and DNA in polyacrylamide gels. *Electrophoresis* **8**:93–99.

Brunet, S., Thibault, P., Gagnon, E., Kearney, P., Bergeron, J.J., and Desjardins, M. (2003). Organelle proteomics: Looking at less to see more. *Trends Cell Boil.* **13**:629–638.

Celis, J.E., Celis, P., Ostergaard, M., Basse, B., Lauridsen, J.B., Ratz, G., Rasmussen, H.H., Orntoft, T.F., Hein, B., Wolf, H., Celis, A. (1999). Proteomics and immunohistochemistry define some of the steps involved in the squamous differentiation of the bladder transitional epithelium: A novel strategy for idenfying metaplastic lesions. *Cancer Res.* **59**(12):3003–3009.

Chervet, J.P., Ursem, M., and Salzmann, J.B. (1996). Instrumental requirements for nanoscale liquid chromatography. *Anal. Chem.* **68**(9):1507–1512.

Chevallet, M., Santoni, V., Poinas, A., Rouquie, D., Fuchs, A., Kieffer, S., Rossignol, M., Lunardi, J., Garin, J., and Rabilloud, T. (1998). New zwitterionic detergents improve the analysis of membrane proteins by two-dimensional electrophoresis. *Electrophoresis* **19**(11):1901–1909.

Clarke, C., Titley, J., Davies, S., and O'Hare, M. (1994). An immunomagnetic separation method using superparamagnetic (Macs) beads for large-scale purification of human mammary luminal and myoepithelial cells. *Epithelial Cell Biol.* **3**(1):38–46.

Comisarow, M.B., and Marshall, A.G. (1974). Fourier transform ion cyclotron resonance spectroscopy. *Chem. Phys. Lett.* **25**:282–283.

Conrods, T.P., Anderson, G.A., Veenstra, T.D., Pasa-Tolic, L., and Smith, R.D. (2000). Utility of accurate mass tags for proteume-wide protein identification. *Anal. Chem.* **72**:3349–3354.

Craig, R., and Beavis, R.C. (2004) TANDEM: Matching proteins with tandem mass spectra. *Bioinformatics* Feb. 19 [Epub ahead of print].

Dai, Y., Whittal, R.M., and Li, L. (1999). Two-layer sample preparation: A method for MALDI-MS analysis of complex peptide and protein mixtures. *Anal. Chem.* **71**:1087–1091.

Davis, M.T., and Lee, T.D. (1998). Rapid protein identification using a microscale electrospray LC/MS system on an ion trap mass spectrometer. *J. Am. Soc. Mass Spectrom.* **9**(3):194–201.

Desprez, T., Amselem, J., Caboche, M., and Hofte, H. (1998). Differential gene expression in Arabidopsis monitored using CDNA arrays. *Plant J.* **14**(5):643–652.

Dovichi, N.J. (1997). DNA sequencing by capillary electrophoresis. *Electrophoresis* **18**(12–13): 2393–2399.

Dunn, M.J., and Corbett, J.M. (1996). In *Two-Dimensional Polyacrylamide Gel Electrophoresis. High Resolution Separation and Analysis of Biological Macromolecules. Part B, Applications.* Karger, B.L., and Hancook, W.S. (Eds.). San Diego, Academic Press, Vol. 271, pp. 177–203.

Edman, P., and Begg, G. (1967). A protein sequenator. *Eur. J. Biochem.* **1**(1):80–91.

Emmert-Buck, M.R., Gillespie, J.W., Paweletz, C.P., Ornstein, D.K., Basrur, V., Appella, E., Wang, Q.H., Huang, J., Hu, N., Taylor, P., Petricoin, E.F. 3rd (2000). An approach to proteomic analysis of human tumors. *Mol. Carcinog.* **27**(3):158–165.

Eng, J., McCormack, A.L., and Yates, J.R. III (1994). An approach to correlate tandem mass spectral data of peptides with amino acid sequences in a protein database. *J. Am. Soc. Mass Spectrom.* **5**:976–989.

Fenn, J.B., Mann, M., and Meng, C.K. (1990). Electrospray ionization—principles and practice. *Mass Spectro. Rev.* **9**(1):37.

Fialka, I., Pasquali, C., Lottspeich, F., Ahorn, H., and Huber, L.A. (1997). Subcellular fractionation of polarized epithelial cells and identification of organelle-specific proteins by two-dimensional gel electrophoresis. *Electrophoresis* **18**(14):2582–2590.

Fichmann, J. (1999). Advantages of immobilized pH gradients. *Methods Mol. Biol.* **112**:173–174.

Field, H.I., Fenyo, D., and Beavis, R.C. (2002). RADARS, a bioinformatics solution that automates proteome mass spectral analysis, optimises protein identification, and archives data in relational database. *Proteomics* **2**:36–47.

Figeys, D., Zhang, Y., and Aebersold, R. (1998a). Optimization of solid phase microextraction–capillary zone electrophoresis–mass spectrometry for high sensitivity protein identification. *Electrophoresis* **19**(13):2338–2347.

Figeys, D., Gygi, S., McKinnon, G., and Aebersold, R. (1998b). An integrated microfluidics–tandem mass spectrometry system for automated protein analysis. *Anal. Chem.* **70**(18):3728–3734.

Figeys, D., Ning, Y., and Aebersold, R. (1997). A microfabricated device for rapid protein identification by microelectrospray ion trap mass spectrometry. *Anal. Chem.* **69**(16):3153–3160.

Futcher, B., Latter, G.I., Monardo, P., McLaughlin, C.S., and Garrels, J.I. (1999). A sampling of the yeast proteome. *Mol. Cell. Biol.* **19**(11):7357–7368.

Garin, J., Diez, R., Kieffer, S., Demine, J.-F., Duclos, S., Gagnon, E., Sadoul, R., Rondeau, C., and Desjardsins, M. (2001). The phagosome proteome: Insight into phagosome functions, *J. Cell Biol.* **152**(1):165–180.

Ghaemmaghami, S., Huh, W.K., Bower, K., Howson, R.W., Belle, A., Dephoure, N., O'Shea, E.K., and Weissman, J.S. (2003). Global analysis of protein expression in year, *Nature* **425**:737–741.

Gibson, B.W., and Biemann, K. (1984). Strategy for the mass spectrometric verification and correction of the primary structures of proteins deduced from their DNA sequences. *Proc. Natl. Acad. Sci. U.S.A.* **81**:1956–1960.

Goldberg, M.E., Expert-Bezancon, N., Vuillard, L., and Rabilloud, T. (1996). Non-detergent sulphobetaines: A new class of molecules that facilitate in vitro protein renaturation. *Fold Des.* **1**(1):21–27.

Gomm, J.J., Browne, P., Coope, R., Liu, Q., Buluwela, L., and Coombes, R. (1995). Isolation of pure populations of epithelial and myoepithelial cells from the normal human mammary gland using immunomagnetic separation with dynabeads. *Anal. Biochem.* **226**(1):91–99.

Gorg, A. (1993). Two-dimensional electrophoresis with immobilized PH gradients: Current state. *Biochem. Soc. Trans.* **21**(1):130–132.

Gorg, A., Postel, W., and Gunther, S. (1988). The current state of two-dimensional electrophoresis with immobilized pH gradients. *Electrophoresis* **9**(9):531–546.

Gygi, S., and Aebersold, R. (1999). Absolute quantitation of 2-D protein spots. *Methods Mol. Biol.* **112**:417–421.

Gygi, S., Corthals, G., Zhang, Y., Rochon, Y., and Aebersold, R. (2000). Evaluation of two-dimensional gel electrophoresis-based proteome analysis technology. *Proc. Nat. Acad. Sci.* **97**(17):9390–9395.

Gygi, S.P., Rist, B., Gerber, S.A., Turecek, F., Gelb, M.H., and Aebersold, R. (1999a). Quantitative analysis of complex protein mixtures using isotope-coded affinity tags. *Nature Biotech.* **17**:994–997.

Gygi, S., Rochon, Y., Franza, B., and Aebersold, R. (1999b). Correlation between protein and mRNA abundance in yeast. *Mol. Cel. Biol.* **19**(3):1720–1730.

Himeda, C.L., Ranish, J.A., Angello, J.C., Maire, P., Aebersold, R., and Hauschka, S.D. (2004). Quantitative proteomic identification of six4 as the trex-binding factor in the muscle creatine kinase enhancer. *Mol. Cell Biol.* **24**:2132–2143.

Hooker, T.F., and Jorgenson, J.W. (1997). A transparent flow gating interface for the coupling of microcolumn LC with CZE in a comprehensive two-dimensional system. *Anal. Chem.* **69**:4134–4142.

Howell, K.E., Devaney, E., and Gruenberg, J. (1989). Subcellular fractionation of tissue culture cells. *Trends Biochem. Sci.* **14**(2):44–47.

Hunt, D.F., Yates, J.R., Shabanowitz, J., Winston, S., and Hauer, C.R. (1986). Protein sequencing by tandem mass spectrometry. *Proc. Natl. Acad. Sci. U.S.A.* **83**:6233–6237.

Hunt, D.F., Bone, W.M., Shabanowitz, J., Rhodes, J., and Ballard, J.M. (1981). Sequence analysis of oliyopeptides by second ion/collision activated dissociation mass sepectrometry. *Anal. Chem.* **53**:1704–1706.

Issaq, H.J., Conrads, T.P., Janini, G.M., and Veenstra, T.D. (2002). Methods for fractionation, separation and profiling of proteins and peptides. *Electrophoresis* **23**:3048–3061.

Jackson, P., Urwin, V.E., and Mackay, C.D. (1988). Rapid imaging, using a cooled charge-coupled-device, of fluorescent two-dimensional polyacrylamide gels produced by labelling proteins in the first-dimensional isoelectric focusing gel with the fluorophore 2-methoxy-2,4-diphenyl-3(2h)furanone. *Electrophoresis* **9**(7):330–339.

Jung, E., Heller, M., Sanchez, J.C., and Hochstrasser, D.F. (2000). Proteomics meets cell biology: The establishment of subcellular proteomes. *Electrophoresis* **21**(16):3369–3377.

Karas, M., and Hillenkamp, F. (1988). Laser desorption ionization of proteins with molecular masses exceding 10,000 daltons. *Anal. Chem.* **60**:2299–2301.

Kern, W., Kohlmann, A., Wuchte, C., Schnittger, S., Schoch, C., Mergenthaler, S., Ratei, R., Ludwig, W.D., Hiddemann, W., and Haferlach, T. (2003). Correlation of protein expression and gene expression in acute leukemia. *Cytometry* **55B**:29–36.

Klose, J., and Kobalz, U. (1995). "Two-dimensional electrophoresis of proteins: An updated protocol and implications for a functional analysis of the genome. *Electrophoresis* **16**(6):1034–1059.

LaCourse, W.R. (2002). Column liquid chromatography: Equipment and instrumentation. *Analy. Chem.* **72**:37R–51R.

Lal, A., Lash, A.E., Altschul, S.F., Velculescu, V., Zhang, L., McLendon, R.E., Marra, M.A., Prange, C., Morin, P.J., Polyak, K., Papadopoulos, N., Vogelstein, B., Kinzler, K.W., Stansberg, R.L., and Riggins, G.L. (1999). A public database for gene expression in human cancers. *Cancer Res* **59**(21):5403–5407.

Lewis, K.C., Opiteck, G.J., Jorgenson, J.W., and Sheeley, D.M. (1997). Comprehensive online RPLC-CZE-MS of peptides. *J. Am. Soc. Mass Spectrom.* **8**(5):495–500.

Li, J., Kelly, J.F., Chernushevich, I., Jed Harrison, D.J., and Thibault, P. (2000). Separation and identification of peptides from gel-isolated membrane proteins using a microfabricated device for combined capillary electrophoresis/nanoelectrospray mass spectrometry. *Anal. Chem.* **72**:599–609.

Li, I., Thibault, P., Bings, N.H., Skinner, C.D., Wang, C., Colyer, C., and Harrison, J. (1999). Integration of microfabricated devices to capillary electrophoresis–electrospray mass spectrometry using a low dead volume connection: Application to rapid analyses of proteolytic digests. *Anal. Chem.* **71**(15):3036–3045.

Loboda, A.V., Krutchinsky, A.N., Bromirski, M., Ens, W., and Standing, K.G. (2000). A tandem quadrupole/time-of-flight mass spectrometer with a matrix-assisted laser desorption/ionization source: Design and performance. *Rapid Commun. Mass Spectrom.* **14**(12):1047–1057.

Lubec, G., Krapfenbauer, K., and Fountoulakis, M. (2003) Proteomics in brain research: Potentials and limitations. *Prog. Neurobiol.* **69**:193–211.

Luche, S., Santoni, V., and Rabilloud, T. (2003). Evaluation of nonionic and zwitterionic detergents as membrane protein solubilizers in two-dimensional electrophoresis. *Proteomics* **3**:249–253.

Madden, S.L., Galella, E.A., Zhu, J.S., Bertelsen, A.H., and Beaudry, G.A. (1997). Sage transcript profiles for P53-dependent growth regulation. *Oncogene* **15**(9):1079–1085.

Mann, M., and Wilm, M. (1994). Error-tolerant identification of peptides in sequence databases by peptide sequence tags. *Anal. Chem.* **66**(24):4390–4399.

Marshall, A., and Hodgson, J. (1998). DNA chips: An array of possibilities. *Nature Biotech.* **16**:27–31.

Martin, S.E., Shabanowitz, J., Hunt, D.F., and Marto, J.A. (2000). Subfemtomole MS and MS/MS peptide sequence analysis using nano-HPLC micro-ESI Fourier transform ion cyclotron resonance mass spectrometry. *Anal. Chem.* **72**:4266–4274.

Matsui, N.M., Smith-Beckerman, D.M., and Epstein, L.B. (1999a). Staining of preparative 2-D gels. Coomassie blue and imidazole-zinc negative staining. *Methods Mol. Biol.* **112**:307–311.

Matsui, N.M., Smith-Beckerman, D.M., Fichmann, J., and Epstein, L.B. (1999b). Running preparative carrier ampholyte and immobilized pH gradient IEF gels for 2D. *Methods Mol. Biol.* **112**:211–219.

Matsumura, H., Nirasawa, S., and Terauchi, R. (1999). Technical advance: Transcript profiling in rice (*Oryza sativa L.*). *Plant J.* **20**(6):719–726.

McCormack, A.L., Schieltz, D.M., Goode, B., Yang, S., Barnes, G., Drubin, D., and Yates, J.R. (1997). Direct analysis and identification of proteins in mixtures by LC/MS/MS and database searching at the low-femtomole level. *Anal. Chem.* **69**(4):767–776.

Moseley, M.A., Deterding, L.J., Tomer, K.B., and Jorgenson, J.W. (1989). Coupling of capillary zone electrophoresis and capillary liquid chromatography with coaxial continuous-flow fast atom bombardment tandem sector mass spectrometry. *J. Chrom.* **480**:197–209.

Munchbach, M., Quadroni, M., Miotto, G., and James, P. (2000). Quantitation and facilitated de novo sequencing of proteins by isotopic N-terminal labeling of peptides with a fragmentation-directing moiety. *Anal. Chem.* **72**:4047–4057.

Neilson, L., Andalibi, A., Kang, D., Coutifaris, C., Strauss, J.F., 3rd, Stanton, J.A., and Green, D.P. (2000). Molecular phenotype of the human oocyte by PCR-SAGE. *Genomics* **63**(1):13–24.

O'Farrell, P.H. (1975). High resolution two-dimensional electrophoresis of proteins. *J. Biol. Chem.* **250**(10):4007–4021.

Opiteck, G.J., Jorgenson, J.W., Moseley III, M.A., and Anderegg, R.J. (1998). Two-dimensional microcolumn HPLC coupled to a single-quadrupole mass spectrometer for the elucidation of sequence tags and peptide mapping. *J. Microcolumn Separations* **10**(4):365–376.

Opiteck, G.J., Lewis, K.C., Jorgenson, J.W., and Anderegg, R.J. (1997). Comprehensive on-line LC/LC/MS of proteins. *Anal. Chem.* **68**:1518–1524.

Ornstein, D.K., Gillespie, J.W., Paweletz, C.P., Duray, P.H., Herring, J., Vocke, C.D., Topalian, S.L., Bostwick, D.G., Linehan, W.M., Petricoin III, E.F., and Emmert-Buck, M.R. (2000). Proteomic analysis of laser capture microdissected human prostate cancer and in vitro prostate cell lines. *Electrophoresis* **21**(11):2235–2242.

Ostergaard, M., Rasmussen, H.H., Nielsen, H.V., Vorum, H., Orntoft, T.F., Wolf, H., and Celis, J.E. (1997). Proteome profiling of bladder squamous cell carcinomas: Identification of markers that define their degree of differentiation. *Cancer Res.* **57**(18):4111–4117.

Page, M.J., Amess, B., Townsend, R.R., Parekh, R., Herath, A., Brusten, L., Zvelebil, M.J., Stein, R.C., Waterfield, M.D., Davies, S.C., and O'Hare, M.J. (1999). Proteomic definition of normal human luminal and myoepithelial breast. *Proc. Natl. Acad. Sci. U.S.A.* **96**(22):12589–12594.

Papayannopoulos, I.A. (1995). The interpretation of collision-induced dissociation tandem mass spectra of peptides. *Mass Spectrom. Rev.* **14**:49–73.

Pappin, D.J.C., Hojrup, P., and Bleasby, A.J. (1993). Rapid identification of proteins by peptide-mass fingerprinting. *Curr. Biol.* **3**(6):327–332.

Patterson, S.D., Thomas, D., and Bradshaw, R.A. (1996). Application of combined mass spectrometry and partial amino acid sequence to the identification of gel-separated proteins. *Electrophoresis* **17**(5):877–891.

Perrin-Cocon, L.A., Marche, P.N., and Villiers, C.L. (1999). Purification of intracellular compartments involved in antigen processing: A new method based on magnetic sorting. *Biochem. J.* **338**:123–130.

Perrot, M., Sagliocco, F., Mini, T., Monribot, C., Schneider, U., Shevchenko, A., Mann, M., Jeno, P., and Boucherie, H. (1999). Two-dimensional gel protein database of *Saccharomyces cerevisiae*. *Electrophoresis* **20**:2280–2298.

Rabilloud, T. (2002). Two-dimensional gel electrophoresis in proteomics: Old, old fashioned, bit it still climbs up the mountains. *Proteomics* **2**:3–10.

Rabilloud, T. (2000). Detecting proteins. Separated by 2-D gel electrophoresis. *Anal. Chem.* **72**(1):48A–55A.

Rabilloud, T. (1999). Silver staining of 2-D electrophoresis gels. *Methods Mol. Biol.* **112**: 297–305.

Rabilloud, T. (1998). Use of thiourea to increase the solubility of membrane proteins in two-dimensional electrophoresis. *Electrophoresis* **19**(5):758–760.

Rabilloud, T. (1996). Solubilization of proteins for electrophoretic analyses. *Electrophoresis* **17**(5):813–829.

Rabilloud, T. (1990). Mechanisms of protein silver staining in polyacrylamide gels: A 10-year synthesis. *Electrophoresis* **11**(10):785–794.

Rabilloud, T., Blisnick, T., Heller, M., Luche, S., Aebersold, R., Lunardi, J., and Braun-Breton, C. (1999). Analysis of membrane proteins by two-dimensional electrophoresis: Comparison of the proteins extracted from normal or *Plasmodium falciparum*-infected erythrocyte ghosts. *Electrophoresis* **20**(18):3603–3610.

Rabilloud, T., Vuillard, L., Gilly, C., and Lawrence, J.J. (1994). Silver-staining of proteins in polyacrylamide gels: A general overview. *Cell Mol. Biol. (Noisy-le-grand)* **40**(1):57–75.

Ramagli, L.S. (1999). Quantifying protein in 2-D page solubilization buffers. *Methods Mol. Biol.* **112**:99–103.

Richert, S., Luche, S., Chevallet, M., Van Dorsselaer, A., Leize-Wagner, E., and Rabilloud, T. (2004). About the mechanism of interference of silver staining with peptide mass spectrometry. *Proteomics* **4**:909–916.

Righetti, P.G., and Bossi, A. (1997a). Isoelectric focusing in immobilized pH gradients: An update. *J. Chromatogr. B Biomed. Sci. App.* **699**(1–2):77–89.

Righetti, P.G., and Bossi, A. (1997b). Isoelectric focusing in immobilized pH gradients: Recent analytical and preparative developments. *Anal. Biochem.* **247**(1):1–10.

Righetti, P.G., and Gianazza, E. (1987). Isoelectric focusing in immobilized pH gradients: Theory and newer methodology. *Methods Biochem. Anal.* **32**:215–278.

Righetti, P.G., Gianazza, E., and Bjellqvist, B. (1983). Modern aspects of isoelectric focusing: Two-dimensional maps and immobilized pH gradients. *J. Biochem. Biophys. Methods* **8**(2): 89–108.

Rosenfeld, J., Capdevielle, J., Guillemot, J.C., and Ferrara, P. (1992). In-gel digestion of proteins for internal sequence analysis after one- or two-dimensional gel electrophoresis. *Anal. Biochem.* **203**(1):173–179.

Ross-Macdonald, P., Coelho, P.S., Roemer, T., Agarwal, S., Kumar, A., Jansen, R., Cheung, K.H., Sheehan, A., Symoniatis, D., Umansky, L., Heidiman, M., Nelson, F.K., Iwasaki, H., Hager, K., Gerstein, M., Miller, P., Roeder, G.S., and Snyder, M. (1999). Large-scale analysis of the yeast genome by transposon tagging and gene disruption. *Nature* **402**(6760):413–418.

Ruan, Y., Gilmore, J., and Conner, T. (1998). Towards arabidopsis genome analysis: Monitoring expression profiles of 1400 genes using cDNA microarrays. *Plant J.* **15**(6):821–833.

Sanchez, J.C., Hochstrasser, D., and Rabilloud, T. (1999). In-gel sample rehydration of immobilized pH gradient. *Methods Mol. Biol.* **112**:221–225.

Santoni, V., Rabilloud, T., Doumas, P., Rouquie, D., Mansion, M., Kieffer, S., Garin, J., and Rossignol, M. (1999). Towards the recovery of hydrophobic proteins on two-dimensional. *Electrophoresis* **20**(4–5):705–711.

Sarto, C., Valsecchi, C., and Mocarelli, P. (2002). Renal cell carcinoma: Handling and treatment. *Proteomics* **2**:1627–1629.

Service, R.F. (1998). Microchip arrays put DNA on the spot. *Science* **282**(5388):396–401.

Shevchenko, A., Loboda, A., Shevchenko, A., Ens, W., and Standing, K.G. (2000). MALDI quadrupole time-of-flight mass spectrometry: A powerful tool for proteomic research. *Anal. Chem.* **72**(9):2132–2141.

Shevchenko, A., Chernushevich, I., Ens, W., Standing, K.G., Thomson, B., Wilm, M., and Mann, M. (1997). Rapid "de novo" peptide sequencing by a combination of nanoelectrospray, isotopic labeling and a quadrupole/time-of-flight mass spectrometer. *Rapid Comm. Mass Spectrom.* **11**(9):1015–1024.

Shevchenko, A., Jensen, O.N., Podtelejnikov, A.V., Sagliocco, F., Wilm, M., Vorm, O., Mortensen, P., Shevchenko, A., Boucherie, H., and Mann, M. (1996a). Linking genome and proteome by mass spectrometry: Large-scale identification of yeast proteins from two dimensional gels. *Proc. Nat. Acad. Sci. U.S.A.* **93**(25):14440–14445.

Shevchenko, A., Wilm, M., Vorm, O., and Mann, M. (1996b). Mass spectrometric sequencing of proteins silver-stained polyacrylamide gels. *Anal. Chem.* **68**(5):850–858.

Sirivatanauksorn, Y., Drury, R., Crnogorac-Jurcevic, T., Sirivatanauksorn, V., and Lemoine, N. (1999). Laser-assisted microdissection: Applications in molecular pathology. *J. Pathol.* **189**(2):150–154.

Smith, B.J. (1994). Quantification of proteins on polyacrylamide gels (nonradioactive). *Methods Mol. Biol.* **32**:107–111.

Stein, W.D., Litman, T., Fojo, T., and Bates, S.E. (2004). A serial analysis of gene expression (SAGE) database analysis of chemosensitivity: Comparing solid tumors with cell lines and comparing solid tumors from different tissue origins. *Cancer Res.* **64**:2805–2816.

Steinberg, T.H., Chernokalskaya, E., Berggren, K., Lopez, M.F., Diwu, Z., Haugland, R.P., and Patton, W.F. (2000). General—ultrasensitive fluorescence protein detection in isoelectric focusing gels using a ruthenium metal chelate stain. *Electrophoresis* **21**(3):486–496.

Steinberg, T.H., White, H.M., and Singer, V.L. (1997). Optimal filter combinations for photographing Sypro orange or Sypro red dye-stained gels. *Anal. Biochem.* **248**(1):168–172.

Steinberg, T.H., Haugland, R.P., and Singer, V.L. (1996a). Applications of Sypro orange and Sypro red protein gel stains. *Anal. Biochem.* **239**(2):238–245.

Steinberg, T.H., Jones, L.J., Haugland, R.P., and Singer, V.L. (1996b). Sypro orange and Sypro red protein gel stains: One-step fluorescent staining of denaturing gels for detection of nanogram levels of protein. *Anal. Biochem.* **239**(2):223–237.

Tan, D., Deeb, G., Wang, J., Slocum, H.K., Winsto, J., Wiseman, S., Beck, A., Sait, S., Anderson, T., Nwogu, C., Ramnath, N., and Loewen, G. (2003). HER-2/neu protein expression and gene alteration in stage I-IIIA non-small-cell lung cancer: A study of 140 cases using a combination of high throughput tissue microarray, immunohistochemistry, and fluorescent in situ hybridization. *Diagn. Mol. Pathol.* **12**:201–211.

Tastet, C., Lescuyer, P., Diemer, H., Luche, S., van Dorsselaer, A., and Rabilloud, T. (2003). A versatile electrophoresis system for the analysis of high- and low-molecular-weight proteins. *Electrophoresis* **24**:1787–1794.

Tonge, R., Shaw, J., Middleton, B., Rowlinson, R., Rayner, S., Young, J., Pognan, F., Hawkins, E., Currie, I., and Davison, M. (2001). Validation and development of fluorescence two-dimensional differential gel electrophoresis proteomics technology. *Proteomics* **1**:377–396.

Unlu, M., Morgan, M., and Minden, J. (1997). Difference gel electrophoresis: A single gel method for detecting changes in protein extracts. *Electrophoresis* **18**(11):2071–2077.

Urwin, V.E., and Jackson, P. (1993). Two-dimensional polyacrylamide gel electrophoresis of proteins labeled with the fluorophore monobromobimane prior to first-dimensional isoelectric focusing: Imaging of the fluorescent protein spot patterns using a cooled charge-coupled device. *Anal. Biochem.* **209**(1):57–62.

Urwin, V., and Jackson, P. (1991). A multiple high-resolution mini two-dimensional polyacrylamide gel electrophoresis system: Imaging two-dimensional gels using a cooled charge-coupled device after staining with silver or labeling with fluorophore. *Anal. Biochem.* **195**(1):30–37.

Velculescu, V.E., Zhang, L., Vogelstein, B., and Kinzler, K.W. (1995). Serial analysis of gene expression. *Science* **270**(5235):484–487.

Wang, H., and Hanash, S. (2003). Multi-dimensional liquid phase based separations in proteomics. *J. Chromatogr. B Analyt Technol. Biomed Life Sci.* **787**:11–18.

Wang, C., Oleschuk, R., Ouchen, F., Li, J., Thibault, P., and Harrison, J. (2000). Integration of immobilized trypsin bead beds for protein digestion within a microfluidic chip incorporating capillary electrophoresis separations and an electrospray ionization interface. *Rap. Comm. Mass Spectrom.* **14**:1377–1383.

Washburn, M.P., Wolters, D., and Yates, J.R.I. (2001). Large-scale analysis of the yeast proteome by multidimensional protein identification technology. *Nature Biotech.* **19**:242–247.

Weeraratna, A.T., Becker, D., Carr, K.M., Duray, P.H., Rosenblatt, K.P., Yang, S., Chen, Y., Bittner, M., Strausberg, R.L., Riggins, G.J., Wagner, U., Kallioniemi, O.P., Trent, J.M., Morin, P.J., and Meltzer, P.S. (2004). Generation and analysis of melanoma SAGE libraries: SAGE advice on the melanoma transcriptome. *Oncogene* **23**:2264–2274.

Wilkins, M.R., Gasteiger, E., Sanchez, J.C., Bairoch, A., and Hochstrasser, D.F. (1998). Two-dimensional gel electrophoresis for proteome projects: The effects of protein hydrophobicity and copy number. *Electrophoresis* **19**(8–9):1501–1505.

Wilkins, M.R., Ou, K., Appel, R.D., Sanchez, J.C., Yan, J.X., Golaz, O., Farnsworth, V., Cartier, P., Hochstrasser, D.F., Williams, K.L., and Gooley, A.A. (1996a). Rapid protein identification using N-terminal "sequence tag" and amino acid analysis. *Biochem. Biophys. Res. Commun.* **221**(3):609–613.

Wilkins, M.R., Hochstrasser, D.F., Sanchez, J.C., Bairoch, A., and Appel, R.D. (1996b). Integrating two dimensional gel databases using the Melanie II software. *Trends Biochem. Sci.* **21**(12):496–497.

Wilm, M., Shevchenko, A., Houthaeve, T., Breit, S., Schweigerer, L., Fotsis, T., and Mann, M. (1996). Femtomole sequencing of proteins from polyacrylamide gels by nano-electrospray mass spectrometry. *Nature* **379**(Feb. 1):466–469.

Wimmer, K., Kuick, R., Thoraval, D., and Hanash, S.M. (1996). Two-dimensional separations of the genome and proteome of neuroblastoma cells. *Electrophoresis* **17**(11):1741–1751.

Yan, J.X., Devenish, A.T., Wait, R., Stone, T., Lewis, S., and Fowler, S. (2002). Fluorescence two-dimensional difference gel electrophoresis and mass spectrometry based proteomic analysis of Escherichia coli. *Proteomics* **2**:1682–1698.

Yates, J.R., McCormack, A.L., Link, A.J., Schieltz, D., Eng, J., and Hays, L. (1996). Future prospects for the analysis of complex biological systems using micro-column liquid chromatography electrospray tandem mass spectrometry. *Analyst* **121**(7):R65–R76.

Yates, J.R. III, Eng, J.K., McCormack, A.L., and Schieltz, D. (1995). Method to correlate tandem mass-spectra of modified peptides to amino-acid-sequences in the protein database. *Anal. Chem.* **67**:1426–1436.

Zhang, W., and Chait, B.T. (2000). Profound: An expert system for protein identification using mass spectrometric peptide mapping information. *Anal. Chem.* **72**:2482–2489.

Zhang, B., Liu, H., Karger, B.L., and Foret, F. (1999). Microfabricated devices for capillary electrophoresis–electrospray mass spectrometry. *Anal. Chem.* **71**(15):3258–3264.

2

MAPPING PROTEIN–PROTEIN INTERACTIONS

Daniel Figeys

Department of Biochemistry, Microbiology, and Immunology, University of Ottawa, Ottawa, Canada

Industrial Proteomics: Applications for Biotechnology and Pharmaceuticals, edited by Daniel Figeys
ISBN 0-471-45714-0

INTRODUCTION

Cellular processes such as cell signaling and regulation are very complex and have multiple input points that can change the direction and magnitude of the signal. It is often simplistically explained as the wiring diagram of the cell, the basis being that the majority of cellular functions are achieved through protein–ligand interactions. Each protein participates in many different interactions throughout its lifetime. In reality, different molecules of the same type of protein can participate in different pathways at the same time. The understanding that proteins are composed of different functional units has increased tremendously over the last 15 years (Pawson and Nash, 2003; Pawson et al., 2002). Proteins are made of domains that infer them some functionality as well as motifs that are implicated in different controls of the protein, such as regulation. These domains and motifs greatly define the role of a protein, and thus the protein's interactions. The complexity of protein interactions can therefore be overwhelming if it is not studied in a systematic manner. For example, in humans, cellular processes involve 30,000 to 60,000 genes, possibly the precursors of up to millions of different proteins in different functional states.

The notion of mapping the human interactome is more science fiction than reality. Simple mathematics would indicate that 60,000 genes represent 1.8 billion possible interactions to verify, that is, if one gene only led to one protein. Considering that genes lead to multiple proteins in different states, this could represent more than 500 billion possible interactions.

More importantly, mapping of protein–protein interactions can be very useful when applied in a disease or a target family-focused approach using relevant model systems. In this chapter we review the mapping of protein–protein interactions using immunopurification coupled with mass spectrometry.

DESCRIPTION OF THE METHOD

The enrichment of interacting partners through affinity purification is a well-established technique (Harlow, 1988). However, it was frequently limited by the lack of information on the interacting proteins. Often, the only information available was the molecular weight of the proteins observed on a gel. These issues were recently reduced through mass spectrometry and bioinformatics, rendering it possible to systematically map protein interactions related to a disease or protein family (Figeys, 2002b, 2003).

The application of mass spectrometry to protein–protein interaction studies has been limited by the number of cells needed to be grown for proper analysis. The initial studies were often performed in organisms for which the starting material was not limiting. Fortunately, the sensitivity of mass spectrometry has continuously improved, thus, significantly reducing the requirement for cellular material. It is now feasible to study protein–protein interactions by mass spectrometry using a single plate of human cell lines. The combination of affinity purification and mass spectrometry generally consists of the selective purification and enrichment of a bait protein and its interactors from cell lysate. The purified proteins are separated by gel electrophoresis, and the isolated protein bands are digested into peptides using trypsin. The peptide mixtures are then analyzed by mass spectrometry to generate information that can be used to identify the proteins present on the gel. Thus, mass spectrometry is able to identify the interactors that are attached to the bait protein. In the following sections, we present in more detail the different steps involved in protein interaction mapping by combining immunopurification and mass spectrometry.

Molecular Biology

We have designed an approach in which three processes occur. First, a clone coding for a tagged bait protein is engineered (Flag tag); second, cells are transfected with the clone and the bait protein expressed; and third the bait protein and its interacting protein partners (preys) are purified using the tag present on the bait protein. One advantage of combining bait tagging with immunopurification/affinity is that the protein of interest is expressed in a relevant cell line, that is, human proteins are expressed in human cells. The epitope tag used is small and does not interfere with the normal protein function, posttranslational modification of proteins, nor their localization. This approach models the normal life cycle of the wild-type protein as closely as possible. A clone must be available for the protein of interest in order to add a small epitope tag that allows the expressed proteins to be readily immunopurified using an antibody. Alternatively, for small genomes it is possible to use homologous recombination approaches to insert a tag directly in the genome. The systematic epitope tagging of large sets of genes allows for high-throughput studies of protein complexes by mass spectrometry.

This immunopurification/mass spectrometry approach based on epitope tagging requires access to clones representative of the genes of interest. Complementary deoxyribonucleic acid (cDNA) clones can be obtained from commercial sources, such as the Invitrogen and the Kazusa collections, as well as from academic endeavors such as the FlexGene repository (*www.hip.harvard.edu*). If they are not available, they can be de novo cloned using standard molecular procedures. The clones can be moved into transfection systems and used to rapidly perform different experiments. We have been using the Gateway system (Walhout et al., 2000b) to rapidly introduce many genes into an entry vector. They can then be easily transferred from their entry vectors to different destination vectors designed for specific experiments. We have designed different destination vectors, in particular, destination vectors to perform epitope tagging at the N or C termini of the protein of interest. This greatly reduces the amount of molecular biology needed to perform these different experiments.

We have also developed a molecular biology protocol to efficiently prepare DNA for transfection (Fig. 2.1). Commercially available clones are often available in a plasmid form. Established polymerase chain reaction (PCR) methodologies can be used to amplify the bait genes from the corresponding parent plasmid DNAs. Those constructs can then be directly transformed in *Escherichia coli* (*E. coli*) followed by a miniprep. The material is then subjected to PCR using primers designed for the BP gateway reaction followed by the BP reaction. The oligonucleotide primers used for PCR introduce an additional nucleotide sequence (29 bp), corresponding to the *att*B

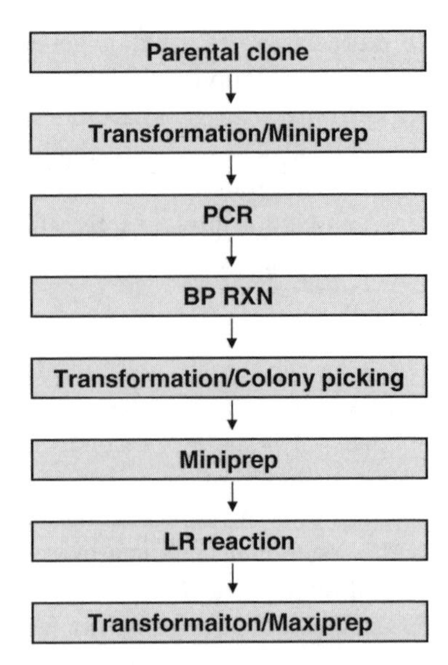

Figure 2.1. Simplified process involved in the molecular biology preparation of clones for protein interaction mapping by immunopurification/mass spectrometry. The Gateway cloning system is used in this process.

recombination sites, onto the ends of the PCR product. The *att*B sites are used as part of the Gateway Cloning Technology* (Invitrogen) (Fig. 2.2). Briefly, a portion of the purified PCR reaction product is added to the BP reaction mixture, which contains a donor vector (encoding *att*P sites) and the BP Clonase mix of recombination proteins. The recombination results in the oriented integration of the *att*B flanked PCR product into the *att*P sites of the donor vector, generating the entry clone in which the bait gene coding region is now flanked with *att*L sites. A portion of the BP reaction is then used to transform *E. coli*, and the entry clone plasmid DNA is purified from selected transformants (antibiotic selection) using plasmid miniprep protocols. The integrity of each entry clone is verified by PCR amplification using gene-specific primers in addition to DNA sequencing of the bait gene to ensure that no mutations had resulted during the PCR reaction or the BP reaction. The constructs that have the correct sequences are subjected to the LR Gateway reaction into a destination vector compatible with *E. coli* transformation and human cell transfection. The LR reaction results in the directional transfer of the bait gene-coding region, flanked by the *att*L sites in the entry clone, to the destination vector through recombination with the *att*R flanked GATERC, generating the expression clone. The final clones are then transformed into *E. coli* and enough DNA prepared by Maxiprep. The FLAG epitope tag(Sigma) has the peptide sequence DYKDDDDK.

Full-Length versus Protein Fragments. There is disagreement over performing protein interactions using full-length proteins or using fragments of proteins. It is

*More detailed information on the Gateway Cloning Technology is available at *http://www.invitrogen.com/GatewayOnline/index.gateway.htm*.

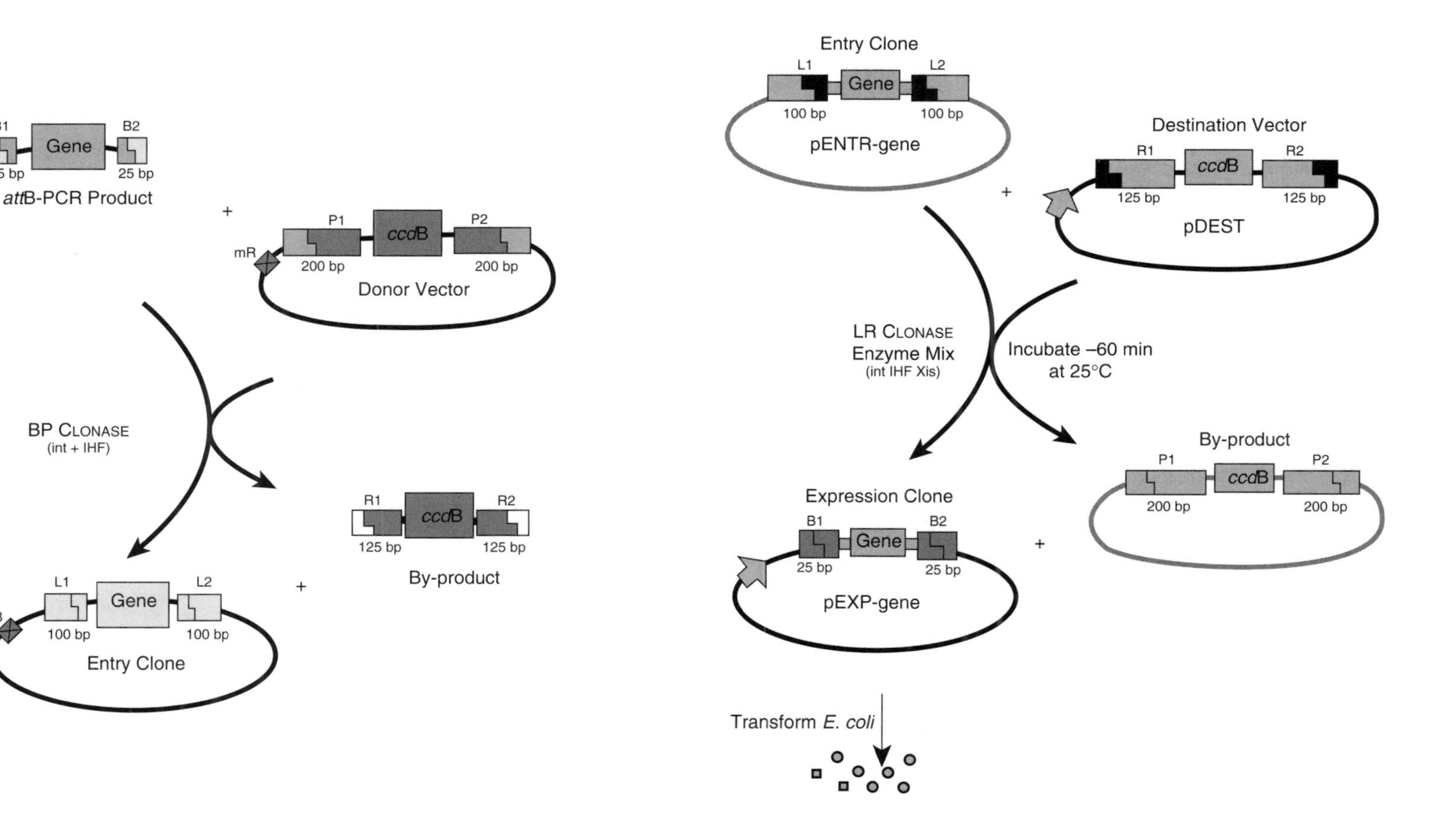

Figure 2.2. (See color insert.) Cartoon view of the Gateway cloning system. Primers are first designed to PCR the coding region of interest and flank it with *att*B sequences that are used with the BP clonase to introduce the gene of interest into the entry clone. The entry clone is then shuffled into the destination vectors using the LR clonase.

important to realize that there are significant differences in terms of working with full-length proteins or protein fragments. A protein function is defined by its interactions as well as by its localization. Protein fragments often do not localize properly and are not involved in all the interactions. Furthermore, it is important to perform a sequence alignment of the protein fragments against all predicted genes to verify the homology level of the fragments to other genes. Basically, the interactions observed for a full-length protein do not equal the interactions observed for its individual fragments. One should see the fragments as new proteins that are not naturally expressed. Thus, the interactions observed with fragments of proteins need to be stringently validated.

It is preferable to focus protein interaction mapping efforts on full-length clones as much as possible. This obviously represents the best-case scenario to ensure valid interactions. In some instances, it is difficult to overexpress the full-length clone. For example, full-length clones of membrane proteins have a lower success rate of over-expression using this approach than cytosolic proteins. There seems to be no hard-set rules to explain the success or failure of membrane proteins as baits. For difficult membrane proteins, cytosolic fragments are often the only options for protein–protein interactions mapping.

Human Cell Transfection

We have been using two different methods for transfecting human cell lines with the constructs. The first is the calcium–phosphate precipitation method. In this instance, approximately 1×10^7 cells are transfected by adding 5 µg of DNA construct in the form of a calcium–phosphate/DNA precipitate. Alternatively, when this method fails, we have the more costly lipofectamine transfection method.

We have developed a fully automated robotic system that performs cell replatting, changing of medium, and transfection either using calcium–phosphate or lipofectamine. The transfection protocol involves many steps, and human errors significantly increase when performing multiple transfection experiments. The advantage of the robotic system is its better consistency and throughput (Fig. 2.3).

Protein Separation

Two days after transfection, the cells are lysed by the addition (1 mL) of lysis buffer [20 mM Tris(Hydroxymethyl) aminomethane Hydrochloride (TrisHCl) pH 7.5, 150 mM NaCl, 1 mM Ethylenediamine Tetra-Acetic Acid (EDTA), 1% NP-40, 0.5% sodium deoxycholate, 10 µg/mL aprotinin, 0.2 mM AEBSF (Calbiochem)], and are clarified by centrifugation for 30 min at 20,000 *g*. Cell lysates and proteins should be maintained at temperatures between 0 and 4°C. The clarified lysate is then subjected to immunoprecipitation by adding 5 µg anti-FLAG monoclonal antibody covalently attached to cross-linked agarose beads (M2, Sigma). The mixture was gently agitated by inversion for 60 min. Immune complexes associated with the insoluble fraction were recovered by centrifugation (1000 *g* for 2 min) and washed by three cycles of resuspension in lysis buffer followed by centrifugation as described above. Immune complexes are eluted from the beads by resuspension in 250 µL 50 mM ammonium bicarbonate (prepared

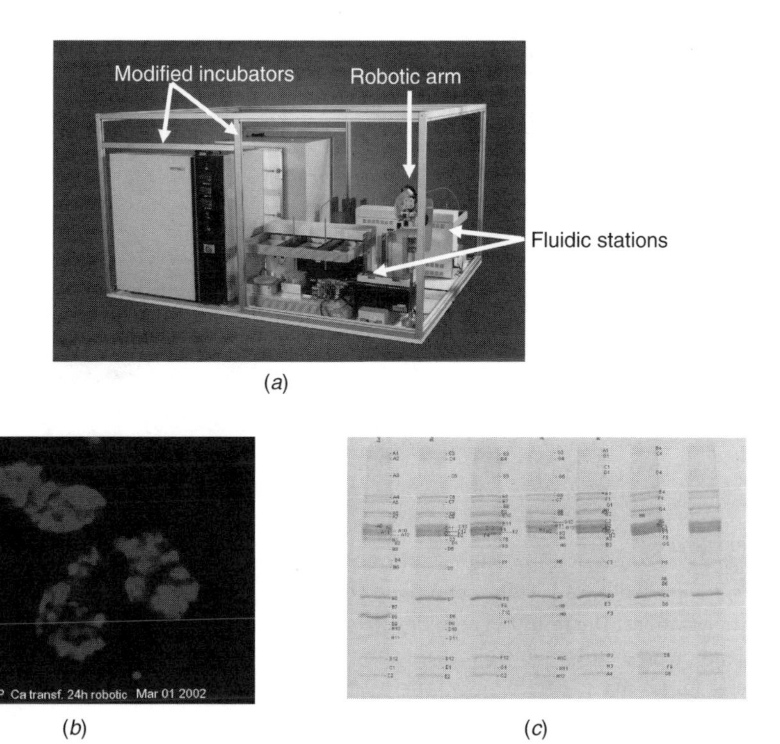

Figure 2.3. (*a*) Robotic system developed for cell plating, culture, and transfection. The robotic arm performs the transport of the cell plates, while the fluidic stations perform all fluid handling necessary to all of the robotic processes. (*b*) Fluorescent imaging of 293°F cells transfected on the robot by calcium/phosphate transfection with GFP protein. (*c*) Illustration of the reproducibility of immunopurification obtained for different cell plates transfected with a human bait protein.

just prior to use) containing 400 μM FLAG peptide. Following a 30-min incubation, beads are removed by centrifugation, and the supernatant containing FLAG peptide as well as the eluted proteins are lyophilized. Proteins are resolved by standard one-dimensional Sodium Dodecyl Sulfate Pulyacrylamide Gel Electrophcresis (SDS-PAGE), and stained with Commassie Blue (GelCode Blue, Pierce) or silver staining. Care must be taken to avoid the introduction of contaminating proteins such as human-skin-derived keratins during the preparation of samples for SDS-PAGE.

Sample Preparation for Mass Spectrometry

In order to facilitate the identification of the recovered immunoprecipitated proteins by mass spectrometry (MS), the stained bands containing one or more protein species are excised from the polyacrylamide gel, digested into polypeptides by treatment in situ with trypsin, and transferred into solutions and concentrations compatible with MS analysis (Fig. 2.4). Techniques for the in-gel processing of proteins have been refined

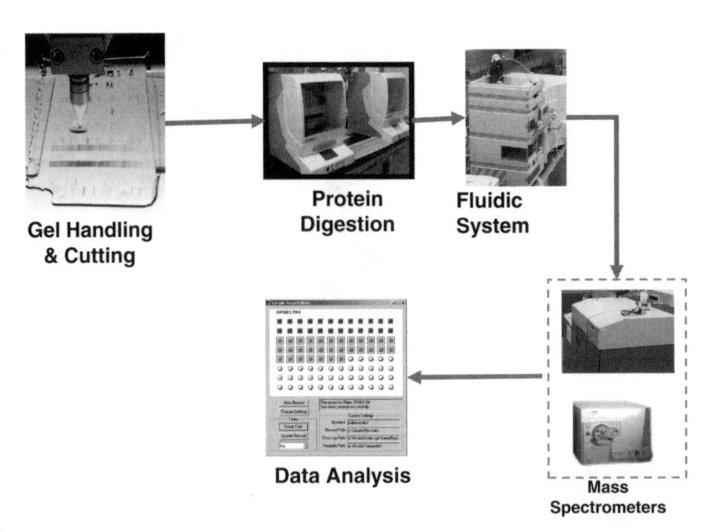

Figure 2.4. Process for the analysis of gel bands obtained from immunopurification. The bands are excised on a custom gel-picking robot and digested in Genomic Solution protein digestion stations. The resulting peptides are separated on a Agilent HPLC/autosampler system coupled only to a custom microbore reverse-phase HPLC column online with an LCQ ion trap or Qstar Pulsar mass spectrometer. The MS/MS data are searched and the results handled through a custom-built data-handling system.

into standardized protocols. The so-called in-gel digestion approach was developed for the enzymatic fragmentation of proteins embedded in gel pieces and the extraction of the resulting peptides (Wilm et al., 1996). Sequencing-grade modified trypsin has been the enzyme of choice for high-throughput identification of proteins. We put a fully automated system in place consisting of an in-house gel picking robot and Genomic Solution digestion system to process proteins. This system delivers the recovered digested proteins in 96 well plates. Each well represents a protein band.

Mass Spectrometers for Protein Identification

Protein identification is achieved by separating the derived tryptic peptides on microbore high-performance liquid chromatography (HPLC) column online with an electrospray ionization (ESI) (Fenn et al., 1989; Whitehouse et al., 1985) tandem mass spectrometer. Briefly, the 96-well plate protein digest is installed on an Agilent microautosample connected to an Agilent microflow HPLC system. The fluid path configuration allows rapid loading of a small reverse-phase (C18) precolumn followed by elution and low-flow separation on a microcapillary column made in a microelectrospray needle. The peptides that elute off of the microcolumn are then electrosprayed into the mass spectrometer. We use LCQ Deca and XP mass spectrometers as well as QStar mass spectrometers for protein identification. Briefly, the mass spectrometers perform one survey scan (MS scan), select the three most abundant ions not already sequenced, and isolate and perform collision-induced dissociation on the three individual ions. This routine is repeated throughout the chromatographic separation. The QStar MS outperforms the LCQ mass spectrometers for the number of and the coverage of the proteins identified per band. We have opted to used the ESI tandem mass spectrum (MS/MS) mass spectrometer instead of the matrix-assisted laser desorption ionization (MALDI) time-of-flight (TOF) mass spectrometer because multiple proteins per bands are present on an immunoprecipitate. Each of our HPLC-ESI-MS/MS systems can perform one gel band analysis every 15 min 24 h a day (i.e., one 96-well plate per day per system). On average, we observe three proteins per band.

Protein Identification Based on Peptide Masses and MS/MS Spectra

Identifying a protein is like identifying an individual in a crowd. The larger the crowd, the more information is required about the individual before achieving a conclusive identification. Similarly, the species of origin of the proteins in question and the quality and quantity of the corresponding available genomic information affects the amount of MS-derived information necessary to unambiguously identify a protein. Generally, MS analyses of peptide mixtures can provide information related to the mass of peptides (MS scan), to their amino acid sequence (MS/MS scan), and sometimes the presence of posttranslational modifications such as phosphoryl groups.

The peptide masses are accurately measured using a MALDI-TOF or a MALDI-Q-Star mass spectrometer down to the low ppm (part per million) precision level. The ensemble of the peptide masses observed in tryptic digests can be used to search protein/DNA databases in a method called peptide mass fingerprinting (Clauser et al., 1995; Cottrell, 1994; Pappin, 1997). In this approach, protein entries in the databases are ranked according to the number of peptide masses that match their predicted tryptic pattern. Some of the commercially available software provides a scoring scheme based on the size of the databases, the number of matching peptides, and the difference peptides. Unambiguous identification can be obtained depending on the number of peptides observed, the accuracy of the measurement, and the size of the genome of the particular species.

The MS/MS spectra are a second set of information that can be used to identify a protein. The MS/MS spectra contain the fragmentation pattern related to the amino acid sequence of specific peptides. The analysis of MS/MS spectra is typically more intensive. The approaches used for the interpretation of these spectra can be classified into three subgroups according to the level of user intervention required.

In the first subgroup no interpretation of the spectra is required. The information contained in the spectra is directly correlated with protein/DNA sequence information contained in databases. Different algorithms have been developed for this specific task. These algorithms automatically search uninterpreted MS/MS spectra against protein and DNA databases and some are freely accessed over the web (for noncommercial entities). Mascot by Matrix Sciences (*www.matrixscience.com*), Sonars by ProteoMetrics (*http://65.219.84.5/service/prowl/sonar.html*), and ProteinProspector from UCSF (*http://prospector.ucsf.edu/*) are the most commonly used web-based MS/MS search engines. On the open-source side, X! tandem is a software that matches tandem mass spectra with peptide sequences (*http://www.proteome.ca/x-bang/tandem/tandem.html*).

Another algorithm that is popular is Sequest (Eng et al., 1994; Yates, 1998; Yates et al., 1995). For every MS/MS spectra submitted, this algorithm searches protein/DNA databases for the top 500 isobaric peptides and generates the corresponding predicted spectra. The predicted spectra are rapidly matched against the measured spectra by multiplication in the frequency domain using a fast-Fourier transform. Correlation parameters, which indicate the quality of the match between predicted and measured spectra are then deduced. A high cross-correlation indicates a good match with the measured spectrum. Furthermore, although protein identification has been performed with as little as one peptide using this algorithm, unambiguous identification of the provenance of a protein is often achieved by the multitude of peptides that match to the same entry in a database.

The approaches in the second subgroup all involve the partial interpretation of the MS/MS spectra and therefore require human intervention. The dominant approach, often called sequence tag (Mann and Wilm, 1994; Patterson et al., 1996; Wilkins et al., 1996), consists of reading the mass spacing between a few specific fragments in an MS/MS spectrum and to generate a short section (tag) of the peptide sequence. Using this tag and the residual mass information, the provenance of the peptide can be ascertained by comparing with the sequence and calculated masses obtained from protein databases for isobaric peptides. Every MS/MS spectrum requires the generation of a tag followed by database searching. Unambiguous identification of the protein is established by the multitude of peptides that matches to the same protein. Over the years, different variations of this approach have been developed to perform database searching using sequence tags. The main limitation of applying the sequence-tag approach in large-scale proteomic centers is the labor requirement and expertise required to manually generate partial interpretation of the MS/MS spectra. Attempts to automate the generation of sequence tags are underway to solve this problem.

The last subgroup, called de novo sequencing of proteins (Papayannopoulos, 1995; Shevchenko et al., 1997), is often used as a last resource when no matching information is available in databases and the quality of the MS/MS spectra is good. The MS/MS spectra of peptides contain ladder-type information, which, in principle, indicates their amino acid sequence. Experienced mass spectrometrists can manually extract the peptide sequence from the CID spectra (de novo sequencing).

Depending on the quality of the data and the complexity of the species under study, a single confidant match between a peptide MS/MS and a protein sequence entry is sufficient to identify a protein, or a family of proteins. The required sequence coverage for unambiguous identification increases for homologous proteins, when the peptide identified is not unique to a protein, when dealing with databases of poor fidelity and/or partial coverage, and to access Single Nucleotide Polymorphisms (SNP) databases. Clearly, every subsequent peptide MS/MS that is matched to the same protein further increases the confidence level of the identification.

In our experience, none of the MS/MS search engines are satisfactory and there is a significant discrepancy between the different results. The search engines perform in a similar manner when dealing with proteins present at high levels. However, the results can be very different when dealing with low-level proteins. We have compared four search engines (Mascot, Pepsea breakpoint, Sonars, and Sequest) for low-level peptide analyses performed on an LCQ and Qstar mass spectrometer. All of the MS/MS spectra were manually verified. Figure 2.5 shows comparisons of the results obtained from different search engines. The red boxes indicate uncertain score for the *y* axis and the yellow boxes indicates uncertain score in the *x* axis. Generally speaking, the correlations are poor. Sequest does not correlate well with either Mascot or Pepsea breakpoint. Mascot, Sonars, and Pepsea breakpoints show trends between the different results. More importantly, a significant number of points fall in the yellow or red area, indicating clear uncertainty in the results. These sequences need to be manually inspected. Similar results are observed for data generated from QStar mass spectrometers, although the scoring ranges are somewhat different for Sonars and Pepsea breakpoints.

The conclusion is clear. None of these search engines can be fully trusted, and, therefore, it is important to incorporate rapid data inspection, especially for proteins with low sequence coverage. We divided data generation from data validation and trained people to ensure that both aspects are adequately covered. We developed a suite of software to rapidly validate all the peptide hits obtained from gel band analyses. In this approach, a technician can validate two 96-well plates per day, which corresponds to 576 proteins per day (we see on average 3 proteins per well).

Bioinformatic Quality Control (QC) of the Data

The bioinformatic analysis as well as QC of the data generated for baits is essential for discerning the true interactors versus the nonspecific interactors. We put a process in place that takes into account the number of repeat immunopurifications, the labeling at the N or C termini, and previous information accumulated in our database.

N- and C-Termini Labeling. We tested the effect of N- and C-termini labeling on numerous baits and the results are somewhat bait dependent. We obtained 82 percent success rate for N-terminus tagging and expression in human cells versus a 67 percent success rate for C-terminus tagging. The difference between the two success rates is statistically significant. The overall success rate of expression of the combined results is 86 percent. Therefore, based on the expression of a bait protein, the C-terminus tagging does not significantly improve the results: 68 percent of the prey proteins are observed with the N-terminus tag protein while 18 percent are observed with the C-terminus tag protein; 14 percent of the proteins are observed with both tags. Figure 2.6 illustrates the reproducibility of immunopurification of an N-terminally tagged bait protein, as well as a comparison of the gel obtained from immunopurification of the same bait tagged either at the N or C termini. It is difficult to predict which of the two taggings will be successful for individual baits. Thus, both tagging approaches should be attempted.

Number of Repeat Immunopurification. The flag-tag approach allows the probing of lower affinity interaction than other methods. The drawback of this approach is that some of these interactions are at the limit of our analytical capabilities. Therefore, it is important to have a process in place to establish the number of repeats that

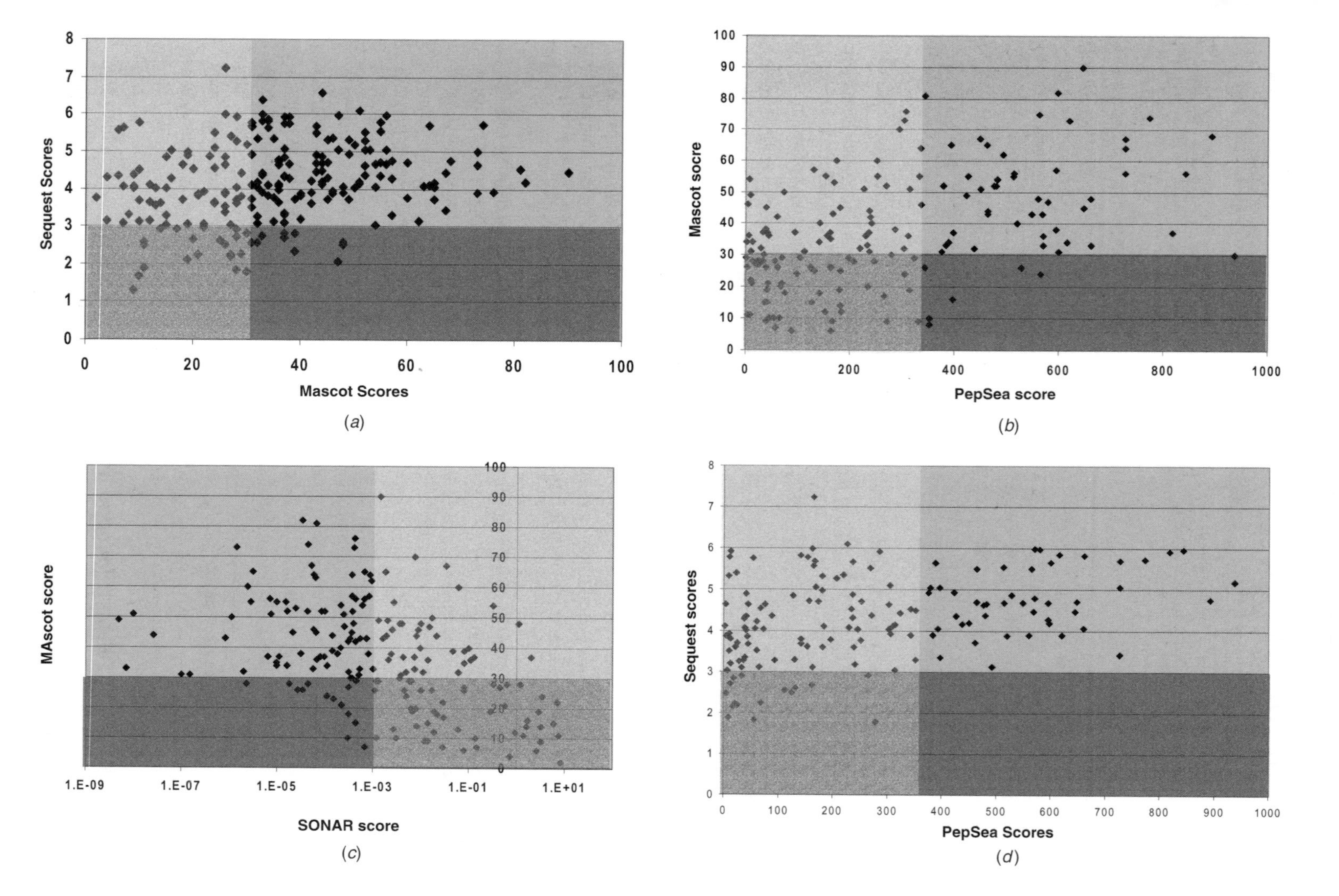

Sequest Scores
Mascot Scores
(a)
Mascot socre
PepSea score
(b)
MAscot score
SONAR score
(c)
Sequest scores
PepSea Scores
(d)

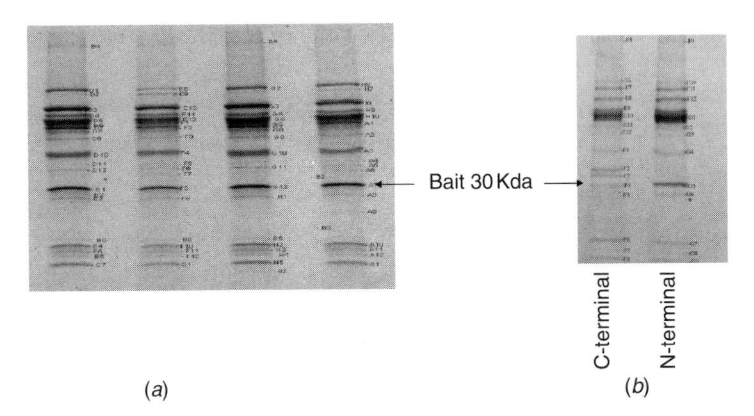

Figure 2.6. (*a*) Reproducibility of N-tagged immunopurification separated by gel electrophoresis (4 repeats). (*b*) Effect of N versus C tagging of the same bait protein.

are necessary to obtain statistically meaningful results. We have subjected a series of independent baits to multiple immunopurification experiments and followed all the hits. Figure 2.7 illustrates the percentage of preys (hits) that are reproducible over all of the runs. We have observed that with four experiments per bait the false positive rate is 15 percent.

Scoring of Interactors. We use several rating systems for the prey proteins that are identified in bait complexes in order to evaluate and prioritize the information. These rating systems are based on an evaluation of frequency, confidence, reproducibility, and biological evaluation of relevance.

We have access to a growing database of human protein interactions that have been discovered by immunopurification/mass spectrometry. It is possible to extract statistical information from this database that can be applied to the data generated. In particular, we have established a quality factor called the prey frequency ($Prey_f$). The prey frequency represents the percentage of baits in our human database for which a prey protein has been observed. The prey frequency is reported for every prey protein entry in the table of interactors for the baits. Prey proteins with high frequencies could be forming nonspecific interactions or are proteins that are involved in common protein processes, that is, chaperone proteins.

Figure 2.5. comparison of MS/MS search engine results for the same set of low-level tryptic peptides obtained from biological samples. (*a*) Sequest versus Mascot, (*b*) Mascot versus Pepsea breakpoint, (*c*) Mascot versus Sonar, and (*d*) Sequest versus Pepsea. The yellow area indicates the lower level of confidence for the *x* axis, while the red area indicates the lower level of confidence for the *y* axis. Overall, a significant number of peptides fall within either the red or yellow area.

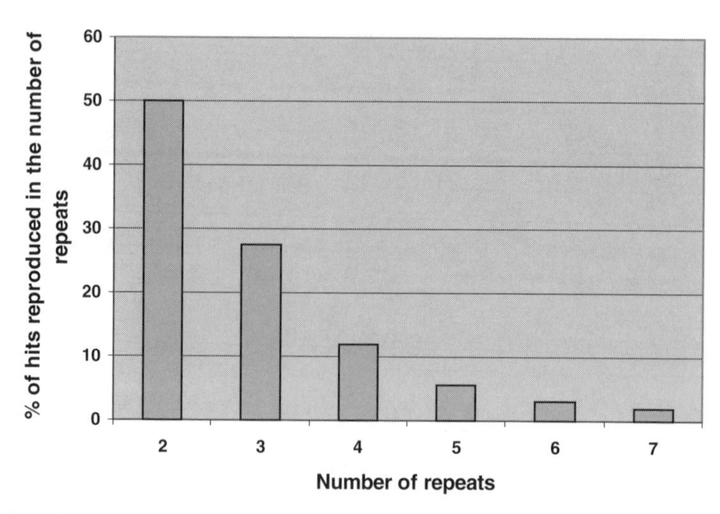

Figure 2.7. Reproducibility of observing prey proteins across multiple immunopurification of the same bait proteins. This illustrates the transient nature of some of the interactions observed by flag tagging. Basically, four repeats are sufficient to obtain less than 15 percent false positive.

We have also established an empirical confidence level that reflects the experimental reliability of the data. The confidence level is calculated from four different variables, including prey reproducibility in different experiments, peptide coverage, search engine rankings, and the correction factor ($Corr_f$) derived from the prey frequency. The confidence level is reported for every interactor as a value from 0 (very low experimental confidence) to 100 (very high experimental confidence). The confidence level is calculated as follows:

$$\text{Confidence level} = (\text{Reproducibility} + \text{Coverage} + \text{Rank})(1 - Corr_f)$$

The prey reproducibility factor provides a score that is dependent on the number of repeats in which the bait has been observed (up to 50 points). The coverage factor (up to 30 points) accounts for the number of peptides that are matched to the prey protein. The rank factor (up to 20 points) reflects that more than one protein per gel band can be seen due to the sensitivity of our mass spectrometers. Higher ranks are given to the more abundant proteins within a band. Finally, a correction factor ($Corr_f$) derived from the prey frequency is applied. Basically, preys that have a high prey frequency get a correction factor closer to 1, and preys that have a low prey frequency get a correction factor closer to 0.

All preys associated with a bait protein have a prey frequency and a confidence level associated with them. This allows the rapid elimination of common interacting proteins as well as irreproducible interactions. In particular, we can setup a cutoff on the prey frequency. For example, if we setup the cutoff at less that 5 percent prey frequency, and we perform more than four repeat experiments, then we achieve a false-positive rate of less than 5 percent.

Bioinformatics on Protein–Protein Interactions

The mapping of protein–protein interactions can generate a significant amount of interaction data that can lead to challenges in viewing and following the different possible interactions. Resources are available over the Internet to access and view protein–protein interactions. The Grid (Breitkreutz et al., 2002, 2003), and Bind (Bader et al., 2003) are examples of such databases and viewers. These approaches are more focused on amassing the data generated by the different public approaches to mapping protein–protein interactions. So far, the public high-throughput mapping of protein–protein interactions is uncoordinated, and the likelihood of achieving cross-validation of the results is low.

The level of complexity greatly increases when dealing with human cells, and the mapping of protein–protein interactions must be approached differently. Thus, the mapping of the interactome using one or a few cell lines will only establish the basal level of protein–protein interactions. In many instances, the results will be irrelevant to the protein in its true state and in the true processes in which it is involved.

Instead of global mining the protein interactions, we focused directly on relevant human protein interactions along the line of target family (kinases and phosphatases) and disease-related proteins. This changes the requirement for databases and viewers and increases the emphasis on annotating of the interacting bait–prey sets using publicly available annotation information and our own in-house statistical information on interactions.

EXPERIMENTAL DESIGN

The most relevant approach for mapping valid protein–protein interactions is to study a subset of the interactome in a hypothesis-driven approach. We have focused our interaction mapping efforts in humans on proteins that show "changes" during disease progression or through stimulation/treatment of specific cells and certain protein families (kinases and phosphatases). Techniques are already in place to measure changes. Gene chip (DeRisi et al., 1996; Schena et al., 1995) and SAGE (Velculescu et al., 1995) can be used to measure changes in gene expression. As well, differential proteomics and differential phosphoproteomics (Ficarro et al., 2002a, 2002b) can be used to measure changes in protein expression and changes in protein phosphorylation. These techniques can provide a list of candidate baits that can be used in immunopurification/mass spectrometry mapping of protein–protein interactions.

The technology for discovering protein–protein interactions requires the experiments to be performed in cell models. The changes that were observed in the tissues can be easily verified to also occur at the cell model level. Focusing on the interactome subset that is changing due to stimulation or disease inherently infers that the upstream pathways and regulations are in place. Thus, this approach increases the likelihood of discovering meaningful interactions.

First-Pass Mapping

The first-pass mapping of protein–protein interactions should be performed in relevant cell lines. In humans, the number of possible interactions based on 30,000 to 60,000 genes is large. Techniques that ask the specific question, "Is protein A interacting with protein B?" are not appropriate for the first pass of high-throughput mapping because they require the possession of all the genes. Instead, the best approach is to rely on techniques that ask the question, "What are the proteins that interact in a specific cell line or tissue with protein A?" We and others have clearly demonstrated that immunopurification/mass spectrometric–based techniques can be used to answer this question (Figeys, 2002a).

We recently published an article on the study of protein–protein interactions by mass spectrometry (Figeys et al., 2001) and demonstrated its application in humans using a flag-tagged chloride conductance regulatory protein (pICLn). Known and novel interactors of chlorine conductance regulatory protein (pICLn) were identified. Large-scale studies of protein–protein interaction in yeast groups were reported by Gavin et al. (Gavin et al., 2002), using the TAP tag approach to study 589 yeast genes, and Ho et al. (2002) using the flag-tag approach to study 725 yeast genes.

So far, no large-scale studies have been published on using these methodologies in a more focused disease or hypothesis-driven approach. We have studied over 500 different human proteins as bait in our in-house discovery efforts. Figure 2.8 illustrates a global view of the interaction network for these 500 human proteins. It has been our experience with human protein interaction mapping that the cell system must be judiciously chosen for the disease of interest or the hypothesis of interest. The results of

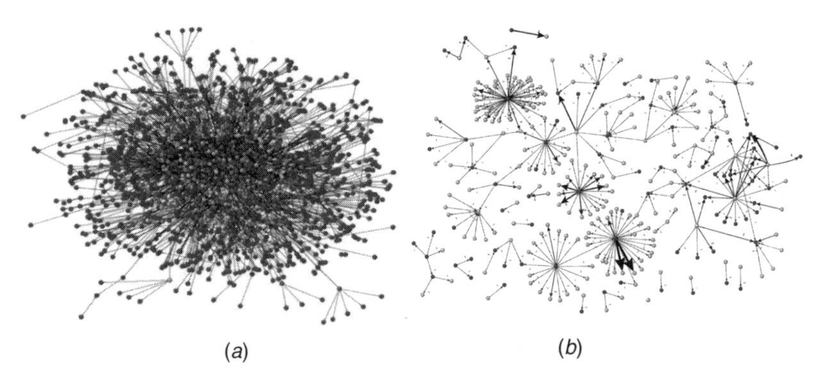

Figure 2.8. (See color insert.) (*a*) Global view of the human protein interactions generated using 500 distinct human bait proteins. (*b*) Interaction view for some of the kinases that are part of the network. (The kinases are in red.)

the MS-based interaction mapping provide a first-pass resolution of the possible interactors related to a gene of interest. The first-pass approach provides a list of potential interactors. A second-stage high-throughput approach is necessary to validate the interactors. Different strategies can be designed to enhance the resolution of the protein–protein interaction map.

Focusing on What Seems to Change

The first-pass interaction mapping generates a list of potentially interacting proteins from different disease states or protein families. In our experience, an average of 5 to 10 proteins is reproducibly associated with particular baits. Fortunately, we can rely on statistics derived from experiments, annotation from the literature, and other "omics" information to select the proteins for the next stage validation. The next step is to validate the protein–protein interactions and to deconstruct immunopurification results into pathways and complexes. The first-pass bait–prey approach does not indicate if the preys are interacting directly with the bait or are interacting with another prey that is itself interacting with the bait. Different techniques are available as secondary approaches to enhance our confidence level in protein–protein interactions.

The 2-Hybrid and Other Systems. The 2-hybrid method was introduced for studying binary protein–protein interactions (Fields and Song, 1989). It is a rapid screen to determine if protein A is binding to protein B. Briefly, in the 2-hybrid, the proteins that are discovered in the first-pass mapping can be expressed in yeast to create two arrays of yeast cells. The first array (the bait array) expresses the different proteins fused with the DNA-binding domain of a transcription factor that lacks the transcription activation domain. The second array (the prey array) expresses the same proteins fused to the transcription activation domain. The two yeast arrays are then mated in order to transfer the coding material and to allow both fusion proteins to be expressed in the same cells. The fusion protein, made of the DNA-binding domain and the protein of interest, binds to the promoter region of the reported gene. If the prey protein interacts with the bait protein, then the activation machinery recognizes the prey-bound activation domain, and the reporter gene is transcribed. The 2-hybrid approach has been previously used to map protein–protein interactions for proteins from *Saccharomyces cerevisiae* (Ito et al., 2000; Uetz et al., 2000), *Helicobacter pylori* (Rain et al., 2001), *Escherichia coli* (Bartel et al., 1996), hepatitis C (Flajolet et al., 2000), vaccinia virus (McCraith et al., 2000), and *Caernorhabditis elegans* (Walhout et al., 2000a).

The 2-hybrid approach is known to have serious limitations. In particular, it is poorly applicable to membrane proteins. Some proteins do not fold properly as hybrid proteins. Yeast 2-hybrid does not handle interactions that require posttranslational modifications. More importantly, the proteins screened through yeast 2-hybrid are taken out of context, and, therefore, the interactions that do not occur in their normal cellular environment can often provide positive yeast 2-hybrid results. Using the 2-hybrid approach to validate the results from the first-pass mapping must be considered carefully. Thus, the results from the 2-hybrid should only be considered if they validate the results from the first pass. However, negative results from 2-hybrid should not invalidate the results obtained in the first-pass immunopurification/mass spectrometry mapping of protein–protein interactions.

Reciprocal MS-Based Protein Interaction Mapping. The immunopurification/mass spectrometry method can be used in a second pass to decipher the protein–protein interactions from the immunopurification results by turning the prey proteins into baits. Figure 2.9 shows an example in yeast of preys turned into baits and the validated interactions. It is not guaranteed that the prey turned into bait proteins will necessarily associate with the first-pass bait protein. This means that either the proteins are only interacting through secondary proteins or that the stochiometry is not favorable. Thus, this approach deciphers interactors but does not necessarily validate them.

Co-localization. Co-localization can be used to validate protein–protein interactions. Recently, a high-throughput approach to determine the localization of green fluorescent protein (GFP)-labeled proteins has been described and applied in yeast (Huh et al., 2003). The large-scale mapping of protein localizations in human cells using this approach has not been reported. However, low-throughput co-localization studies are routinely performed in human cells. The overlapping of co-localization data with protein interactions and gene expression as well as other genomics data sets can help reinforce the validity of protein–protein interactions.

CONCLUSION

The mapping of protein–protein interactions in human cells requires an experimental design directed to answer specific questions such as the interactions of disease associated genes. Different "omics" approaches can be incorporated in the experimental design to reinforce the validity of the interactions.

There is scientific value in trying to map the interactions of all the genes associated with a disease, for example, all the genes that showed changes in expression during disease progression. The applications of protein–protein interaction mapping in industry tend to be more focused on specific sets of proteins that are well associated with a disease and have good annotation. There should be more academic efforts to apply immunopurification/mass spectrometry for the mapping of disease interactome. Most of the interaction mapping has been done so far in smaller genome organisms. In those organisms it is likely that an unstructured approach by different labs to map protein–protein interactions will lead to a level of cross-validation. However, considering the number of genes, gene products, and cells in human, it is unlikely that cross-validation can be achieved in a reasonable time in an unstructured approach. Therefore, academic approaches to map specific human interactomes, such as disease interactomes, should be done in concerted efforts.

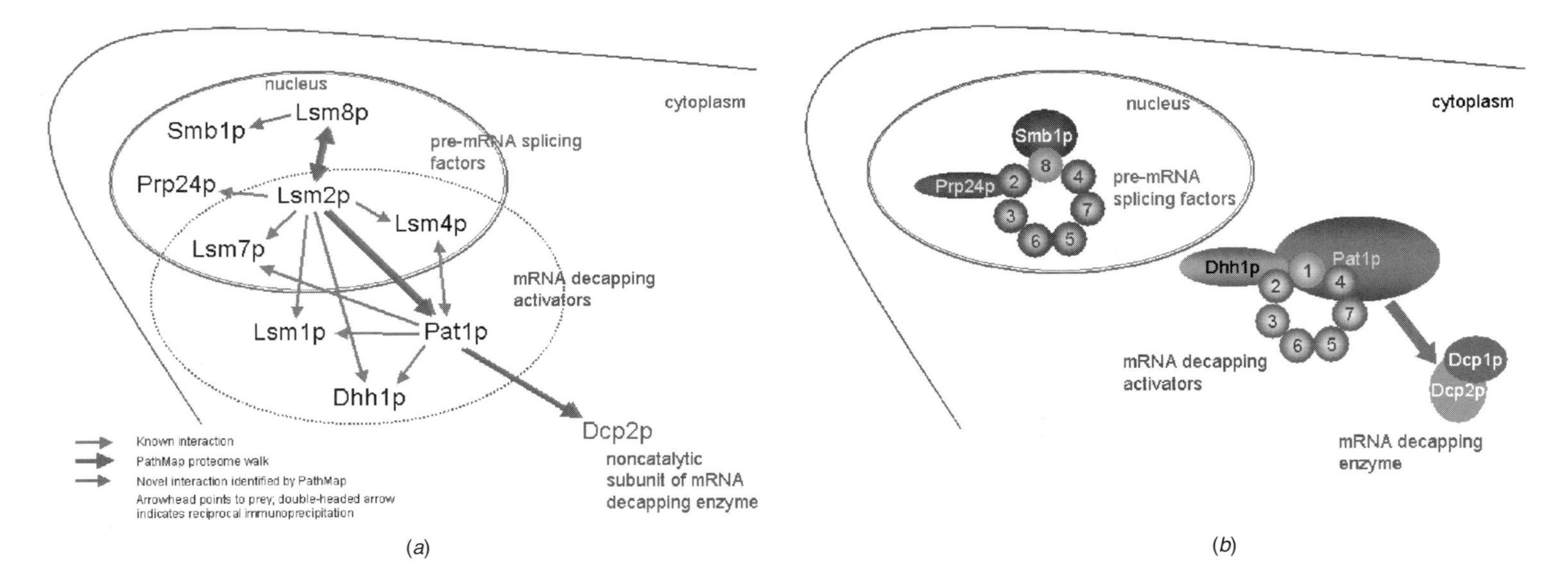

Figure 2.9. Second-pass validation using immunopurification/mass spectrometry approach. (*a*) Observed interactions in yeast by first- and second-pass immunopurification/mass spectrometry approach starting with Lsm8p and walking to Dcp2p. (*b*) Known interactions from the literature.

REFERENCES

Bader, G.D., Betel, D., and Hogue, C.W. (2003). BIND: The Biomolecular Interaction Network Database. *Nucleic Acids Res.* **31**(1):248–250.

Bartel, P.L., Roecklein, J.A., SenGupta, D., and Fields, S. (1996). A protein linkage map of *Escherichia coli* bacteriophage T7. *Nat. Genet.* **12**(1):72–77.

Breitkreutz, B.J., Stark, C., and Tyers, M. (2003). "The GRID: The General Repository for Interaction Datasets." *Genome Biol.* **4**(3):R23.

Breitkreutz, B.J., Stark, C., and Tyers, M. (2002). Osprey: A network visualization system. *Genome Biol.*

Clauser, K.R., Hall, S.C., Smith, D.M., Webb, J.W., Andrews, L.E., Tran, H.M., Epstein, L.B., and Burlingame, A.L. (1995). Rapid mass spectrometric peptide sequencing and mass matching for characterization of human melanoma proteins isolated by two-dimensional PAGE. *Proc. Natl. Acad. Sci. USA* **92**(11):5072–5076.

Cottrell, J.S. (1994). Protein identification by peptide mass fingerprinting. *Pept. Res.* **7**(3):115–124.

DeRisi, J., Penland, L., Brown, P.O., Bittner, M.L., Meltzer, P.S., Ray, M., Chen, Y., Su, Y.A., and Trent, J.M. (1996). Use of a cDNA microarray to analyse gene expression patterns in human cancer. *Nat. Genet.* **14**(4):457–460.

Eng, J., McCormack, A.L., and Yates, J.R.I. (1994). An approach to correlate tandem mass spectral data of peptides with amino acid sequences in a protein database. *J. Am. Soc. Mass Spectrom.* **5**:976–989.

Fenn, J.B., Mann, M., Meng, C.K., Wong, S.F., and Whitehouse, C.M. (1989). Electrospray ionization for mass spectrometry of large biomolecules. *Science* **246**(4926):64–71.

Ficarro, S., Chertihin, O., Westbrook, V.A., White, F., Jayes, F., Kalab, P., Marto, J.A., Shabanowitz, J., Herr, J.C., Hunt, D., and Visconti, P.E. (2002a). Phosphoproteome analysis of capacitated human sperm. Evidence of tyrosine phosphorylation of AKAP 3 and valosin containing protein/P97 during capacitation. *J. Biol. Chem.*

Ficarro, S.B., McCleland, M.L., Todd Stukenberg, P., Burke, D.J., Ross, M.M., Shabanowitz, J., Hunt, D.F., and White, F.M. (2002b). Phosphoproteome analysis by mass spectrometry: Application to *S. cerevisiae*. *Nat. Biotechnol.* **20**(3):301–305.

Fields, S., and Song, O. (1989). A novel genetic system to detect protein-protein interactions. *Nature* **340**(6230):245–246.

Figeys, D. (2003). Proteomics in 2002: A year of technical development and wide-ranging applications. *Anal. Chem.* **75**(12):2891–2905.

Figeys, D. (2002a). Functional proteomics: Mapping protein-protein interactions and pathways." *Curr. Opin. Mol. Therap.* **4**(3):210–215.

Figeys, D. (2002b). Proteomics approaches in drug discovery. *Anal. Chem.* **74**(15):412A–419A.

Figeys, D., McBroom, L.D., and Moran, M.F. (2001). Mass spectrometry for the study of protein-protein interactions. *Methods* **24**(3):230–239.

Flajolet, M., Rotondo, G., Daviet, L., Bergametti, F., Inchauspe, G., Tiollais, P., Transy, C., and Legrain, P. (2000). A genomic approach of the hepatitis C virus generates a protein interaction map. *Gene* **242**(1–2):369–379.

Gavin, A.-C., Bösche, M., Krause, R., Grandi, P., Marzioch, M., Bauer, A., Schultz, J., Rick, J.M., Michon, A.-M., Cruciat, C.-M., Remor, M., Höfert, C., Schelder, M., Brajenovic, M.,

Ruffner, H., Merino, A., Klein, K., Hudak, M., Dickson, D., Rudi, T., Gnau, V., Bauch, A., Bastuck, S., Huhse, B., Leutwein, C., Heurtier, M.-A., Copley, R.R., Edelmann, A., Querfurth, E., Rybin, V., Drewes, G., Raida, M., Bouwmeester, T., Bork, P., Seraphin, B., Kuster, B., Neubauer, G., and Superti-Furga, G. (2002). Functional organization of the yeast proteome by systematic analysis of protein complexes. *Nature* **415**:141–147.

Harlow, E., and Lane, D. (1988). *Antibodies: A Laboratory Manual*. Cold Spring Harbor, NY, Cold Spring Harbor Press.

Ho, Y., Gruhler, A., Heilbut, A., Bader, G.D., Moore, L., Adams, S.-L., Millar, A., Taylor, P., Bennett, K., Boutilier, K., Yang, L., Wolting, C., Donaldson, I., Schandorff, S., Shewnarane, J., Vo, M., Taggart, J., Goudreault, M., Muskat, B., Alfarano, C., Dewar, D., Lin, Z., Michalickova, K., Willems, A.R., Sassi, H., Nielsen, P.A., Rasmussen, K.J., Andersen, J.R., Johansen, L.E., Hansen, L.H., Jespersen, H., Podtelejnikov, A., Nielsen, E., Crawford, J., Poulsen, V., Sùrensen, B.D., Matthiesen, J., Hendrickson, R.C., Gleeson, F., Pawson, T., Moran, M.F., Durocher, D., Mann, M., Hogue, C.W.V., Figeys, D., and Tyers, M. (2002). Systematic identification of protein complexes in *Saccharomyces cerevisiae* by mass spectrometry. *Nature* **415**:180–183.

Huh, W.-K., Falvo, J.V., Gerke, L.C., Carroll, A.S., Howson, R.W., Weissman, J.S., and O'Shea, E.K. (2003). Global analysis of protein localization in budding yeast. *Nature* **425**:686–691.

Ito, T., Tashiro, K., Muta, S., Ozawa, R., Chiba, T., Nishizawa, M., Yamamoto, K., Kuhara, S., and Sakaki, Y. (2000). Toward a protein-protein interaction map of the budding yeast: A comprehensive system to examine two-hybrid interactions in all possible combinations between the yeast proteins. *Proc. Natl. Acad. Sci. U.S.A.* **97**(3):1143–1147.

Mann, M., and Wilm, M. (1994). Error-tolerant identification of peptides in sequence databases by peptide sequence tags. *Anal. Chem.* **66**(24):4390–4399.

McCraith, S., Holtzman, T., Moss, B., and Fields, S. (2000). Genome-wide analysis of vaccinia virus protein-protein interactions. *Proc. Natl. Acad. Sci. U.S.A.* **97**(9):4879–4884.

Papayannopoulos, I.A. (1995). The interpretation of collision-induced dissociation tandem mass spectra of peptides. *Mass. Spectr. Rev.* **14**:49–73.

Pappin, D.J. (1997). Peptide mass fingerprinting using MALDI-TOF mass spectrometry. *Methods Mol. Biol.* **64**:165–173.

Patterson, S.D., Thomas, D., and Bradshaw, R.A. (1996). Application of combined mass spectrometry and partial amino acid sequence to the identification of gel-separated proteins. *Electrophoresis* **17**(5):877–891.

Pawson, T., and Nash, P. (2003). Assembly of cell regulatory systems through protein interaction domains. *Science* **300**(5618):445–452.

Pawson, T., Raina, M., and Nash, P. (2002). Interaction domains: From simple binding events to complex cellular behavior. *FEBS Lett.* **513**(1):2–10.

Rain, J.C., Selig, L., De Reuse, H., Battaglia, V., Reverdy, C., Simon, S., Lenzen, G., Petel, F., Wojcik, J., Schachter, V., Chemama, Y., Labigne, A., and Legrain, P. (2001). The protein-protein interaction map of *Helicobacter pylori*. *Nature* **409**:211–215.

Schena, M., Shalon, D., Davis, R.W., and Brown, P.O. (1995). Quantitative monitoring of gene expression patterns with a complementary DNA microarray. *Science* **270**(5235):467–470.

Shevchenko, A., Chernushevich, I., Ens, W., Standing, K.G., Thomson, B., Wilm, M., and Mann, M. (1997). Rapid "de novo" peptide sequencing by a combination of nanoelectrospray, isotopic labeling and a quadrupole/time-of-flight mass spectrometer. *Rapid Commun. Mass Spectrom.* **11**(9):1015–1024.

Uetz, P., Giot, L., Cagney, G., Mansfield, T.A., Judson, R.S., Knight, J.R., Lockshon, D., Narayan, V., Srinivasan, M., Pochart, P., Qureshi-Emili, A., Li, Y., Godwin, B., Conover, D., Kalbfleisch, T., Vijayadamodar, G., Yang, M., Johnston, M., Fields, S., and Rothberg, J.M. (2000). A comprehensive analysis of protein-protein interactions in *Saccharomyces cerevisiae*. *Nature* **403**:623–627.

Velculescu, V.E., Zhang, L., Vogelstein, B., and Kinzler, K.W. (1995). Serial analysis of gene expression. *Science* **270**(5235):484–487.

Walhout, A.J., Sordella, R., Lu, X., Hartley, J.L., Temple, G.F., Brasch, M.A., Thierry-Mieg, N., and Vidal, M. (2000a). Protein interaction mapping in *C. elegans* using proteins involved in vulval development. *Science* **287**(5450):116–122.

Walhout, A.J., Temple, G.F., Brasch, M.A., Hartley, J.L., Lorson, M.A., van den Heuvel, S., and Vidal, M. (2000b). GATEWAY recombinational cloning: application to the cloning of large numbers of open reading frames or ORFeomes. *Methods Enzymol.* **328**:575–592.

Whitehouse, C.M., Dreyer, R.N., Yamashita, M., and Fenn, J.B. (1985). Electrospray interface for liquid chromatographs and mass spectrometers. *Anal. Chem.* **57**(3):675–679.

Wilkins, M.R., Ou, K., Appel, R.D., Sanchez, J.C., Yan, J.X., Golaz, O., Farnsworth, V., Cartier, P., Hochstrasser, D.F., Williams K.L., and Gooley, A.A. (1996). Rapid protein identification using N-terminal sequence tag and amino acid analysis. *Biochem. Biophys. Res. Commun.* **221**(3):609–613.

Wilm, M., Shevchenko, A., Houthaeve, T., Breit, S., Schweigerer, L., Fotsis, T., and Mann, M. (1996). Femtomole sequencing of proteins from polyacrylamide gels by nano-electrospray mass spectrometry. *Nature* **379**:466–469.

Yates, J.R.I. (1998). Peptide sequencing by tandem mass spectrometry. *Cell Biology: A Laboratory Handbook*, Vol. 4. San Diego, Academic, pp. 529–538.

Yates, J.R.I., Eng, J.K., McCormack, A.L., and Schieltz, D. (1995). Method to correlate tandem mass-spectra of modified peptides to amino-acid-sequences in the protein database. *Anal. Chem.* **67**:1426–1436.

3

PROTEIN POSTTRANSLATIONAL MODIFICATIONS: PHOSPHORYLATION SITE ANALYSIS USING MASS SPECTROMETRY

Roland S. Annan and Francesca Zappacosta

Proteomics and Biological Mass Spectrometry, Department of Computational, Analytical, and Structural Sciences, GlaxoSmithKline, King of Prussia, Pennsylvania

Industrial Proteomics: Applications for Biotechnology and Pharmaceuticals, edited by Daniel Figeys
ISBN 0-471-45714-0

INTRODUCTION

If one includes the proteolytic processing of the precursor sequence encoded in the messenger ribonucleic acid (mRNA), it is very likely true that all proteins are posttranslationally modified (PTM) to some extent (Wold, 1981). After the assembly of the predicted amino sequence on the mRNA–ribosome complex, the released polypeptide chain is processed and delivered to its proper compartment. Following this, a protein may be further modified as a means of regulating its function. One of the most common regulatory PTMs is reversible protein phosphorylation.

Protein kinases, the enzymes responsible for phosphorylation of amino acid side chains represent a very large gene family. In the human genome approximately 450 protein kinases have been identified. While this number is somewhat smaller than predicted prior to the complete sequencing of the human genome (Cohen, 2000), the total number of kinases (580) still represents 1.7 percent of all human genes (Manning et al., 2002). Not surprising given the large number of protein kinases, as many as 30 percent of all the proteins in a cell are thought to be phosphorylated at any given time (Cohen, 2000). A single protein can be phosphorylated by several protein kinases on the same or at different sites, and each of these sites may be dephosphorylated by protein phosphatases at different times.

Much of the activity in the cellular proteome is under the control of reversible protein phosphorylation. Phosphorylation-dependent signaling regulates differentiation of cells, triggers progression of the cell cycle, and controls metabolism, transcription, apoptosis, and cytoskeletal rearrangements. Signaling via reversible phosphorylation also plays a critical role in intracellular communication and the immune response. Phosphorylation can function as either a positive or negative switch, activating or inactivating enzymes. It can serve as a recognition element, targeting a substrate for ubiquitin-dependent proteolysis, or serve as a docking site to recruit other proteins into multiprotein complexes. Phosphorylation can trigger a change in the three-dimensional structure of a protein or initiate translocation of the protein to another compartment of the cell (Hanada and Yoshimura, 2002; Mustelin et al., 2002; Henneke et al., 2003; Rane and Reddy, 2002; Angers-Loustau, 1999). Disruption of normal cellular phosphorylation events is responsible for a large number of human diseases (Hunter, 1998; Cohen, 2001; Zhu et al., 2002).

Knowledge of the specific amino acids phosphorylated on a protein is a critical component in assembling a complete understanding of how a signaling pathway works. Within the last 10 years mass spectrometry (MS) has emerged as a key technology in this difficult task (Wilm et al., 1996; Zhang et al., 1998; Posewitz and Tempst, 1999; Steen et al., 2001; Zappacosta et al., 2002). Isolating phosphopeptides preferentially from all of the other peptides present in a protein digest remains the most challenging aspect of phosphosite mapping and still relies, in most cases, on labeling the protein beforehand with radioactive [^{32}P]-phosphate and Cherenkov counting of high-performance liquid chromatography (HPLC) fractions from the protein digest. Efforts to isolate phosphopeptides that circumvent this bothersome protocol have focused mainly on two approaches: immobilized metal-ion affinity chromatography both offline with matrix-assisted laser desorption ionization (MALDI) (Posewitz and Tempst, 1999) or electrospray (ES) (Kocher et al., 2003) and online coupled to an ES mass spectrometer (Watts et al., 1994; Cao and Stults, 1999), and selective detection of MS techniques that are based on the unique fragmentation behavior of phosphopeptides (Huddleston et al., 1993; Carr et al., 1996). By using any of several different MS-based techniques it is now clearly possible to map in vivo derived phosphorylation sites at the femtomole level, without resorting to the use of radioactivity. Currently, the primary challenge for phosphosite mapping is to ensure that even low stoichiometry sites have not been overlooked.

Multisite phosphorylation of individual proteins appears to be quite common and may be more the rule than the exception (Cohen, 2000), thus it is important to understand which phosphorylation sites modulate or are active in a given biological pathway. Adding to the complexity of this problem is the fact that phosphorylation-dependent function may not depend on activity at a single site but rather be dependent upon serial activation of several sites, and that multiple phosphorylation sites on a given protein may control multiple functions. In order to unravel this structure–function relationship, a thorough quantitative analysis of the phosphorylation profile is necessary.

In this chapter we describe technologies used in our laboratory to map phosphorylation sites on proteins and how we and others are attempting to understand phosphorylation-dependent regulation of protein function by quantitating protein phosphorylation profiles.

EXPERIMENTAL

Phosphopeptide-Specific Liquid Chromatography-Electrospray Ionization Mass Spectrometry (LC-ESMS)

The experimental setup for the first dimension of analysis has been described in detail (Zappacosta et al., 2002). Protein digests were concentrated on a C18 trap cartridge used in place of the sample loop on the HPLC injector. After washing, the peptides were back flushed off of the cartridge at 4 µL/min with an acetonitrile/water gradient onto an LC-Packings PepMap C18 capillary column (180 µm × 15 cm, 3-µm particles) fitted directly into the injector. HPLC mobile phases contained 0.1 percent formic acid and 0.02 percent trifluoro acetic acid (TFA). Solvent flow from the HPLC pumps to the injector was regulated by an LC-Packings Accurate microflow splitter. The column outlet was connected to a Micromass nanoflow ion source (Manchester, UK). Column flow was split prior to the MS by means of a 0.15-mm i.d. microvolume Valco tee insert (Houston, TX), which directed 0.4 to 0.6 µL/min to a 20-µm i.d. tapered fused silica electrospray tip from New Objectives (Cambridge, MA). The remainder of the flow was sent to the prep line for manual collection into polymerase chain reaction (PCR) tubes. MS spectra were recorded on a PE-Sciex API III+ (Concord, Ontario, Canada) triple quadrupole mass spectrometer equipped with a Micromass nanoflow ion source as described above. Negative ion LC-ESMS using single-ion monitoring (SIM) for the phosphate-specific marker ions *m/z* 63 (PO_2-) and *m/z* 79 (PO_3-) was performed as previously described (Huddleston, 1993). Fractions were immediately stored at –70°C until needed for further analysis.

Phosphopeptide Selective Precursor Ion Scanning

Precursor ion spectra for *m/z* 79 were recorded on a Sciex API 3000 triple quadrupole mass spectrometer equipped with a nanoelectrospray source. One-half of each HPLC fraction was made basic by adding two volumes of 50/50 methanol/water containing 10 percent ammonium hydroxide (30 wt %) and 1.5 µL was loaded into the nanospray needle for analysis. Spectra were acquired in the MCA mode as described previously (Carr et al., 1996).

MALDI Analysis of Phosphorylated Peptides

The MALDI mass spectra were recorded on a Micromass TofSpec SE reflectron time-of-flight mass spectrometer equipped with a time-lag-focusing source. Spectra were calibrated externally using two peptide standards. Samples were prepared in α-cyano-4-hydroxycinnamic acid (10 mg/mL in 50:50 ethanol/acetonitrile) or in 2,5-dihydrobenzoic (DHB) acid (saturated solution in 50:50 ethanol/acetonitrile). An aliquot of 0.2 µL from each HPLC fraction was mixed with 0.2 µL of matrix solution and 0.2 µL spotted on the target. Post-source decay (PSD) spectra were recorded in a single segment, and 100 to 200 laser shots were averaged in a single spectrum.

Phosphatase Treatment of Phosphorylated Peptides

Peptides were dephosphorylated by using calf intestinal phosphatase (CIP) (New England Biolabs). For CIP treatment, 8 µL of NH_4HCO_3 50 mM and 1 µL CIP (diluted 1:20) were added to 1/10 of samples (1 µL). Samples were incubated at 37°C for 30 min, dried down, and dissolved in 2 µL acetonitrile 25 percent; 0.4 µL were used for MALDI/MS analysis in DHB.

Targeted Phosphopeptide Sequencing and Phosphosite Mass Spectrometry Western Analysis by LC-ES-MS/MS

Phosphorylated peptides were sequenced by LC-ES-MS/MS on a Micromass QTOF equipped with a nanoflow ion source. Protein digests or individual fractions were loaded on a C18 trap cartridge and were back-flushed onto an LC-Packings PepMap C18 capillary column (75 µm × 15 cm, 3-µm particles) at 0.3 µL/min. HPLC mobile phases contained 0.1 percent formic acid and 0.02 percent TFA. MS/MS data were collected on a single precursor or on a set of predefined precursors using a 2-s scan with a precursor selection window of +3 Da. Whenever possible different precursors were analyzed at different times during the LC separation according to their retention time. For closely eluting peptides, the instrument was set to alternate between precursors after each scan.

Phosphopeptide Sequencing by NanoES MS/MS

Alternatively, peptides were sequenced by nanoelectrospray tandem mass spectrometry on a Micromass QTOF. Individual fractions were dried and resuspended in 3 µL of 60/40 methanol/water containing 5 percent formic acid. One-half of the sample (1.5 µL) was loaded into the nanospray needle.

N-Terminal Isotope Tagging (NIT) Labeling of Protein Digests

The NIT labeling of peptide mixtures was carried out using a modification of the protocol previously described (Zhang et al., 2002; Zappacosta and Annan, 2003). Briefly, one volume of 2 M O-methylisourea in 100 mM $NaHCO_3$ pH 11.0 was added to the protein digest. Samples were incubated at 37°C for 3 h. Acylation with propionic-d_0 or $-d_{10}$ anhydride was carried out for 30 min at 37°C after the addition of 1 µL reagent per 30 pmol of protein. After the incubation the d_0 and d_5 samples were pooled and the excess reagent was removed by desalting on a C18 ZipTip (Millipore) or a MicroTrap (Michrom BioResources, Inc.). Peptides were eluted with 5 µL of 70 percent acetonitrile in 0.1 percent TFA. Ninety microliters of 1 M NH_4OH (pH 11.0) were added to the labeled peptide mixture to promote deacylation of the tyrosine residues. After 1 h incubation at 37°C the samples were acidified by the addition of an appropriate volume of 10 percent TFA and stored at −20°C until further analysis.

RESULTS

ES-MS-Based Phosphorylation Site Mapping of Pho4

The yeast transcription factor Pho4 is regulated in response to changes in the availability of phosphate. This regulation is through the activity of the cyclin-cyclin-dependent kinase (cdk) complex Pho80–Pho85. Pho4 contains six potential cdk phosphorylation sites (Table 3.1). Distinct and separable functions have been assigned to five of the six sites (Komeili and O'Shea, 1999). Other than the specific analysis of the above described cdk sites, there are no published reports on the overall phosphorylation status of Pho4.

A comprehensive phosphopeptide mapping protocol must be able to detect and identify both low-abundance and low-stoichiometry phosphorylation sites in the presence of a complex mixture of highly abundant nonphosphorylated peptides (Annan et al., 2001; Zappacosta et al., 2002). The strategy employed in our laboratory uses two different, highly selective MS techniques to isolate and identify phosphorylated peptides (see Fig. 3.1). These are followed by targeted tandem mass spectrometry to locate the specific site of modification. In the first dimension of the process, phosphopeptides present in the proteolytic digest of a protein are selectively detected and collected into fractions by monitoring for the phosphopeptide specific marker ion PO_{3^-} (*m/z* 79) produced in the ion source during online LC-ESMS (Huddleston et al., 1993). The inclusion of a chromatographic step in the overall strategy partitions the total phosphopeptide pool, reducing the complexity of the nonphosphorylated background and providing a

TABLE 3.1. In Vivo Phosphorylation of SDS-PAGE Derived PHO4

Cyclin/cdk Site	M_r Found	Peptide	Sequence	% P[a]
S1, Ser^{100}	ND[b]	ND	ND	ND
S2, Ser^{114}	3474.8	111–143	LLY**S**PLIHTQSAVPVTISPNLVATATSTTSANK	5
S3, Ser^{128}	3474.8	111–143	LLYSPLIHTQSAVPVTI**S**PNLVATATSTTSANK	5
S4, Ser^{152}	1088.5	149–157	SNS**S**PYLNK	26
S5, Ser^{204}	936.5	202–209	RV**S**PVTAK	15
S5, Ser^{223}	2615.5	210–235	TSSSAEGVVVASE**S**PVIAPHGSSHSR	60
Noncyclin/ cdk Site				
Ser^{6}	1823.8	4–19	TT**S**EGIHGFVDDLEPK	15
Ser^{166}	3107.7	160–188	GKPGPD**S**ATSLFELPDSVIPTPKPKPKPK	48
Ser^{236}	825.5	236–241	**S**LSKRR	ND
Ser^{243}	1241.6	242–252	S**S**GALVDDDKR	60

[a] Percentage of phosphorylation as determined by LC-ESMS by comparing the phosphorylated vs. non-phosphorylated peak intensities.
[b] Not detected.

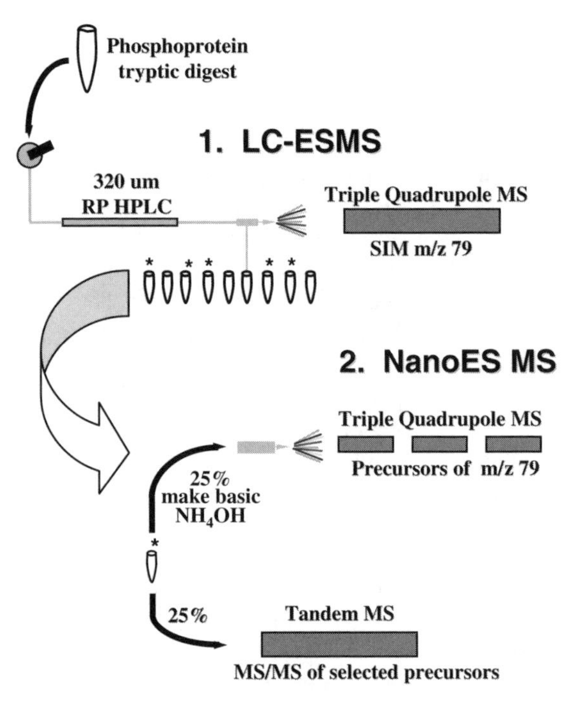

Figure 3.1. Schematic diagram of the multidimensional ESMS phosphopeptide mapping strategy employed in our laboratory to detect and identify phosphorylated peptides.

substantial measure of sample cleanup, all of which facilitate the identification and sequencing of the specific phosphopeptides. In the second dimension of the analysis, the molecular weights of the phosphopeptides in the selected fractions are determined by precursor ion scans for *m/z* 79 in the negative ion mode. A tentative assignment of the amino acid sequence for some of the peptides can usually be made at this point; however, direct sequencing is almost always necessary. In the final step of analysis phosphopeptides are sequenced by positive-ion tandem MS using either nanoelectrospray or targeted LC-MS/MS of selected precursors. This strategy, which has been developed and refined in our laboratory over the last few years, is particularly well suited to phosphoproteins containing many sites of phosphorylation and variable stoichiometry at individual sites.

To establish the overall state of Pho4 phosphorylation, we purified Pho4 as a GST-Pho4 fusion protein from yeast grown in phosphate-rich medium. Following SDS-PAGE, we excised the Pho4 band, digested it with trypsin, and mapped the phosphorylation sites using the multidimensional strategy outlined above. The overall phosphorylation profile of Pho4 from yeast grown in phosphate-rich medium is shown in Fig. 3.2*b*. The LC-MS *m/z* 79 trace for phosphopeptides is greatly simplified compared to the corresponding full-scan trace collected in a separate LC-MS run (Fig. 3.2*a*). Peaks labeled 1 to 9 were manually collected and the molecular weight of the phosphopeptides present in each fraction were determined by analyzing an aliquot of each fraction by nanoES using a precursor ion scan for *m/z* 79. By way of example, we show the *m/z* 79 precursor ion spectrum from fraction 5 in Figure 3.2*c*. A single phosphopeptide with an apparent average molecular mass of 2615.5 is represented by the 2^-, 3^-, and 4^- charge states. From this data we were able to tentatively assign this peptide to a Pho4 tryptic peptide covering amino acids 210–235 plus one mole of phosphate. This peptide contained one of the six consensus cdk sites at Ser^{223} (S6), however, it also contains seven other serines and one threonine. In order to confirm the sequence and conclusively assign the site of phosphorylation, we sequenced the triply charged ion (monoisotopic $[M+3H]^{3+}$ 872.4) for this 26 amino acid peptide using nanoES tandem MS. The resulting product ion spectrum was transformed by the instrument software to produce the singly charged spectrum shown in Figure 3.2*d*. Although the sequence data from the C-terminus of the peptide is somewhat sparse, the abundant y_8 ion arising from cleavage at Pro^{228} and the absence of any satellite peaks representing y_8-98 show clearly that Ser^{231}, Ser^{232}, and Ser^{234} are not phosphorylated. On the other hand a nearly complete set of y_n ions from y_8 to y_{24} show conclusively that phosphorylation occurs on the S6 cdk site, Ser^{223}. The mass difference between y_{12} and y_{13} is 167 Da indicating phosphoserine at the y_{13} position and each subsequent y_n ion is 80 Da higher in mass than a fragment ion that contains unmodified Ser^{223}. In this way we identified and sequenced the phosphopeptides in each of the fractions shown in Figure 3.2*b*.

In the cyclin–cdk complex, Pho80/Pho85 has been shown to phosphorylate Pho4 in vitro on five of the six consensus cdk sites (O'Neill et al., 1996), S1, S2, S3, S4, and S6 (see Table 3.1). In our phosphorylation analysis we identified four of these sites as being utilized in vivo, failing, however, to find any evidence for phosphorylation at S1. This may not be surprising since $Serine^{100}$ in the S1 site was found previously to be a

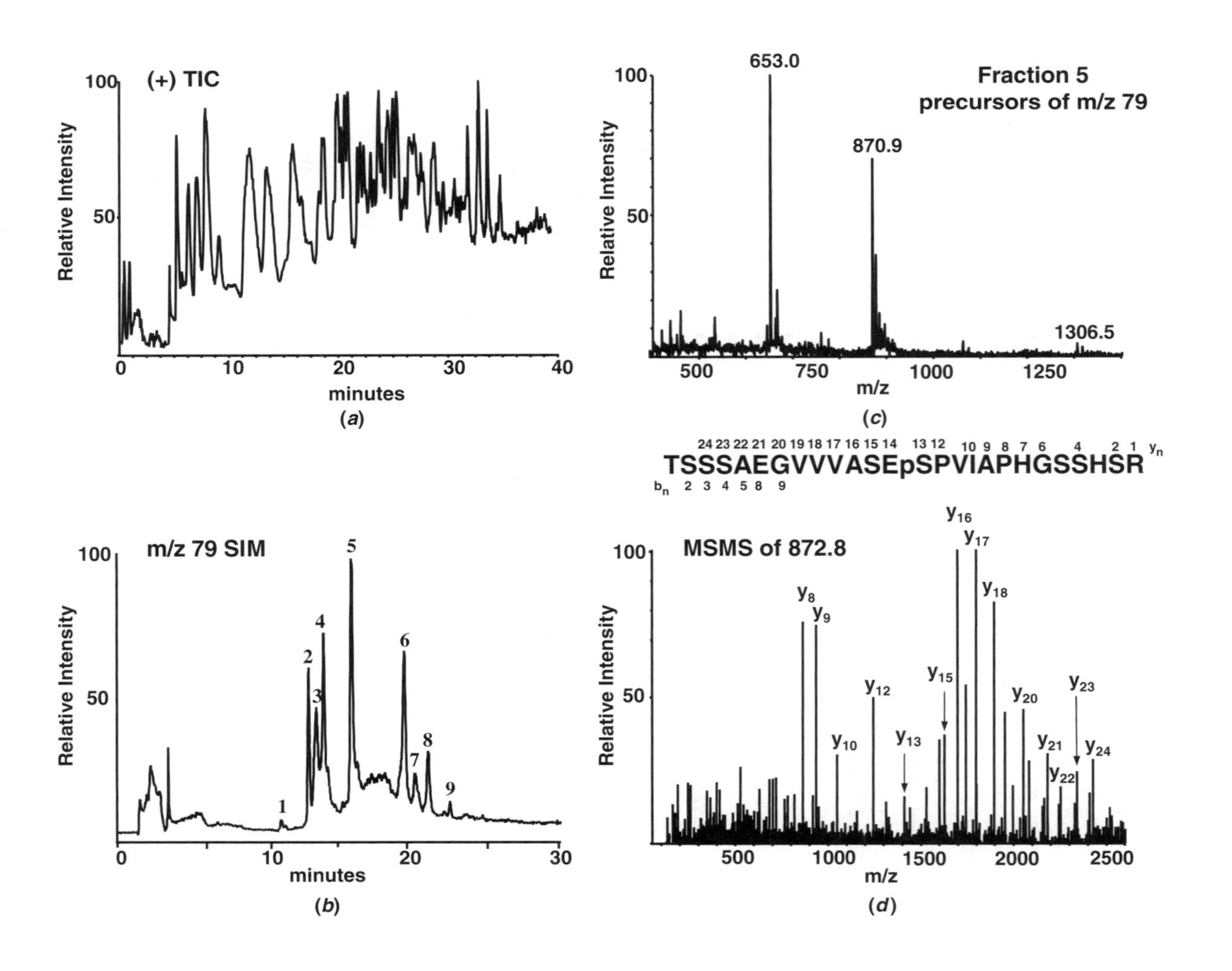
(+) TIC
Relative Intensity
100
50
0
10
20
30
40
minutes
(a)
m/z 79 SIM
1
2
3
4
5
6
7
8
9
minutes
(b)
Fraction 5
precursors of m/z 79
653.0
870.9
1306.5
500
750
1000
1250
m/z
(c)
24 23 22 21 20 19 18 17 16 15 14 13 12 10 9 8 7 6 4 2 1 yn
TSSSAEGVVVASEpSPVIAPHGSSHSR
bn 2 3 4 5 8 9
MSMS of 872.8
y8
y9
y10
y12
y13
y15
y16
y17
y18
y20
y21
y22
y23
y24
1500
2000
2500
(d)

poor in vitro substrate for Pho80/Pho85 (Komeili, 1999). On the other hand our analysis did turn up phosphorylation at the S5 site that had previously been reported to not undergo phosphorylation in vitro in response to Pho80/Pho85 treatment. Of course, it is possible that this site is utilized by another kinase in vivo and has a role independent of the four Pho80/Pho85 sites. Indeed in addition to the five cdk sites identified, we identified seven other peptides that contained four (Ser^{6}, Ser^{166}, Ser^{236}, Ser^{243}) novel phosphorylation sites in a variety of sequence motifs (see Table 3.1).

The great advantage of the method described here is the impressive selectivity of the first two dimensions of the analysis and the ability of the method to separate the phosphopeptides into small pools. In this analysis we identified 12 phosphopeptides representing 9 different sites. A crude estimate of the stoichiometry at the each of the 9 sites shows in vivo utilization ranging from 5 to 65 percent (Table 3.1). A mixture of this complexity, with such a range of stoichiometry, would severely challenge any attempt to analyze it as an unfractionated mixture or using a simple one-dimensional online strategy.

Recently, we have shifted from nano-ES to targeted LC-MS/MS for our peptide sequencing. By targeting a single precursor *m/z* (or several in a time-dependent manner), we are able to integrate the MS/MS signal over the entire elution profile of the peptide, thus obtaining some of the nano-ES signal averaging advantage while taking advantage of the tremendous concentration advantage provided by 75-μm (i.d.) capillary LC columns. The use of the chromatographic step also significantly reduces interference from chemical noise and other peptides of similar *m/z*.

MALDI-Based Approaches to Mapping Phosphorylation Sites on Proteins

Matrix-assisted laser desorption/ionization on a time-of-flight mass spectrometer (MALDI-TOF) is a technique that is well suited to the analysis of protein digests. It is relatively easy to perform, fairly tolerant of sample composition and sample preparation, and copes well with relatively complex mixtures (less than 100 peptides). For peptide analysis, modern MALDI-TOF instrumentation has low femtomole sensitivity, resolution on the order of 10,000 to 20,000 full width at half maximum (FWHM) and a mass accuracy better than 20 ppm. A further advantage of MALDI is that the sample

Figure 3.2. Phosphopeptide mapping of in vivo derived yeast Pho4. (*a*) Total ion current for full-scan LC-ESMS analysis of the SDS-PAGE purified Pho4 tryptic digest. (*b*) Single-ion monitoring LC-ESMS trace for *m/z* 79 and 63. Fractions numbered 1 to 9 were manually collected for further analysis. (*c*) Negative ion nano-ES with *m/z* 79 precursor scan of fraction 5. The spectrum shows 2^-, 3^-, and 4^- charge states for a phosphopeptide with average mass of 2615.5. (*d*) Positive ion nano-ES CID product ion spectrum of the triply charged ion (*m/z* 872.8) for the 2615 peptide found in fraction 5. The y_n ion series unambiguously determines the site of phosphorylation as Ser^{223}. For clarity not all ions have been labeled. Actual coverage is indicated on the peptide sequence.

is frozen in time on the MALDI target. However, the challenge of phosphopeptide analysis by MALDI is to find the phosphopeptides in the presence of other, probably more abundant nonphosphorylated peptides. Unfortunately, phosphopeptides generally seem to ionize less efficiently than nonphosphorylated peptides and thus tend to be minor components of most mixtures. Furthermore, the scan techniques described above are not available on MALDI-TOF instruments. Two analytical strategies described below when coupled with MALDI TOF as the readout have proven to be useful for phosphopeptide mapping.

A number of broad-specificity protein phosphatases are available commercially that will efficiently remove phosphate from the side chains of serine, threonine, and tyrosine residues. MALDI mass spectra of a phosphoprotein tryptic digest recorded before (–) and after (+) treatment with phosphatase will generate a difference map that can be used to identify candidate phosphopeptides. The phosphopeptides are identified based on the disappearance of a peak present in the before (–) sample and the simultaneous appearance or increased intensity of another peak 80 Da (or multiple of 80 Da) lower in the after (+) sample. Figure 3.3 shows a region of the MALDI mass spectrum from an unfractionated tryptic digest of myosin-V tail protein (amino acids 1442 to 1853 from mouse myosin-Va) before (top) and after (bottom) treatment with calf intes-

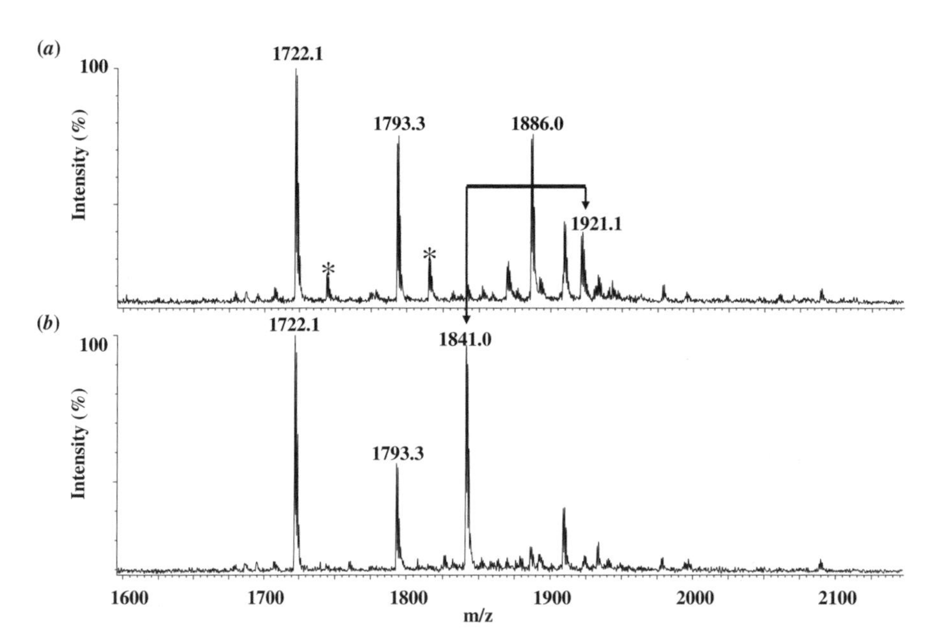

Figure 3.3. Determination of phosphorylation sites using phosphatase treatment and MALDI analysis. (*a*) Partial MALDI mass spectrum of myosin-V tryptic digest. Peaks labeled with an asterisk are Na adducts. (*b*) Partial MALDI mass spectrum of myosin-V tryptic digest after treatment with calf intestinal phosphatase (CIP). A peak at *m/z* 1921.1, present in the before CIP sample has disappeared in the after CIP sample and a new peak 80 Da lower at *m/z* 1841.0 has appeared in the after CIP sample.

tinal phosphatase. A peak at *m/z* 1921.1, present in the before sample has disappeared in the after sample and a new peak 80 Da lower in mass has appeared in the after sample at *m/z* 1841.1. This data suggests that the peptide at *m/z* 1921, is a candidate phosphopeptide spanning amino acids 1648 to 1664 containing one mol of phosphate. This assignment was confirmed by sequencing the *m/z* 1921 peptide by nano-ES, which also identified the site of phosphorylation as Ser^{1650} (Karcher et al., 2001). There is a second peak at *m/z* 1886 in the top spectrum that also disappears after treatment with phosphatase. However, unlike the *m/z* 1921 peptide, no peak 80 Da lower in mass (or multiples of 80 Da) appears in the after spectrum and further analysis revealed that this peak was not a phosphopeptide. It is not clear why this peptide disappears following phosphatase treatment, but it underscores the need to follow up any preliminary analysis with direct peptide sequencing.

An alternative to identify the phosphorylated peptides directly in the sample is to isolate the phosphopeptides from the bulk of the sample. Fortunately, the phosphate group provides a convenient handle for the isolation or enrichment of phosphopeptides from protein digests via immobilized metal affinity chromatography (IMAC) (Porath, 1992; Cao and Stults, 1999; Posewitz and Tempst, 1999). IMAC using immobilized Fe(III) was originally shown by Porath and co-workers to bind phosphoamino acids, phosphopeptides, and phosphoproteins (Porath, 1988; Muszynska et al., 1992). Figure 3.4 shows the MALDI mass spectrum of an unfractionated protein digest that contains

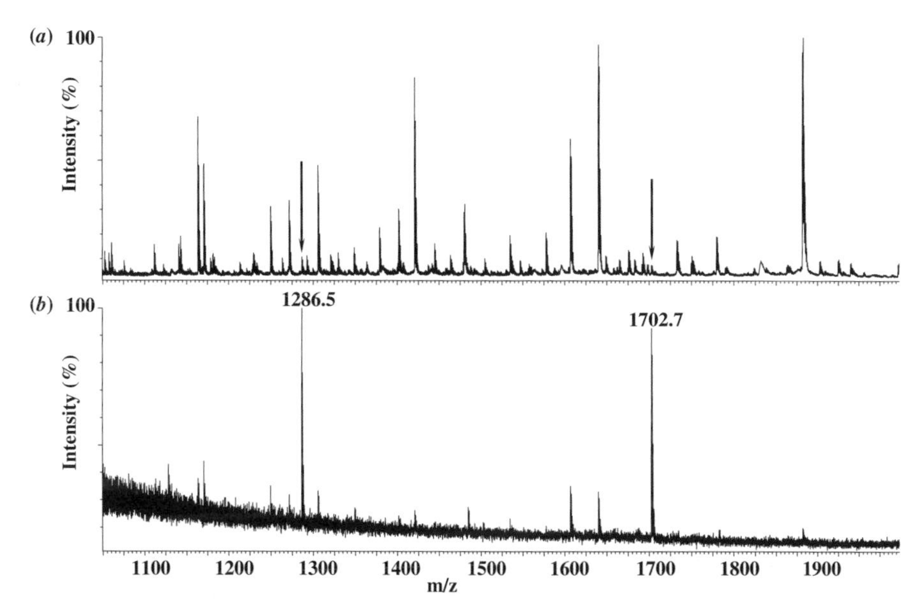

Figure 3.4. Phosphopeptide enrichment using immobilized metal affinity chromatography (IMAC). (*a*) MALDI spectrum of an unfractionated protein digest spiked with two phosphopeptides (MH+ 1286.4 and 1702.8) present at approximately 10 percent stoichiometry. (*b*) After IMAC the two phosphopeptides are the most abundant peaks in the mass spectrum.

two phosphopeptides (MH+ 1286.4 and 1702.8) present at approximately 10 percent stoichiometry. The top panel shows the protein digest prior to IMAC and the bottom panel after enrichment on IMAC. In the top panel the presence of the two phosphopeptides is not obvious and none are among the top 50 most abundant peaks. After IMAC (bottom panel) the two phosphopeptides are the most abundant peaks in the spectrum. It is important to remember that the IMAC resins are not selective for phosphopeptides, rather they have a high affinity for acidic peptides. Peptides with multiple aspartic and/or glutamic acid residues are a major source of competition for binding to the resin. It is, however, becoming increasingly clear that acidic peptides are not the only source of unwanted binding (Jensen, unpublished data).

Given that phosphorylation stiochiometry on in vivo derived proteins is generally much less than 100 percent and that peptides containing Glu and Asp residues are very common, IMAC should be considered an enrichment tool, not a purification device. Thus the question remains how to identify the phosphopeptides in the enriched pool. One possibility is to use the phosphatase reaction described above and interrogate the MALDI mass spectra of the IMAC-enriched sample for peptides that disappear. Another and perhaps simpler approach is to look for peptides that undergo a neutral loss of 98 or 80 Da upon metastable decomposition. In the flight tube of a MALDI-TOF mass spectrometer phosphopeptides readily undergo metastable decomposition to produce abundant $[MH\text{-}H_3PO_4]^+$ and $[MH\text{-}HPO_3]^+$ ions (Annan and Carr, 1996). In a single-stage reflectron instrument or a TOF/TOF-type instrument (Backlund et al., 2003), using a rapid postsource decay (PSD) (Kaufmann et al., 1993) experiment, these fragment ions can be properly focused and detected at their correct masses corresponding to M-98 and M-80, respectively. This simple experiment, which can be done in a batch mode on the top *n* peptides, easily distinguishes phosphorylated peptides from nonphosphorylated peptides. By way of illustration, we determined that from the top ten most abundant peptides shown in Figure 3.5*a*, MH^+ 1640.9, 1702.9, and 1862.9 were phosphorylated (e.g., see Fig. 3.5*b*). In a second pass of the next five most abundant, we found that MH^+ 1782.9 was also phosphorylated. We have shown previously that by using sufficiently high laser energies this experiment has a sensitivity on the order of 20 fmol on target (Chen et al., 1999). Because of the high sensitivity, ease, and speed of doing this experiment, it is an attractive approach to identifying phosphopeptides in simple mixtures such as an HPLC or IMAC-enriched fraction (Loughrey-Chen et al., 2002). It would also be useful for confirming phosphopeptides identified by the +/– phosphatase strategy. In either case no additional sample is required. The original sample spot on the MALDI target is sufficient for the entire PSD experiment.

Quantitative Analysis of Protein Phosphorylation

Many phosphoproteins are phosphorylated at more than one site; yet the biological implications of this multisite phosphorylation are not clear. The challenge that follows phosphosite mapping is to understand which phosphorylation sites modulate a given biological activity or activate a step in a signaling pathway. In some cases phosphorylation or dephosphorylation at a single site can be sufficient to confer function (Karcher et al., 2001). On proteins with multiple phosphorylation sites, individual sites may each

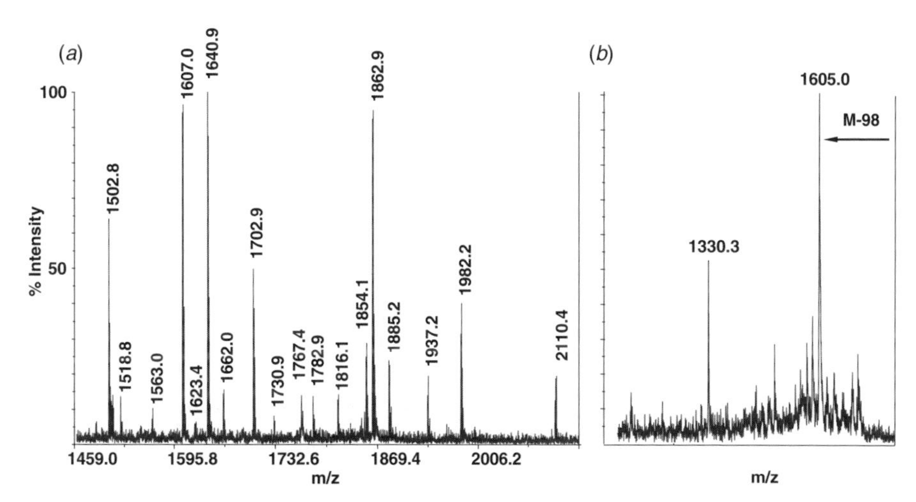

Figure 3.5. MALDI and MALDI-PSD analysis for phosphopeptide detection. (*a*) Partial MALDI mass spectrum of peptide and phosphopeptide mixture ca. 100 fmol. (*b*) MALDI-PSD spectrum of MH+ 1702.9 from top 10 analysis showing neutral loss of H_3PO_4 (M-98).

regulate a different function or activity (Komeili and O'Shea, 1999), and in other cases a single phosphorylation-dependent function may depend upon processive or distributive phosphorylation of multiple sites (Verma et al., 1997). Recently, it has been shown that multisite phosphorylation can act to set a threshold for regulation of protein activity (Nash et al., 2001). In order to unravel this structure–function relationship a thorough quantitative analysis of the phosphorylation profile is necessary. Changes in the stoichiometry at specific phosphorylation sites in response to stimulation or a change in the local environment speak to the functional significance of those particular phosphorylation sites.

The most common approach to measuring changes in phosphorylation at one or a few sites is detection of the protein with an antibody against the specific phosphoepitope. While this readout is fast and sensitive, raising phosphosite-specific antibodies can be both a long and frustrating process. An alternative to the analysis of epitopes that prove refractory to the production of antibodies is the "mass spec western" (Arnott et al., 2002). In this case the mass spectrometer is set up to selectively detect one or two peptides that are specific for the protein of interest. If those peptides are phosphorylated, then a phosphosite-specific mass spec western assay can be established to quantitate phosphorylation at one or several different epitopes. On an ion trap or the quadrupole-TOF type of instruments, this is done using a full-scan LC-MS/MS experiment and targeting specific precursor ions for the phosphorylated peptides. Figure 3.6*a* shows the total ion current (TIC) for two doubly charged precursors representing the phosphorylated and nonphosphorylated forms of a tryptic peptide derived from 5 ng of a 70-kDa protein cut blindly from an SDS-PAGE gel. This data was acquired by setting up specific time domains for each precursor and monitoring for a single precursor in each time window. Because the experiment is set up to target both the phosphorylated

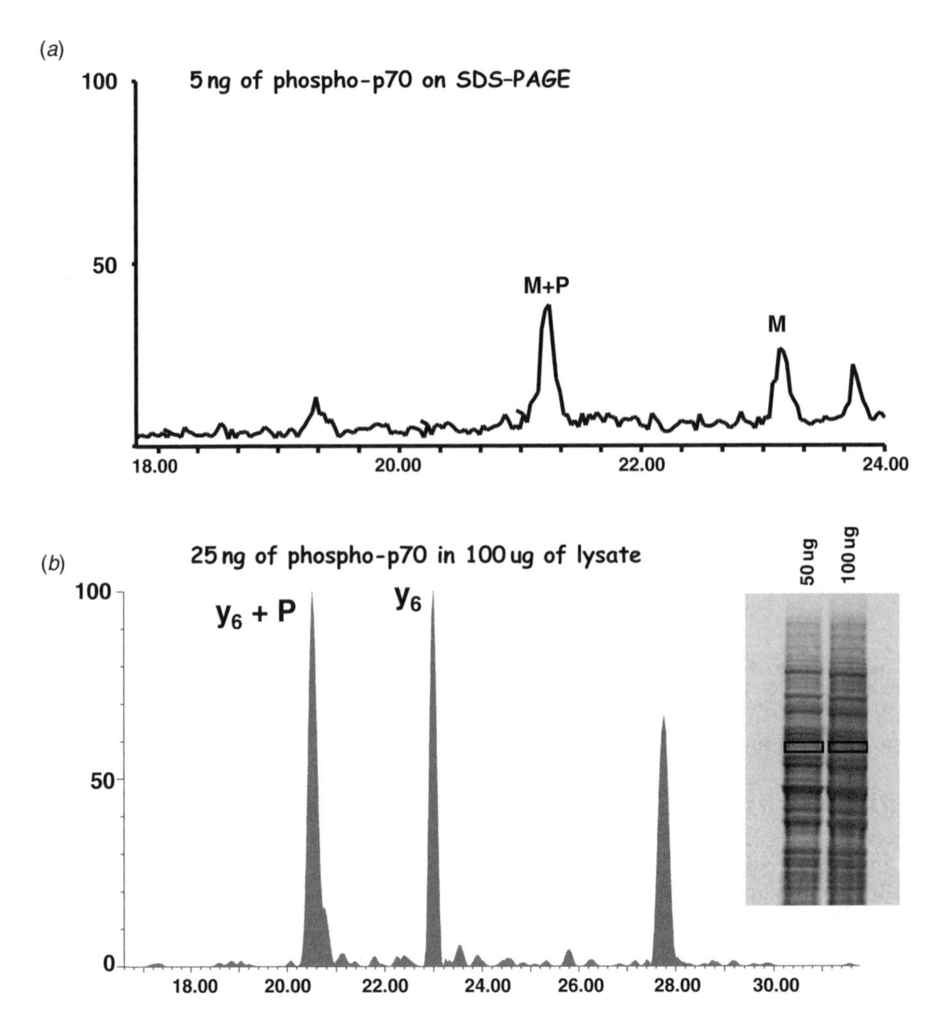

Figure 3.6. Quantitation of phosphorylation using phosphosite-specific mass spec western assay. (*a*) TIC for two doubly charged precursors representing the phosphorylated and nonphosphorylated forms of a tryptic peptide derived from 5 ng of protein cut blindly from an SDS-PAGE gel. (*b*) Twenty-five nanograms of the same protein were spiked into 100 μg of yeast lysate and separated on a SDS-PAGE; a band corresponding to the protein MW was cut (inset). The same two precursor as in panel (*a*) were monitored by full-time LC-ESMSMS. In this panel the extracted ion chromatogram of two abundant y_n ions diagnostic for the phosphorylated and nonphosphorylated forms of the peptide are shown.

and the nonphosphorylated forms of a particular sequence, a measure of the absolute stoichiometry at the specific site can be derived. Although the stoichiometry measured in this way might have an error associated with the possibly different ionization efficiency of the nonphosphorylated peptide and its phosphorylated counterpart, it is a convenient measure of stoichiometry and can be used to accurately determine changes in phosphorylation stoichiometry.

In the analysis of low-abundance proteins from cell lysates the issue of background signal from highly abundant proteins in the lysate must be addressed. Because experiments of this type record a complete MS/MS spectrum with every scan, diagnostic sequence ion chromatograms can be reconstructed during data analysis to reduce the background in the precursor ion TIC and add specificity to the experiment. Figure 3.6*b* shows the analysis of the same precursors described in Figure 3.6*a*, but using 25 ng of the protein spiked into 100 μg of yeast lysate. The background from yeast proteins migrating at the same molecular weight on the gel is now significant as judged by the TIC for the data-dependent LC-MS/MS analysis of the sample collected in a separate LC-MS experiment (data not shown). However, by extracting the signal for two abundant y_n ions diagnostic for the phosphorylated and nonphosphorylated forms of the peptide (dark trace), a relatively clean signal is derived, and it is possible to determine the phosphorylation stoichiometry at this site (ca. 60 percent). In this case the change in phosphorylation at any given site is determined directly from a change in the apparent stoichiometry for each sample, and it is not necessary to normalize the measurements for protein load. If the nonphosphorylated peptide is not available for analysis, another nonphosphorylated peptide from the protein can be used as the normalization factor (Ruse et al., 2002); however, in this case the changes in site-specific phosphorylation will be based on relative measurements. If a measure of absolute quantitation is necessary, then either a set of external standards (Lee et al., 2002) or stable isotope-labeled internal standards (Barr et al., 1996; Ruse et al., 2002; Gerber et al., 2003) with carefully prepared calibration curves are necessary. Using approaches similar to that described, the kinetics of phosphorylation at five different sites on the C-terminus of rhodopsin, a highly abundant light-sensing protein from the rod cells of the retina were studied (Lee et al., 2002). Via targeted LC-MS/MS of selected precursors, the authors were able to determine the relative amount of each monophosphorylated species for the three primary sites of phosphorylation even though they exist on the same peptide. These authors used synthetic rhodopsin phosphopeptides as external standards to establish response factors and calibration curves for the various phosphorylated forms of the C-terminal rhodopsin peptides.

An alternative to the full-scan targeted LC-MS/MS approach for peptide quantitation described above is the use of multiple reaction monitoring (MRM) experiments on a triple quadrupole mass spectrometer. In this experiment, the MS targets and monitors a single transition (or several) for a precursor of interest, that is, a certain precursor ion fragments to yield a certain product ion. Using a triple quadrupole MS the first quad (Q1) selects the precursor, the second quadrupole (Q2) acting as a collision cell fragments it via CID, and the third quadrupole (Q3) detects a single product ion. Because this experiment utilizes two dimensions of selectivity and the analyzer (Q3) is not scanning, but operating in a single-ion monitoring (SIM) mode, it is perhaps the most sensitive way to detect a single species in a complex mixture. It has been used in pharmacology, drug metabolism, and clinical chemistry for nearly 20 years (Hunt et al., 1982, Schweer et al., 1986; Edlund et al., 1989) to detect small molecules and their metabolites in complex biological matrices. More recently the concept of quantitation via MRM has been applied to the analysis of protein expression (Barr et al., 1996; Barnidge et al., 2003; Gerber et al., 2003) and changes in phosphorylation stoichiometry (Stemman et al., 2001). The advantages of the MRM method on the triple quadrupole mass spectrometer are clearly demonstrated in Gerber et al. (2003) where 300 amol of myoglobin can be detected in a band excised from 50 μg of yeast whole-cell lysate via MRM for two different myoglobin peptides. Using the MRM strategy with stable isotope-labeled internal standards, the same authors quantitated cell-cycle-dependent changes in phosphorylation on Ser1126 of the human separase protein. In comparing results from targeted LC-MS/MS on an ion trap and MRM on a triple quadrupole MS, Gerber et al. report that 25 times more total protein was necessary to perform the analysis on the ion trap.

Measuring changes in phosphorylation stoichiometry on many sites simultaneously or on more than one protein at a time presents a much greater challenge. There are no technologies available to selectively capture and/or detect both the phosphorylated and the nonphosphorylated forms of any sequence; thus, any attempts to measure changes in phosphorylation over a large number of sites will have to rely on measuring relative changes. Currently, the most accurate way to accomplish this is to perform a pairwise analysis of two samples that are differentially labeled with a stable isotope tag. This can be accomplished by chemical labeling (Oda et al., 2001; Zhou et al., 2001; Goshe et al., 2002), enzymatic labeling (Bonenfant et al., 2003; Zhang et al., 2002), or metabolic labeling (Oda et al., 1999). However, quantitative analysis of more than just a few phosphorylation sites also requires a detection method less specific than the targeted LC-MS/MS and MRM techniques described above. The most common solution to this has been to couple a phosphopeptide enrichment strategy with routine MS detection of all isotopically labeled species (Salomon et al., 2003). The success of this approach is obviously dependent on the selectivity and efficiency of the enrichment protocol. An alternative strategy is to use phosphopeptide-selective precursor ion scanning to specifically detect isotopically labeled phosphopeptides in unfractionated mixtures (Zappacosta and Annan, 2003). Figure 3.7 shows the quantitative phosphopeptide analysis of two samples representing the same protein from two different sources. Both samples are chemically labeled with a single isotope tag on the N-terminus (Zhang et al., 2002), source 1 containing a d_0 tag and source 2 containing a d_5 tag. The samples are pooled and analyzed by precursor ion scanning for m/z 79 (Carr et al., 1996; Wilm et al., 1996). No enrichment was used prior to analysis. The negative ion mass spectrum shown in Figure 3.7*a* illustrates the actual complexity of the sample. Figure 3.7*b* demonstrates the selectivity of the precursor ion scan. Three scenarios present themselves in this spectrum. A singlet labeled with d_0 represents a phosphorylation site utilized only in source 1. A singlet labeled with d_5 represents a phosphorylation site utilized only in source 2. Which of these two possibilities the singlet actually represents is determined by direct sequencing. The third scenario shown in the expansion (Fig. 3.7*c*) is differential phosphorylation between the two sources. In this case the ratio of d_0/d_5 indicates a fourfold increase in phosphorylation at this site. Using this strategy, multiple sites on a single protein or in fractions from affinity-captured protein complexes can be monitored simultaneously. Of course, those phosphorylation sites that have been shown by the experiment to be differentially regulated must be identified by direct

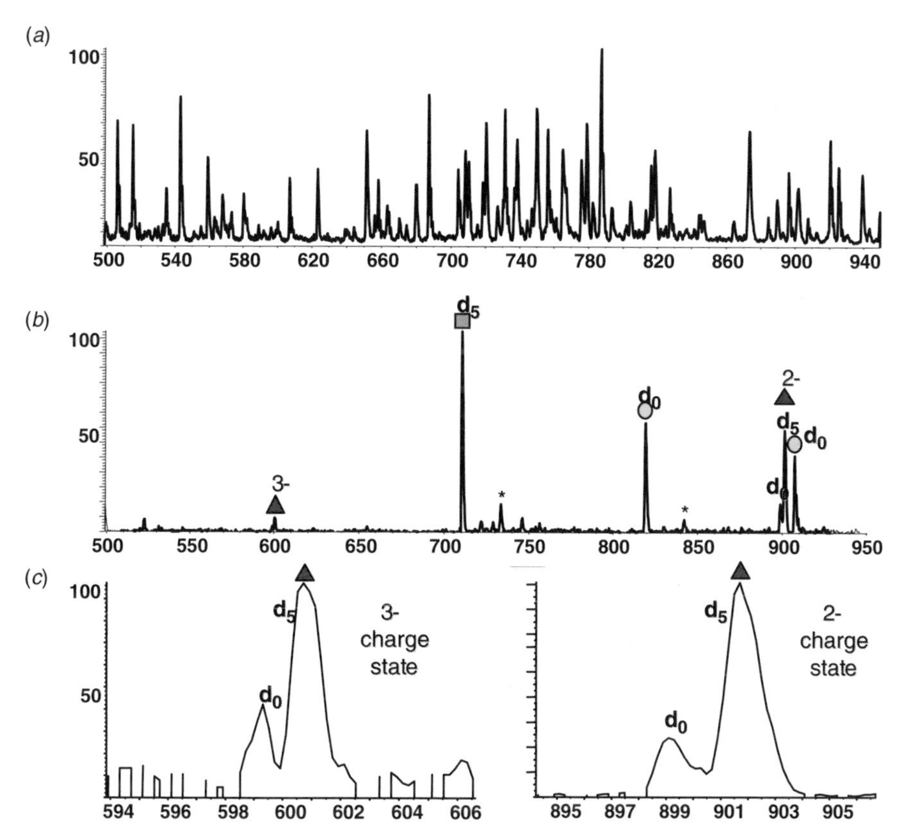

Figure 3.7. Quantitation of phosphorylation stoichiometry using N-terminal isotope tagging (NIT) and phosphopeptide-selective precursor ion scanning. (*a*) nano-ES full-scan (–) ion MS spectrum of the unfractionated protein digest. (*b*) Precursor ion scan for *m/z* 79. Single peaks labeled with a circle represent phosphorylation sites utilized only in source 1 (d_0 tag). A singlet labeled with a square represents phosphorylation sites utilized only in source 2 (d_5 tag). The case of differential phosphorylation between the two sources is shown by peak pairs (d_0/d_5) labeled with a triangle. (*c*) Expansion of the peak pairs for the differentially phosphorylated phosphopeptide showing two different charge states.

sequencing, which is easily accomplished via targeted MS/MS. Since only the differentially regulated phosphopeptides need to be identified, this approach can be described as a results-driven strategy (Griffin et al., 2003).

Several groups have attempted an analysis of the so-called phosphoproteome, that is, the entire phosphopeptide content of a cell. Most of this work has used IMAC, either alone (Ficarro et al., 2002) or in combination with another form of preliminary enrichment (Salomon et al., 2003) to isolate the phosphoproteome prior to LC-MS/MS identification. As an alternative to IMAC, Oda et al. (2001) and Goshe et al. (2002) have chosen to chemically replace the phosphate groups on the side chains of phosphoserine and phosphothreonine residues with an isotope-encoded biotinylated affinity tag. Generally, however, the number of phosphopeptides identified by these approaches has been somewhat disappointing (about 200 to 400 phosphopeptides mostly from the more abundant proteins in a cell). Thus, quantitating large numbers of phosphoepitopes efficiently remains a significant challenge.

CONCLUSIONS

Reversible protein phosphorylation is a critical component of the machinery that controls cellular function. Techniques to map phosphorylation sites, while not perfect, are these days fairly reliable. Although radioactive Edman sequencing remains the most sensitive technique for direct identification of phosphorylated residues, mass spectrometry has emerged as the method of choice, especially for in vivo derived proteins, due to its flexibility, selectivity, sensitivity, and nonreliance on radioactivity. That said it is perfectly clear that no single MS-based method is capable of identifying all of the sites on a hyperphosphorylated protein in a single experiment (Loughrey-Chen et al., 2002). As the recent literature bears witness to, the fact that multisite phosphorylation is quite common and that utilization at any given site is often substoichiometric (Verma et al., 1997), the comprehensive analysis of low-abundance phosphoproteins remains a challenge. A further challenge is to understand the functional relevance of any given phosphorylation site or set of sites on a protein or in a pathway. Quantitation of phosphorylation stoichiometry and how it changes in response to stimulus remains the most direct means to achieve this. Again, the selectivity, sensitivity, and flexibility of the mass spectrometer make it a valuable tool to help address this point. The ability to quantitate a few sites on a single protein is currently a reality, even for low-abundance proteins derived directly from cells. However, the quantitation of many sites, such as those in a signaling cascade or a signaling complex, in a single set of experiments is still a challenge.

Acknowledgments

We are indebted to current and former members of the Proteomics and Biological Mass Spectrometry Group at GSK, Susan Loughrey Chen, Michael Huddleston, Dean McNulty, Therese Sterner, Steven Carr, and Xiaolong Zhang for their experimental and intellectual contributions to the work presented in this chapter.

REFERENCES

Angers-Loustau, A., Cote, J.F., and Tremblay, M.L. (1999). Roles of protein tyrosine phosphatases in cell migration and adhesion. *Biochem. Cell Biol.* **77**:493–505.

Annan, R.S., and Carr, S.A. (1996). Phosphopeptide analysis by matrix-assisted laser desorption time-of-flight mass spectrometry. *Anal. Chem.* **68**:3413–3421.

Annan, R.S., Huddleston, M.J., Verma, R., Deshaies, R.J., and Carr, S.A. (2001). A multidimensional electrospray MS-based approach to phosphopeptide mapping. *Anal. Chem.* **73**:393–404.

Arnott, D., Kishiyama, A., Luis, E.A., Ludlum, S.G., Marsters, J.C., Jr., and Stults, J.T. (2002). Selective detection of membrane proteins without antibodies: A mass spectrometric version of the Western blot. *Mol. Cell Proteomics* **1**:148–156.

Backlund, P.S., Annan, R.S., Zappacosta, F., Sterner, T., Kowalak, J.A., and Yergey, A.L. (2003). Rapid identification and mapping of phosphopeptides by combined immobilized metal ion chromatography and MALDI TOF-TOF analysis, Proceedings of the 51th Conference on Mass Spectrometry and Allied Topics. Montréal.

Barnidge, D.R., Dratz, E.A., Martin, T., Bonilla, L.E., Moran, L.B., and Lindall, A. (2003). Absolute quantification of the G protein-coupled receptor rhodopsin by LC/MS/MS using proteolysis product peptides and synthetic peptide standards. *Anal. Chem.* **75**:445–451.

Barr, J.R., Maggio V.L., Patterson, D.G., Jr., Cooper, G.R., Henderson, L.O., Turner, W.E., Smith, S.J., Hannon, W.H., Needham, L.L., and Sampson, E.J. (1996). Isotope dilution-mass spectrometric quantification of specific proteins: model application with apolipoprotein A-I. *Clin. Chem.* **42**:1676–1682.

Bonenfant, D., Schmelzle, T., Jacinto, E., Crespo, J.L., Mini, T., Hall, M.N., and Jenoe, P. (2003). Quantitation of changes in protein phosphorylation: A simple method based on stable isotope-labeling and mass spectrometry. *Proc. Natl. Acad. Sci. U. S. A.* **100**:880–885.

Cao, P., and Stults, J.T. (1999). Phosphopeptide analysis by online immobilized metal-ion affinity chromatography–capillary electrophoresis–electrospray ionization mass spectrometry. *J. Chromatogr. A.* **853**:225–235.

Carr, S.A., Huddleston, M.J., and Annan, R.S. (1996). Selective detection and sequencing of phosphopeptides at the femtomole level by mass spectrometry. *Anal. Biochem.* **239**:180–192.

Chen, S.L., Annan, R.S., and Carr, S.A. (1999). Isolation and identification of phosphopeptides using Ga-IMAC and selective, MS or biochemical approaches, Proceedings of the 47th Conference on, Mass Spectrometry and Allied Topics.

Cohen, P. (2001). The role of protein phosphorylation in human health and disease. *Eur. J. Biochem.* **268**:5001–5010.

Cohen, P. (2000). The regulation of protein function by multisite phosphorylation—a 25 year update. *Trends Biochem. Sci.* **25**:596–601.

Edlund, O., Bowers, L., Henion, J., and Covey, T.R. (1989). Rapid determination of methandrostenolone in equine urine by isotope dilution liquid chromatography-tandem mass spectrometry. *J. Chromatogr.* **497**:49–57.

Ficarro, S.B., McCleland, M.L., Stukenberg, P.T., Burke, D.J., Ross, M.M., Shabanowitz, J., Hunt, D.F., and White, F.M. (2002). Phosphoproteome analysis by mass spectrometry and its application to *Saccharomyces cerevisiae*. *Nat. Biotechnol.* **20**:301–305.

Gerber, S.A., Rush, J., Stemman, O., Kirschner, M.W., and Gygi, S.P. (2003). Absolute quantification of proteins and phosphoproteins from cell lysates by tandem MS. *Proc. Natl. Acad. Sci. U. S. A.* **100**:6940–6945.

Goshe, M.B., Veenstra, T.D., Panisko, E.A., Conrads, T.P., Angell, N.H., and Smith, R.D. (2002). Phosphoprotein isotope-coded affinity tags: Application to the enrichment and identification of low-abundance phosphoproteins. *Anal. Chem.* **74**:607–616.

Griffin, T.J., Lock, C.M., Li, X.J., Patel, A., Chervetsova, I., Lee, H., Wright, M.E., Ranish, J.A., Chen, S.S., and Aebersold, R. (2003). Abundance ratio-dependent proteomic analysis by mass spectrometry. *Anal. Chem.* **75**:867–874.

Hanada, T., and Yoshimura, A. (2002). Regulation of cytokine signaling and inflammation. *Cytokine Growth Factor Rev.* **13**:413–421.

Henneke, G., Koundrioukoff, S., and Hubscher, U. (2003). Multiple roles for kinases in DNA replication. *EMBO Rep.* **4**:252–256.

Huddleston, M.J., , Annan, R.S., Bean, M.F., and Carr, S.A. (1993). Selective detection of phosphopeptides in complex mixtures by electrospray liquid chromatography/mass spectrometry. *J. Am. Soc. Mass Spectrom.* **4**:*710*–715.

Hunt, D.F., Giordani, A.B., Rhodes, G., and Herold, D.A. (1982). Mixture analysis by triple-quadrupole mass spectrometry: Metabolic profiling of urinary carboxylic acids. *Clin. Chem.* **28**:2387–2392.

Hunter, T. (1998). The phosphorylation of proteins on, tyrosine: its role in cell growth and disease. *Philos. Trans. R. Soc. Lond. B Biol. Sci.* **353**:583–605.

Karcher, R.L., Roland, J.T., Zappacosta, F., Huddleston, M.J., Annan, R.S., Carr, S.A., and Gelfand, V.I. (2001). Cell cycle regulation of myosin-V by calcium/calmodulin-dependent protein kinase, II. *Science* **293**:1317–1320.

Kaufmann, R., Spengler, B., and Lutzenkirchen, F. (1993). Mass spectrometric sequencing of linear peptides by product-ion analysis in a reflectron time-of-flight mass spectrometer using matrix-assisted laser desorption ionization. *Rapid Commun. Mass Spectrom.* **7**:902–910.

Kocher, T., Allmaier, G., and Wilm, M. (2003). Nanoelectrospray-based detection and sequencing of substoichiometric amounts of phosphopeptides in complex mixtures. *J. Mass Spectrom.* **38**:131–137.

Komeili, A., and O'Shea, E.K. (1999). Roles of phosphorylation sites in regulating activity of the transcription factor Pho4. *Science* **284**:977–980.

Lee, K.A., Craven, K.B., Niemi, G.A., and Hurley, J.B. (2002). Mass spectrometric analysis of the kinetics of in vivo rhodopsin phosphorylation. *Protein Sci.* **11**:862–874.

Loughrey-Chen, S., Huddleston, M.J., Shou, W., Deshaies, R.J., Annan, R.S., and Carr, S.A. (2002). Mass spectrometry-based methods for phosphorylation site mapping of hyper-phosphorylated proteins applied to Net1, a regulator of exit from mitosis in yeast. *Mol. Cell Proteomics* **1**:186–196.

Manning, G., Whyte, D.B., Martinez, R., Hunter, T., and Sudarsanam, S. (2002). The protein kinase complement of the human genome. *Science* **298**:1912–1934.

Mustelin, T., Abraham, R.T., Rudd, C.E., Alonso, A., and Merlo, J.J. (2002). Protein tyrosine phosphorylation in T cell signaling. *Front Biosci.* **7**:d918–d969.

Muszynska, G., Dobrowolska, G., Medin, A., Ekman, P., and Porath, J.O. (1992). Model studies on iron(III) ion affinity chromatography. II. Interaction of immobilized iron(III) ions with phosphorylated amino acids, peptides and proteins. *J. Chromatogr.* **604**:19–28.

Nash, P., Tang, X., Orlicky, S., Chen, Q., Gertler, F.B., Mendenhall, M.D., Sicheri, F., Pawson, T., and Tyers, M. (2001). Multisite phosphorylation of a CDK inhibitor sets a threshold for the onset of DNA replication. *Nature* **414**:514–521.

Oda, Y., Nagasu, T., and Chait, B.T. (2001). Enrichment analysis of phosphorylated proteins as a tool for probing the phosphoproteome. *Nat. Biotechnol.* **19**:379–382.

Oda, Y., Huang, K., Cross, F.R., Cowburn, D., and Chait, B.T. (1999). Accurate quantitation of protein expression and site-specific phosphorylation. *Proc. Natl. Acad. Sci. U. S. A.* **96**:6591–6596.

O'Neill, E.M., Kaffman, A., Jolly, E.R., and O'Shea, E.K. (1996). Regulation of PHO4 nuclear localization by the PHO80-PHO85 cyclin-CDK complex. *Science* **271**:209–212.

Porath, J. (1992). Immobilised metal ion affinity chromatography. *Protein Expr. Purif.* **3**:263–281.

Porath, J. (1988). High-performance immobilized-metal-ion affinity chromatography of peptides and proteins. *J. Chromatogr.* **443**:3–11.

Posewitz, M.C., and Tempst, P. (1999). Immobilised gallium(III) affinity chromatography of phosphopeptides. *Anal. Chem.* **71**:2883–2892.

Rane, S.G., and Reddy, E.P. (2002). JAKs, STATs and Src kinases in hematopoiesis. *Oncogene* **21**:3334–3358.

Ruse, C.I., Willard, B., Jin, J.P., Haas, T., Kinter, M., and Bond, M. (2002). Quantitative dynamics of site-specific protein phosphorylation determined using liquid chromatography electrospray ionization mass spectrometry. *Anal. Chem.* **74**:1658–1664.

Salomon, A.R., Ficarro, S.B., Brill, L.M., Brinker, A., Phung, Q.T., Ericson, C., Sauer, K., Brock, A., Horn, D.M., Schultz, P.G., and Peters, E.C. (2003). Profiling of tyrosine phosphorylation pathways in human cells using mass spectrometry. *Proc. Natl. Acad. Sci. U. S. A.* **100**:443–448.

Schweer, H., Seyberth, H.W., and Schubert, R. (1986). Determination of prostaglandin E2, prostaglandin F2 alpha and 6-oxo-prostaglandin F1 alpha in urine by gas chromatography/mass spectrometry and gas chromatography/tandem mass spectrometry: a comparison. *Biomed. Environ. Mass Spectrom.* **13**:611–619.

Steen, H., Kuster, B., Fernandez, M., Pandey, A., and Mann, M. (2001). Detection of tyrosine phosphorylated peptides by precursor ion scanning quadrupole TOF mass spectrometry in positive ion mode. *Anal. Chem.* **73**:1440–1448.

Stemmann, O., Zou, H., Gerber, S.A., Gygi, S.P., and Kirschner, M.W. (2001). Dual inhibition of sister chromatid separation at metaphase. *Cell* **107**:715–726.

Verma, R., Annan, R.S., Huddleston, M.J., Carr, S.A., Reynard, G., and Deshaies, R.J. (1997). Phosphorylation of Sic1p by G1 Cdk required for its degradation and entry into S phase. *Science* **278**:455–460.

Watts, J.D., Affolter, M., Krebs, D.L., Wange, R.L., Samelson, L.E., and Aebersold, R. (1994). Identification by electrospray ionization mass spectrometry of the sites of tyrosine phosphorylation induced in activated Jurkat T cells on the protein tyrosine kinase ZAP-70. *J. Biol. Chem.* **269**:29520–29529.

Wilm, M., Neubauer, G., and Mann, M. (1996). Parent ion scans of unseparated peptide mixtures. *Anal. Chem.* **68**:527–533.

Wold, F. (1981). In vivo chemical modification of proteins (post-translational modification). *Annu. Rev. Biochem.* **50**:783–814.

Zappacosta, F., and Annan, R.S. (2003). Relative quantitation of phosphorylation stoichiometry and protein expression using an N-terminal isotope tagging strategy. Proceedings of the 51th Conference on, Mass Spectrometry and Allied Topics. Montréal.

Zappacosta, F., Huddleston, M.J., Karcher, R.L., Gelfand, V.I., Carr, S.A., and Annan, R.S. (2002). Improved sensitivity for phosphopeptide mapping using capillary column HPLC and microionspray mass spectrometry: Comparative phosphorylation site mapping from gel-derived proteins. *Anal. Chem.* **74**:3221–3231.

Zhang, X., Jin, Q.K., Carr, S.A., and Annan, R.S. (2002). N-Terminal peptide labeling strategy for incorporation of isotopic tags: A method for the determination of site-specific absolute phosphorylation stoichiometry. *Rapid Commun. Mass Spectrom.* **16**:2325–2332.

Zhang, X., Herring, C.J., Romano, P.R., Szczepanowska, J., Brzeska, H., Hinnebusch, A.G., and Qin, J. (1998). Identification of phosphorylation sites in proteins separated by polyacrylamide gel electrophoresis. *Anal. Chem.* **70**:2050–2059.

Zhou, H., Watts, J.D., and Aebersold, R. (2001). A systematic approach to the analysis of protein phosphorylation. *Nat. Biotechnol.* **19**:375–378.

Zhu, X., Lee, H.G., Raina, A.K., Perry, G., and Smith, M.A. (2002). The role of mitogen-activated protein kinase pathways in Alzheimer's disease. *Neurosignals* **11**:270–281.

4

USE OF HIGH-THROUGHPUT CRYSTALLOGRAPHY AND IN SILICO METHODS FOR STRUCTURE-BASED DRUG DESIGN

Leslie W. Tari and Duncan E. McRee

ActiveSight, San Diego, California

Andy J. Jennings

Syrrx, Inc., San Diego, California

Industrial Proteomics: Applications for Biotechnology and Pharmaceuticals, edited by Daniel Figeys
ISBN 0-471-45714-0

INTRODUCTION

X-ray crystallography, which is used to elucidate the three-dimensional structures of biological macromolecules at atomic resolution, has had a profound impact on biology and medicine over the last 40 years. Knowledge of the three-dimensional structures of biological macromolecules provides unparalleled insights into the functions of individual macromolecules and macromolecular complexes, the exact nature of protein–ligand interactions, and the catalytic mechanisms of enzymes. From an applied standpoint, information about the structural and chemical landscape of a macromolecular drug target can be exploited to develop small-molecule drugs using structure-based drug design (SBDD), which is an iterative method that uses protein structural information to develop and optimize small-molecule inhibitors. By determining crystal structures of protein targets with natural substrates, natural product inhibitors, compounds derived from libraries or de novo designed scaffolds, which pinpoint key interactions, medicinal chemists and molecular modelers are able to devise strategies for making rational modifications to molecules that improve compound affinity and selectivity, as well as compound pharmacokinetics. Through iterative cycles of rational compound design/modification, target-based screening, and co-crystal structure determination, SBDD is able to play a key role in the development of drug candidates with novel scaffolds and can dramatically accelerate the rate at which small-molecule candidates enter clinical testing. An ever-increasing number of drugs on the market are developed using structure-based methodologies, including the human immunodeficiency virus (HIV) protease inhibitor Agenerase (Kim et al., 1995) (Vertex, Kissei and Glaxo Welcome) and the neuraminidase inhibitors Relenza (von Itzstein et al., 1993) (Biota and Glaxo Welcome) and Tamiflu (Kim et al., 1997) (Gilead Sciences and Roche).

While there is no contesting the utility of X-ray crystallography in drug discovery, protein structure determination by this method has traditionally been slow, labor intensive, and expensive, due to serious bottlenecks in a number of steps in the process. However, key technological breakthroughs over the last 20 years have dramatically reduced the time, difficulty, and expense in obtaining protein structures by X-ray crystallography, so that structure determination has moved into the *high-throughput* realm. The major bottleneck of obtaining suitable protein samples for crystallization was removed with the advent of molecular biology tools developed in the 1980s and 1990s that allow facile recombinant expression of target proteins, so that methods for parallel expression and purification of large numbers of gene products now exist (Ding et al., 2002; Claverie et al., 2002; Edwards et al., 2000; Brizuela et al., 2001; Lesley, 2001; Holz et al., 2002). Such parallel methods have allowed scientists to explore diverse arrays of multiple constructs, homologs, and variants for specific protein targets, greatly increasing the chances of obtaining crystallizable protein samples. The development of robotic systems for the setup and imaging of crystallization experiments has also led to exponential increases in throughput of crystallization experiments, as well as increased reproducibility of results. Concurrently, advances in the structure determination process, including structure solution by multiple-wavelength anomalous dispersion (MAD) using selenomethionine incorporated overexpressed protein (Hendrickson et al., 1989, 1990), rapid diffraction data collection at synchrotron beamlines using flash-cooling (Garman, 1999), beamline sample-mounting robotics (Rupp et al., 2002; Karain, 2002; Muchmore, 2000), and automated structure solution methods (Rupp et al., 2002; Lamzin and Pemakis, 2000; Diller et al., 1999; Terwilliger and Berendzen, 1999) have provided the complimentary elements enabling high-throughput structural biology (Rupp et al., 2002; Yasutake et al., 2002; Sugahara and Miyano, 2002; Stewart et al., 2002; Stevens and Wilson, 2001; Stevens, 2000; Schmid, 2002; Kuhn et al., 2002; Krupka et al., 2002; Buchanan, 2002; Burley and Bonanno, 2002; Blundell et al., 2002).

The impact of a structure-based design program on drug discovery is directly proportional to the rate at which co-crystal structures can be determined, and the speed with which that information can feed back to the next compound design cycle. Thus, to be an effective component of a drug discovery program, a cost-efficient parallel approach that simultaneously evaluates the effect of construct, expression, purification, and the chemical space explored during crystallization on the structure determination process is essential.

HIGH-THROUGHPUT CONSTRUCT SELECTION, CLONING, AND MICROSCREENING

Some of the most important variables in obtaining crystallizable protein samples are the correct selection of gene boundaries of the protein of interest as well as the location and type of purification tag. Based on scanning a large number of structure determinations in the literature, a change in length of two amino acids or a mutation of a single amino acid in an expression construct can make the difference in obtaining diffraction quality crystals. Combinatorial and knowledge-based construct design followed by small-scale protein expression have been used to quickly identify useful variants suitable for structural analysis. Success can often be had from samples that have had structurally heterogeneous elements, usually at the ends of the proteins, removed. Identification of heterogeneous regions has been accomplished by systematic truncations from the N- and C-terminal ends of the protein of interest, knowledge-based methods, where the protein of interest is compared with homologs, or a low-resolution crystal structure is used to identify heterogeneous regions, and limited proteolytic digestion. Other variables, including the location and nature of affinity purification and secretion tags, the inclusion of proteolytic sites for tag removal, internal sequence truncations (to remove structurally heterogeneous or hydrophobic loops), bicistronic expression of partner proteins, and site-specific mutagenesis to enhance sample solubility and monodispersity, or to prevent or mimic a posttranslational modification (e.g., serine to aspartate mutations to mimic phosphorylation), are all commonly used strategies by crystallography groups. It has long been thought that the most prized biophysical property of a protein that portends crystallizability is sample monodispersity, as measured by light-scattering and gel filtration techniques (D'Arcy, 1994).

PROTEIN EXPRESSION

Systems have been developed for rapid preparative protein expression in *Escherichia coli* for prokaryotic proteins and soluble eukaryotic proteins, and less rapid systems for intractable eukaryotic proteins as well as for proteins that undergo posttranslational modifications, in insect cells using baculovirus (Possee, 1997), yeast, and mammalian cells. In a system developed at the Genomics Institute for the Novartis Research Foundation (GNF) a high-throughput preparative-scale protein expression system in *E. coli* uses a proprietary 96-tube fermentation system (utilizing 100 mL tubes) (Fig. 4.1) that permits cultures to reach high cell densities (OD_{600} in the range 20 to 30) in 8 h (Ding et al., 2002, U.S. Published Patent Application, No. 2002/0146818). The system is reported to provide tens of milligrams of protein per tube (Stevens, 2000). Scalability is facile since multiple cultures (up to 96) can be dedicated to those proteins that are expressed at low levels. For proteins that express at even moderate levels, this fermentation system makes it possible to generate sufficient quantities of biomass for the processing of several hundred clones per week. This capability has been used by the Joint Center for Structural Genomics to scan through the entire soluble genome of *Thermatoga maritima* for crystal hits (Page et al., 2003).

For eukaryotic proteins and proteins that are intractable in *E. coli*, baculovirus-based expression in insect cells is often employed. In terms of throughput and cost,

(*a*) (*b*)

Figure 4.1. Syrrx protein expression systems. (*a*) The 96-well high-throughput *E. coli* fermentation system for protein expression. The 100-mL tubes can yield up to 20 mg of raw protein per tube. (*b*) The "wave" baculovirus fermentation system for parallel expression of eukaryotic proteins.

baculovirus expression systems can be at least one and sometimes two orders of magnitude slower and more expensive to implement than *E. coli* expression systems. This presents a large bottleneck to eukaryotic structural proteomics. However, in SBDD, the number of realistic targets at this point is relatively few and the throughput envisioned in structural proteomics projects is not needed, making these systems suitable.

HIGH-THROUGHPUT ROBOTIC NANOVOLUME CRYSTALLIZATION AND IMAGING

Historically, the setup of crystallization experiments has been a serious bottleneck in obtaining macromolecular structures. A competent technician is hard pressed to setup 200 crystallization experiments by hand in a day. At GNF/JCSG, this bottleneck was removed by developing robotics to process crystallization trials in an accelerated fashion. Other groups are building automated crystallization systems and some are becoming available commercially such as the RoboDesign Crystalmation system, the Syrrx/RTS system, and a system from Data Centric Automation recently installed at GlaxoSmithKline (Fig. 4.2) (Stevens, 2003). These new systems go beyond liquid handling systems such as those available from Gilson, to couple a database and laboratory information management system (LIMS) system to the process to track large numbers of samples with barcoding. Automated crystallization has a number of advantages over conventional methods, with the most obvious benefits being speed and reproducibility. In addition, robotic systems allow for the implementation of higher-density drop plating configurations for a more condensed experimentation scale, minimizing

(*a*) (*b*)

Figure 4.2. Robotic crystallization systems. (*a*) The Syrrx Agencourt crystallization system. It consists of a number of stations arranged around a central robotic arm, which moves plates between stations and also positions the plates. The stations are a well-filling station that can put any combination of 96 solutions into each well as needed, a mixing station, a well dispensing station (left rear), a protein dispensing station (left), a taping station to seal the plates, and a stacker for finished plates (right). (*b*) The Agencourt protein dispense station. A crystal plate is shown on the dispense block, ready to receive protein. The proteins are kept in a 96-well plate on cooling blocks in the rear of the station.

plate storage space requirements. In addition, an automated crystallization system facilitates efficient tracking of experimental data, a key requirement when working in any high-throughput environment.

One of the major advantages of the GNF robotic crystallization system, now commercially available from Syrrx/RTS, is the use of submicroliter (or nanoliter) crystallization volumes (U.S. Patent 6,296,673). By using volumes 20 to 50 times lower than those routinely employed in standard microliter volume experiments, submicroliter volume techniques dramatically lessen the amount of sample required to comprehensively sample crystallization space (Santarsiero et al., 2002). An extensive exploration of crystallization space at multiple temperatures can be carried out with less than a milligram of protein. The reduced material requirements allow small volumes of large numbers of samples to be screened in parallel, accelerating assessments of construct, tag design, buffer components, and purification method by orders of magnitude. Using conventional methods, these experiments would be executed serially over many weeks. However, with the biomass requirements decreased by a factor of 20 to 50 for growth and purification, tens of experimental protocols for a given protein target can be tested per week (Page et al., 2003). Nanovolume crystallization droplets have much smaller surface-area-to-volume ratios than microliter droplets, increasing the equilibration rate in vapor diffusion experiments. The faster turnaround time for nanovolume crystallization experiments facilitates the rapid assessment and optimization of experimental parameters that impact crystal growth and structure determination (Hosfield et al., 2003). Nanovolume crystallization tends to produce crystals that are smaller than microliter volume methods, which at first glance would seem to be a disadvantage. However, advances in detector technology coupled with the availability of intense in-house and synchrotron X-ray sources allows for routine data collection without losses in diffraction resolution. Additionally, smaller crystals exhibit an increased surface-area-to-volume ratio that provides easier diffusion of cryoprotectants into the crystal lattice, minimizing damage due to osmotic shock during cryoprotection, reducing crystal mosaicity (Goodwill et al., 2001). Lastly, it has been theorized that a reduction in the volume of a crystallization experiment mimics the effects of microgravity by reducing convective flow (Carotenuto et al., 2002; Ng et al., 2002; Vekilov, 1999).

A new crystallization system that uses nanovolumes and is based on a free-interface diffusion technique and may offer orthogonal advantages over the more traditional vapor diffusion methods is the Topaz Crystallizer available from Fluidigm (Hansen et al., 2002). A small volume of protein is pumped through the chip to small chambers where it is allowed to slowly mix with a number of precipitants in separate chambers through small channels. Air pressure is used to control small pressure valves throughout the chip to open and close channels and to pump liquids. The chip is based on fluidic microprocessor technology developed by Stephen Quake and colleagues at Caltech (Thorsen et al., 2002) that has potential for automating and miniaturizing a large number of steps in the proteomics laboratory.

The ability to produce large numbers of crystallization trials leads to a new bottleneck at the inspection step since all of these drops need to be scanned every few days to observe crystal growth. This problem has been solved by a number of groups by

(*a*) (*b*)

Figure 4.3. Protein crystal incubators and imagers. (*a*) The Syrrx gantry imaging system. Stacks of plates are stored on either side of a central gantry arm. Each stack can be taken to one of two imagers by the central arm to be imaged on a number of different schedules. A typical schedule would be days 1, 3, 7, 14, and 21. Plates are typically discarded after 28 days. The inset shows the screen of the imager while imaging a plate. (*b*) RoboDesign RoboMicroscope III crystal incubation and imaging system. The incubators, which can be set at different temperatures, are positioned on either side of a temperature-controlled CCD imager. The incubators allow for random access of crystallization plates.

developing automated imaging systems that combine a storage system with a motorized microscope and change coupled device (CCD) camera. A recent review lists 10 crystal storage and imaging systems available commercially (Stevens, 2003). These systems store the images as computer files that can be scanned quickly by the crystallographer at his leisure and examined for crystal growth. The crystallization plate storage developed at GNF and deployed at Syrrx involves a gantry configuration crystal image and storage system (Fig. 4.3) that can automatically capture and process 138,000 images per day at both 4 and 20°C (Stevens, 2003).

To find the crystals in the large number of images that are generated by these systems, automated crystal recognition software is actively being developed by a number of groups including GNF/JCSG (Spraggon et al., 2002). However, due to the complex and irregular nature of precipitates in crystallization droplets, refractive "skins" that can form on the surfaces of droplets and other complicating factors, the software sometimes misses crystals, or more frequently, characterizes precipitates and skins as crystals. While the software can be "tuned" to detect more than 95 percent of successful crystallization experiments (i.e., to reduce the chance of missing a crystal), this accuracy comes at a cost of a very high rate of false-positive indications (Spraggon et al., 2002). This situation has proven acceptable for scanning easy-to-crystallize prokaryotic targets but may be less suitable for drug discovery applications where a smaller number of high value, very difficult targets with a low probability of crystallizing are being pursued.

HIGH-THROUGHPUT X-RAY DATA COLLECTION

With the high-throughput (HT) paradigm scientists are able to generate several hundreds of crystals/month suitable for diffraction analysis. In such a context, tens to hundreds of crystals must be screened prior to data collection, and data collection throughput must be high, so that complete data sets are collected on 5 to 10 crystals in a single day. To achieve these levels of throughput, data collection at intense synchrotron radiation sources is essential to rapidly obtain high-resolution diffraction patterns on the small crystals generated from automated crystallization experiments. Recent developments in rotating anode X-ray sources and optics by Rigaku/MSC have led to the development of in-house X-ray sources suitable for HT applications such as the FR-E superbright generator that provides flux at the sample equivalent to second-generation synchrotron X-ray beamlines by focusing the X-rays onto a small area. There is also the need for an automated way to load samples into and from the X-ray beam. The manual procedure incurs a heavy time penalty, particularly at synchrotron sources, where entering and exiting the experimental hutch is a time-consuming procedure. An automated system can be operated remotely and run unsupervised on a semicontinual basis when immediate judgment calls are not essential (i.e., diffraction screening). Of equal importance, sample tracking and data capture from diffraction experiments can be automated.

The first automated system was described by Abbott Laboratories (Muchmore et al., 2000) and is now being manufactured by Rigaku/MSC as the ACTOR system (Fig. 4.4). Automated systems are in place at the Stanford Synchrotron Radiation Lab

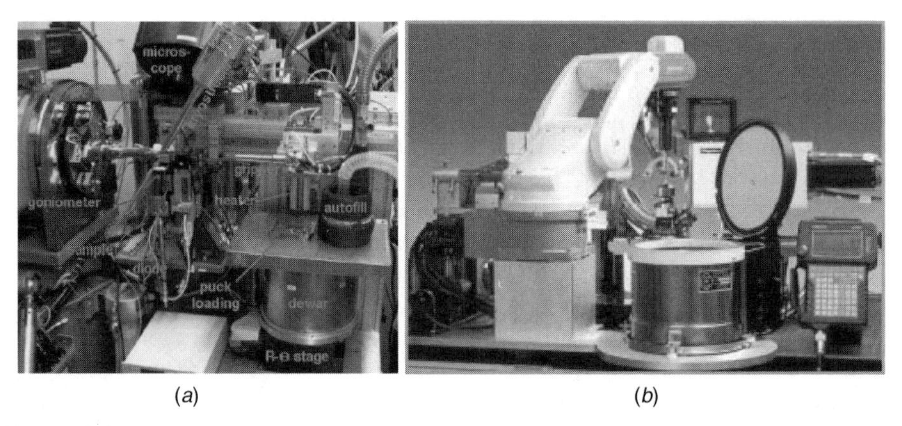

Figure 4.4. Robotic crystal mounting systems. (*a*) The RoboHutch robot used at the Syrrx beamline 5.0.3 at the Advanced Light Source. (*b*) The ACTOR 6-axis mounting robot sold by Rigaku/MSC. The dewars in both systems can hold 60 or more crystal samples in magnetic cassettes. Both systems are equipped with high-resolution CCD cameras and motorized *x*-*y*-*z* goniometers to facilitate automated crystal mounting, centering, diffraction screening, and data collection on multiple samples.

(SSRL) (Abola et al., 2000) and at the Advanced Light Source (ALS) in Berkeley (Snell et al., 2002). Other beamlines throughout the world are actively building automated sampling handling and many systems should be in place by the time this is published. These systems robot utilizes a robotic arm that retrieves conventionally flash-frozen crystals mounted in rayon loops on magnetic bases from storage and places them on a motorized goniometer while keeping them frozen at liquid nitrogen temperatures; both storage and goniometer positions are flexible. The samples point upward from a disk-shaped magnetic base while immersed in liquid nitrogen, and a gripper closes down from above, rotates to a horizontal position, and mounts the crystals on a horizontal goniostat. The use of the mounting robot in the hutch has had a dramatic impact on throughput, such that 100 crystals can be screened with X-rays in 3 to 5 h (Snell et al., 2002).

STRUCTURE DETERMINATION AND REFINEMENT

Structure determination, the process of turning the integrated X-ray intensities into atomic coordinates, has becomes a streamlined process in most cases. In particular, the perfection of the multiple anomalous-dispersion (Se-MAD) method pioneered by Hendrickson, Pähler, and Smith, has greatly lessened the time between initial crystallization of a protein and the time the first model is available (Hendrickson et al., 1989). In this method, the methionines in a protein are replaced by seleno-methionine when the protein is expressed in the cell (Hendrickson et al., 1990). Since selenium absorbs X-rays differently than the rest of the atoms in the protein at wavelengths around the selenium absorption edge, the difference in the intensities that result by taking data at these wavelengths can be treated as isomorphous derivatives. The advantage over the traditional heavy-atom method of solving protein structures is that the data is fully isomorphous and can be obtained from a single crystal minimizing experimental errors. Initially, this method required the use of a methionine auxotroph and was limited to *E. coli* protein expression. Procedures have been worked out for inserting seleno-methionine in yeast, insect, and mammalian cell expression systems by adding exogenous seleno-methionine while simultaneously down-regulating internal methionine synthesis (Bushnell et al., 2001). This has allowed the use of the Se-MAD method on more difficult human drug target proteins.

Phasing of Se-MAD data can be rapidly accomplished using a number of available computer programs (Rupp et al., 2002; Lamzin and Perrakis, 2000; Diller et al., 1999; Terwilliger and Berendzen 1999). If these fail to work straightforwardly, it is usually more successful to grow a new crystal and recollect the MAD data than to continue with a difficult structure determination. Model coordinates can be rapidly obtained by using the automated tracing and refinement procedures of Perrakis and Lamzin as coded in the computer program ARP/wARP (Lamzin and Perrakis, 2000).

The rapid increase in the number of solved structures has made the method of determining structures by molecular replacement increasingly successful. For this method to work, a previously solved structure with sequence identity of about 30 percent or greater needs to be available. However, as the structural proteomics projects now in progress turn out an ever-increasing number of structures, this probability increases greatly. In difficult cases of molecular replacement the challenge is often selecting the correct solution out of a panel of likely candidate solutions. In these cases, using ARP/wARP to refine and rebuild the models of all the solutions will often show the correct solution by a rapid drop in refinement statistics, particularly R-free. The incorrect solutions will initially drop but then become hung up at high values of R-free and not proceed further. Doing this rapidly and efficiently is greatly enhanced by having available a LINUX farm so that several instances of refinement can be run in parallel.

Once an initial structure determination is completed, it is worth stopping and carefully considering whether this crystal form is the ideal one to continue with for SBDD work. It is rare that a protein will crystallize in only one form, and Murphy's law seems to ensure that this is not the first form found. The ideal crystal for SBDD would diffract to better than 2.1 Å resolution, can have compounds bound by soaking into the crystal, and would be grown in conditions without molar concentrations of salt, as this will make it difficult to dissolve compounds. Often the model will have some mobile ends or loops with poor or no density, and a common strategy is to trim these parts in the construct and rescreen for crystals. Other strategies include replacing a few surface residues with another type. Of course, care must be taken not to disrupt the active site of the protein or to adversely affect the activity of the protein. Carefully improving the freezing conditions has also led to crystals with improved diffraction characteristics. In this regard it is worth seeing if the crystal can be grown in the presence of a cryoprotectant so that adding it later is not needed.

DATA TRACKING

The vast amount of experimental data generated from a highly parallel crystallization platform absolutely requires a fully integrated information system. For each construct that enters the platform all data relevant to gene cloning, gene expression, protein purification, protein biophysical characterization, crystallization conditions, crystal annotations, crystal harvesting, diffraction, and structure solution must be captured. Such an informatics platform should provide scientists with real-time access to any experimental data pertaining to a specific project and tracks all the physical materials that progress through the system. Most importantly, the information system provides the basis for efficient bidirectional data sharing that increases platform efficiency. By providing a means for instant access to all the experimental data, the informatics platform enables rapid assessment of key experimental parameters in the gene to structure process to ultimately accelerate the speed in which diffraction-quality crystals of traditionally difficult targets are obtained.

STRUCTURE ANALYSIS

The structure of a protein gives rise to all of its properties, whether it is the mechanism of an enzyme, the recognition of a ligand by its receptor, or the orientation of the protein in its environment. Here we are concerned with the secondary, tertiary, and quaternary structure of a protein. Structure analysis at these levels enables us to place the protein into its correct class, to identify its function, and to speculate upon how we might moderate its activity. A single protein structure provides a wealth of information, but it is the situation where we have more than one structure of a protein that enables us to gain as much insight as possible into the protein in question. Here we shall consider both scenarios.

Single Structure

A single protein structure of high quality allows us to place it within its protein class and family. There are a number of structural databases that classify proteins into various groups (structural classification of proteins (SCOP) (Murzin et al., 1995), class(C) architecture(A) topology(T) homologous superfamily(H) (CATH) (Orengo et al., 1997)), but all share a common concept: Proteins of similar structure have similar function. This is key to any analysis of protein structures for it allows the information gained from any member of that protein group to be applied to any other member. This pooling of data is at the heart of any initial structural analysis. So, by comparing the three-dimensional arrangement of any helices, strands, and turns in a protein with those in the database, one may very rapidly determine which group it belongs to, its function, any homologous proteins, and perhaps any natural or synthetic ligands.

One may often arrive at very similar answers by primary sequence comparisons with sequence databases [e.g., Swiss-Prot (Bairoch and Apweiler, 1996)] except in the case of convergent evolution: the process by which proteins from different ancestral sources mutate independently to perform the same function. These proteins may share so little sequence similarity that it is impossible to find similarly functioning proteins in this way. A structural comparison with always enable those proteins with the same function to be identified [structure alignment of multiple proteins (STAMP) (Russell and Barton, 1993)]. In those cases where the structure is the first in its family to be solved, it will still enable much of the mechanism and function to be determined.

Probably the most important area of interest in the protein will be its binding pocket (also known as an active site), the region of the protein that recognizes any ligands. These ligands moderate the activity in both receptors and enzymes. Receptors can be activated (by an agonist), deactivated (by an inverse antagonist), or rendered unable to be activated (by an antagonist). Enzymes may bind molecules upon which they carry out some reaction (a substrate) or those that the enzyme is unable to consume (an inhibitor).

The most valuable protein structure is one that has been solved with a ligand in the active site. These co-complex structures enable us to identify and quantify the interactions between protein and ligand, so allowing us to see how the protein recognizes the ligand as well as how the ligand performs its function. The knowledge gained from this can lead to the design of ligands that perform some desired function and is the basis of structure-based drug design.

Once the amino acids of interest have been identified in the active site, other aspects of binding can be examined, for example, the presence of other substances in the active site. These can be metals (metalloproteins), waters, or other ligands. Metals may be present in a protein structure for two reasons: structural (Kabsch et al., 1990) or mechanistic (Pavletich et al., 1999). The former plays no direct part in the function of the protein and merely serves to hold the structure in the desired conformation. The latter is often present to carry out a redox reaction (Holmgren, 1989) or to chelate a ligand. Probably the best-known example of the latter is hemoglobin where the iron functions to carry oxygen around the body.

If the protein structure obtained has no ligand bound (an apo structure), one may still be able to identify the important aspects of the active site. One will not, however, be able to analyze to any great degree of confidence the relative contributions of each amino acid within the active site.

We are not only concerned with the recognition of a ligand by the protein but also by the interactions between the protein and other macromolecules. Proteins may interact with other proteins or with nucleic acids. The protein–protein or protein–nucleic acid interface can be analyzed if such a quaternary structure has been solved. This will rarely provide information of use in drug design (unless a ligand is sought that will interfere with the dimerisation of two proteins) but is key in the overall understanding of the protein system in question.

Multiple Structures

When one is fortunate enough to have more than one structure of a protein, even more information can be gained. Two or more structures with different ligands bound will show how different molecules are recognized and bound. By comparing common features in these binding modes we can arrive at a pharmacophore that may itself be used to identify other potential ligands. It is very difficult to predict the movement of backbone atoms in a protein upon ligand binding (Najmanovich et al., 2000), and hence multiple structures provide insight into the malleability of the site that would be impossible to obtain by other methods. Multiple structures also inform us about the mobility of a protein. We may see a region well resolved in one structure but disordered or ordered differently in another protein. A single structure might mislead us as to the relative mobility of a region.

Regardless of how many structures one has, any containing a ligand will provide additional insight into the whole system. Computational techniques have progressed to the point where the conformation of small molecules may be predicted with a high degree of accuracy. These calculations may be carried out in vacuo or, better, with solvation to simulate the behavior of the ligand in the extracellular medium. Upon binding, a ligand has its entropy drastically reduced and is desolvated. With a protein–ligand structure in hand we can compare the bound conformation with the unbound conformation. Proteins often bind ligands in one of their energetically minimal conformations, but there may be some energetic penalty to pay to achieve the correct geometry to fit the site. It is possible to calculate these differences in energy and gain insight into the design of better molecules, for instance, by producing conformationally restrained analog that do not pay the penalty of entropy loss upon binding.

A protein may also make use of water to achieve the correct recognition site for a particular ligand [class I/II transfer ribonucleic acid (tRNA) synthetases (Serre et al., 2001)]. If such water molecules are found in the crystal structure (so-called crystallographic waters), it is possible to replace them with ligand atoms and hence increase the quality of the ligand (García-Sosa et al., 2003). This is a technique commonly employed by medicinal chemists to achieve better drug candidates.

DOCKING

With the crystal structure of a protein of interest in hand, what use is it? In addition to carrying out structure-based drug design one may perform docking experiments. Docking is the computational technique of virtually screening ligands in silico. We may use this technique to:

- Prioritize purchase and screening of external compounds
- Prioritize screening of compound banks
- Prioritize planned synthetic chemistry work
- Attempt to explain structure–activity relationship (SAR)
- Attempt to explain selectivity
- Attempt to explain metabolic data

Prioritization in the above examples is important because resources in a drug discovery project are invariably finite. One would ideally purchase every compound available, make all those not commercially available, and then test all in the assay of interest. Obviously, this is impossible, and therefore a method to select the best compounds for biological screening is required. Docking is one such technique and may be thought of as something akin to a child's blocks-and-holes toy—but in many dimensions.

All docking approaches fall into three classes: high throughput, medium throughput, and low throughput. The distinction between the three classes is a gray one and often depends strongly upon the size of the problem and the computing power available.

High Throughput

The high-throughput approach is the highest speed method and the lowest quality in terms of theory. Rather than attempting to look at a realistic representation of the protein–ligand system, high-throughput techniques use a static representation of the protein. The ligands to be docked are flexible, but conformational sampling is not done with a full mathematical force field—rather a simpler, rule-based system is often employed. The measure of "goodness of fit" so obtained is merely a number bearing no relation to binding energy: It is simply a prioritization score. Even the ordering of the compounds is not accurate enough to be called a ranking. Rather, one may expect poor binders to appear in the bottom half of the results and good binders in the top.

The cutoff chosen is a very subjective one and is more often dictated by the resources available. Nevertheless, good results have been obtained with this methodology, as it is effective at down-prioritizing those compounds that could never fit the active site. This work is often referred to as an enrichment experiment due to the gross categorization into good and poor ligands that results.

The algorithms often work purely in integer space for speed, and this lends itself to easy parallelization across a multiprocessor machine such as a cluster. Such docking clusters are becoming more common within drug discovery groups as their speed and relative cost make them a very economical tool.

- Pros of high-throughput docking (HTD)
  - Fastest docking protocol available
  - Trivial to set up experiments
- Cons of HTD
  - Low level of theory
  - Rigid protein
  - Often requires knowledge of computing system architecture to implement optimal parallelisation
  - Results will not usually reflect any subtle SAR, only good binders and poor binders
  - Cannot simulate solvent, polarization, or quantum mechanics (QM) interactions (pi-stacks)

With a technique this fast, 3 million compounds per day can be docked by some of the clusters currently available. It is possible to take a large subset of every compound commercially available and screen them against a protein target. One may actually derive selectivity information across a panel of targets in only a few days. The large combinatorial arrays that are now possible to synthesize rapidly can also be prioritized or even gross SARs derived if the biological data is known. A general rule of thumb here would be to look at the capacity of the biological screen and select the top *N* compounds to fill the screen.

Medium Throughput

If one desires to look at a more accurate approximation of the protein–ligand system, one may employ medium-throughput techniques. As noted earlier, there is no hard-and-fast rule that dictates which techniques fall into this category, but usually the inclusion of protein flexibility is a good indicator. The complexity of simulating any protein flexibility will decrease the throughput of a docking technique in a manner that is exponential with the number of protein residues allowed to move or sampled. The assumption of a static protein in HTD is a very poor one, and allowing some flexibility adds a great deal of quality to the results of the docking experiment. These results are still more a prioritization than an accurate ranking of binding energies but don't suffer the big problem of static protein techniques: The ligands you find will look very similar to the ligand you removed from the protein structure prior to beginning your high-throughput docking. Merely moving a moderately sized amino acid side chain into one of its alternate conformers can provide enough space to accommodate a group as large as a phenyl ring—a very significant result when attempting to find new chemotypes for the protein target of interest.

- Pros of medium-throughput docking (MTD)
  - Faster than low-throughput docking (LTD)
  - Results are of higher quality than HTD
  - Flexibility of some protein atoms
  - Can explain some SAR
  - Can simulate some solvent by implicit methods
- Cons of MTD
  - Slower than HTD
  - Cannot incorporate relatively large changes in protein structure
  - The results do not correlate at all with binding energies
  - Cannot simulate polarization, QM interactions, or explicit solvent

Medium-throughput docking is sometimes employed as a refinement step after HTD. The top *N* compounds from HTD are subjected to a further round of more accurate docking. The decrease in speed of the technique is offset by a reduction in the number of ligands to be assessed.

Low Throughput

The final category of docking experiments refers to those methods considered low throughput. These are the highest quality calculations of binding affinity and use the most accurate depiction of the protein–ligand system. The increase in theory adversely affects the speed of the techniques and makes it possible to perform these experiments on relatively few compounds.

- Pros of LTD
  - Best quality method, so should be the most accurate
  - Can incorporate full flexibility in the system
  - Full description force field
  - Can incorporate some quantum mechanics/molecular mechanics (QM/MM) methods
  - Binding energies, not just scores
  - Can simulate explicit solvent molecules
- Cons of LTD
  - The slowest of all the categories of technique

- ◦ Very easy to get trapped in ligand local minima as conformational sampling may become a problem
- ◦ Can be tricky to set up the correct charge models for complex systems, e.g., P450s, metalloproteins, etc.

However, one generally would only undertake this work on compounds that are already known to be good binders: The purpose of the work is to explain/alter the observed binding or selectivity of the compounds in question. Many techniques fall into this category, from the molecular mechanics simulation of large proteins with ligands to the quantum mechanical calculation of subtle electronic effects such as polarization.

DRUG DESIGN

The analysis of protein structure is often performed with one goal in mind: to design a ligand capable of binding to it and moderating its activity. The use of protein structure to design these compounds, which may eventually become drugs, is known as structure-based drug design (SBDD). Strictly speaking, the technique would be more accurately termed structure-based ligand design as the factors that determine if a compound is a suitable drug candidate are often not related to the affinity of the compound for the protein target.

We have already touched upon all the points important for SBDD. To recap, these are:

- Important active site residues
- Known ligands and what determines how they interact with the protein
- Presence of metals or water molecules bound to the active site
- Protein flexibility

The starting point for SBDD can vary. One might have a protein structure with no ligands bound and so necessitates de novo drug design. A structure may contain a natural ligand with undesired properties, for instance, an agonist bound where one desires an antagonist or a substrate where one desires an inhibitor. Lastly, a synthetic lead compound may be crystallized in the structure and the information used to increase the affinity, increase the selectivity, or lower the metabolic liability. Each scenario requires different approaches.

Example 1: A Protein Structure Devoid of Any Ligands

An apo structure may be used for de novo ligand design if the active site can be identified. Protein mutation work will often be available to indicate those amino acids that affect ligand binding and/or activity. Once identified, active site amino acids can be selected as partners for protein–ligand interactions. Many packages exist to design ligands from scratch [e.g., SPROUT (Gillet et al., 1993), LeapFrog (Tripos, Inc.), LUDI

(Böhm, 1992)] and approach the task in differing ways. Some attempt to dock chemical groups and subsequently join those groups that appear to bind well to the active site. Others dock a starting group or substructure, then add groups to this starting point, in effect "evolving" a ligand in the active site. Another technique has already been discussed—that of HTD. We may use HTD to identify a promising starting structure and make changes to this molecule, both deleting as well as adding portions. No technique appears to be superior, the quality of the results dictated by the skill of the user rather than the method employed.

Example 2: Structure Containing the Natural Ligand

This represents an excellent starting point for drug design, and one may expect a high probability of success. The literature contains many examples of how an endogenous agonist has been used to design an antagonist. Antagonists are generally larger than their agonist counterparts, presumably the antagonist possessing good affinity for the active site but not the subtle characteristics required to cause an activation of the system. When presented with a structure containing a substrate, it is often enough to know the catalytic mechanism to design an inhibitor. For instance, if the protein in question is a protease, one may replace the labile amide group with an unreactive replacement (for instance an N-methyl amide).

Example 3: Leading Drug Candidate in the Structure

Drug companies often attempt to crystallize lead structures in the protein target in order to enable further design work and explain SAR. A careful examination of such a co-complex will highlight any areas where suboptimal interactions are present or where selectivity may be affected. As part of an iterative process of analysis → crystallization analysis → design, one may achieve impressive results (e.g., the design of cathepsin B inhibitors, which will be discussed later).

STRUCTURE PREDICTION

If a protein structure is unavailable, there is still much that computational techniques can offer to the medicinal chemist or crystallographer. It is possible to derive models of protein structures that are of high enough quality to be used as the basis for SBDD or molecular replacement work. SBDD using a protein model is identical to SBDD using an experimentally determined structure, albeit with the knowledge that the model is certainly wrong in some aspects. Protein structure models may also assist the obtaining of an experimental structure. Many factors can prevent a structure from being obtained experimentally, for instance:

- Insoluble protein: One can find no conditions under which the protein will dissolve prior to crystallization experiments.

- Aggregated protein: The protein obtained will not form an ordered crystal lattice and prefers to "clump" in a seemingly random manner.
- Flexible domains: A very mobile section of the protein will prevent crystallization as it perturbs the crystal lattice as it forms.
- Inability to solve crystal diffraction data: The data cannot be solved due to the lack of a suitable reference structure.

Protein models produced using modern techniques can be good enough to assist the experimentalist to overcome those problems listed above. A good model can:

- Indicate areas of protein structure that play no direct role in ligand recognition or protein–protein interaction—amino acids in these regions can be mutated to assist solubility of the protein.
- Indicate patches of surface residues that are highly polar or highly lipophilic—it is often these patches that drive aggregation and mutants can be designed to improve this behavior without affecting important protein characteristics.
- Indicate flexible regions of protein that play no role in ligand recognition and may therefore be removed to assist the protein crystal packing.
- Provide a template for molecular replacement work to solve crystal diffraction data.

Of course, the key here is to obtain as high a quality, and complete a model, as possible. Two widely used techniques are threading (Miyazawa and Jernigan, 1999) and homology modeling.

Threading algorithms attempt to map the query protein sequence onto a database of example structures or folds. A scoring function of some form evaluates all possible solutions and those with the best scores are output. The Critical Assessment of techniques for protein Structure Prediction (CASP) competitions (Venclovas et al., 2001) have shown that threading methods can find the correct solution among the top N results but seldom score the correct solution as the best one. Threading is an important technique but cannot approach the quality of model obtained when homology modeling is possible.

Homology modeling refers to the use of a suitable, related protein as the template for the query protein sequence. This may at first glance seem to be analogous to threading, but, whereas that method makes no assumption of protein family membership nor sequence similarity between the query and the template, homology modeling makes both of these assumptions. We have introduced the concept of similar structure bestowing similar function on a protein. Another concept is that protein structure is dictated by protein sequence. Therefore, if it is possible to place the query protein within the correct family, one may use the closest experimentally determined member of that family as a template for a model of the query. The name of the technique, therefore, stems from the use of this related, or homologous, protein. Unlike threading, the query sequence is matched to the template structure by the use of a sequence alignment rather than the evaluation of all possible structural solutions of the two. Hence, this technique is unsuitable when the sequence similarity between the query and template protein is low.

CATHEPSIN B INHIBITOR DESIGN

To illustrate some of the principles discussed we shall look at the published work detailing the design of reversible cathepsin B inhibitors (Greenspan et al., 2001). Cathepsin B is a lysosomal cysteine protease, a member of the papain superfamily. It is implicated in the pathology of numerous diseases such as rheumatoid arthritis, cancer, and neurodegenerative disorders. The crystal structure of the rat enzyme was available that possesses only two point mutations in active site with respect to the human enzyme. This structure was a complex of the enzyme (one THE) Protein Domain Database (PDB) Entry called 1THE with an irreversible inhibitor, Cbz-Arg-Ser(OBn)-CMK, which binds via the reaction of its chloromethyl ketone moiety and an enzyme serine residue. This structure was used as the template for an homology model of the human enzyme.

Using a known ligand (Cbz-Phe-$NHCH_2CN$, $IC_{50} = 62\,\mu M$) as the starting point, these workers (Greenspan et al., 2001) examined a series of dipeptidyl nitriles. They performed computational docking experiments around the binding site of the homology model, allowing some residues to be flexible (Q23, C29, Y75, E122, E245, H199). The result of these experiments showed that the predicted lowest energy conformation of Cbz-Phe-$NHCH_2CN$ was consistent with the crystal structure. From this, they were able to observe that both the S_2 and S_3 subsites were suboptimally occupied by this ligand. The crystal structure of the human enzyme complexed with Cbz-Phe-$NHCH_2CN$ was obtained subsequently and showed excellent agreement with the modeled pose.

High-resolution X-ray crystallographic data and molecular modeling were used to optimize the P_1, P_2, and P_3 substituents of this template. Cathepsin B is unique in its class in that it contains a carboxylate recognition site in the S_2' pocket of the active site. Inhibitor potency and selectivity were enhanced by tethering a carboxylate functionality from the carbon α to the nitrile to interact with this region of the enzyme. This resulted in the identification of a 7-nM inhibitor of cathepsin B, with excellent selectivity over other cysteine cathepsins.

REFERENCES

Abola, E., Kuhn, P., Earnest, T., and Stevens, R.C. (2002). Automation of X-ray crystallography. *Nat. Struct. Biol.* **7** (suppl):973–977.

Bairoch, A., and Apweiler, R. (1996). The Swiss-Port protein sequence data bank and its new supplement TREMBL. *Nucleic Acids Res.* **24**:21–25.

Blundell, T.L., Jhoti, H., and Abell, C. (2002). High-throughput crystallography for lead discovery in drug design. *Nat. Rev. Drug Discov.* **1**(1):45–54.

Böhm, H.J. (1992). The computer program LUDI: A new method for the de novo design of enzyme inhibitors. *J. Comp. Aided Molec. Design* **6**:69.

Brizuela, L., Braun, P., and LaBaer, J. (2001). FLEXGene repository: From sequenced genomes to gene repositories for high-throughput functional biology and proteomics. *Mol. Biochem. Parasitol.* **118**(2):155–165.

Claverie, J.M., Monchuis, V., Audic, S., Poirul, O., Abergel, C. (2002). In search of new antibacterial target genes: A comparative/structural genomics approach. *Comb. Chem. High Throughput Screen.* **5**(7):511–522.

D'Arcy, A. (1994). Crystallizing proteins—a rational approach? *Acta Crystallogr. D Biol. Crystallogr.* **D50**:469–471.

Buchanan, S.G. (2002). Structural genomics: Bridging functional genomics and structure-based drug design. *Curr. Opin. Drug Discov. Devel.* **5**(3):367–381.

Burley, S.K., and Bonanno, J.B. (2002). Structuring the universe of proteins. *Annu. Rev. Genomics Hum. Genet.* **3**:243–262.

Bushnell, D.A., Kramer, P., and Kornberg, R.D. (2001). Selenomethionine incorporationin *Saccharomyces cerevisiae* RNA polymerase II. *Structure (Camb)* **9**(1):R11–14.

Carotenuto, L., Cartwright, J.H., Castagnolo, D., Garcia Ruiz, J.M., Otalura, F. (2002). Theory and simulation of buoyancy-driven convection around growing protein crystals in microgravity. *Microgravity Sci. Technol.* **13**(3):14–21.

Diller, D.J., Redinbo, M.R., Pohl, E., Hol, W.G. (1999). A database method for automated map interpretation in protein crystallography. *Proteins* **36**(4):526–541.

Ding, H.T., Ren, H., Chen, O., Fang, G., Li, L.F., Wang, Z., Jia, X.Y., Liang, Y.H., Hu, M.H., Li, Y., Luo, J.C., Gu, X.C., Su, X.D., Luo, M., Lu, S.Y. (2002) Parallel cloning, expression, purification and crystallization of human proteins for structural genomics. *Acta Crystallogr. D Biol. Crystallogr.* **58**(Pt 12):2102–2108.

Edwards, A.M., Arrowsmith, C.H., Chaistendat, D., Dharamsi, A., Friesen, J.D., Greenblatt, J.F., Veladim (2000). Protein production: Feeding the crystallographers and NMR spectroscopists. *Nat. Struct. Biol.* **7**(Suppl):970–972.

Finnin, M.S., Donigian, J.R., Cuhen, A., Richon, V.M., Rifkind, R.A., Marks, P.A., Breslow, R., Pauletich, N.P. (1999). Structures of a histone deacetylase homologue bound to the TSA and SAHA inhibitors. *Nature* **401**:188–193.

García-Sosa, A.T., Mancera, R.L., and Dean, P.M. (2003). WaterScore: A novel method for distinguishing between bound and displaceable water molecules in the crystal structure of the binding site of protein-ligand complexes. *J. Molec. Model.* **9**(3):172–182.

Garman, E. (1999). Cool data: Quantity AND quality. *Acta Crystallogr. D Biol. Crystallogr.* **55**(Pt 10):1641–1653.

Gillet, V.J., Johnson, A.P., Mata, P., Sike, S., and Williams, P. (1993). SPROUT—a program for structure generation. *J. Comput.-Aided Mol. Des.* **7**:127–153.

Goodwill, K.E., Tennant, M.G., and Stevens, R.C. (2001). High-throughput X-ray crystallography for structure-based drug design. *Drug Discovery Today (Genomics Suppl)* **6**:S113–118.

Greenspan, P.D., Clark, K.L., Tommasi, R.A., Cowen, S.D., McQuire, L.W., Farley, D.L., Van Duzer, J.H., Goldbery, R.L., Zhou, H., Du, Z., Fitt, J.J., Coppa, D.E., Fang, Z., Macchia, W., Zhu, L., Capparelli, M.P., Goldstein, R., Wigg, A.M., Doughty, J.R., Bohacek, R.S., Knap, A.K. (2001). Identification of dipeptidyl nitriles as potent and selective inhibitors of cathepsin B through structure-based drug design. *J. Med. Chem.* **44**(26):4524–4534.

Hansen, C.L., Skordalakes, E., Berger, J.M., and Quake, S.R. (2002). A Robust and scalable microfluidic metering method that allows protein crystal growth by free interface diffusion. *Proc. Natl. Acad. Sci.* **99**(26):16531–16536.

Hendrickson, W.A., Horton, J.R., and LeMaster, D.M. (1990). Selenomethionyl proteins produced for analysis by multiwavelength anomalous diffraction (MAD): A vehicle for direct determination of three-dimensional structure. *Embo J.* **9**(5):1665–1672.

Hendrickson, W.A., Horton, J.R., Murthy, H.M., Pahler, A., Smith, J.L. (1989). Multiwavelength anomalous diffraction as a direct phasing vehicle in macromolecular crystallography. *Basic Life Sci.* **51**:317–324.

Holmgren, A. (1989). Thioredoxin and glutaredoxin systems. Minireview. *J. Biol. Chem.* **264**:13963–13966.

Holz, C., Hesse, O., Bolotina, N., Stahl, U., Lang, C. (2002). A micro-scale process for high-throughput expression of cDNAs in the yeast *Saccharomyces cerevisiae. Protein Expr. Purif.* **25**(3):372–378.

Hosfield, D., Palan, J., Hilgers, M., Scheibe, D., McRee, D.E., and Stevens, R.C. (2003). A fully integrated protein crystallization platform for small-molecule drug discovery. *J. Struct. Biol.* **142**(1):207–217.

Kabsch, W., Mannherz, H.G., Suck, D., Pai, E.F., and Holmes, K.C. (1990). *Nature* **347**:37–44.

Karain, W.I., Bourenkou, G.P., Blume, H., Bartunik, H.D. (2002). Automated mounting, centering and screening of crystals for high-throughput protein crystallography. *Acta Crystallogr. D Biol. Crystallogr.* **58**(Pt 10 Pt 1):1519–1522.

Kim, C.U., Lew, W., Williams, M.A., Liu, H., Zhang, L., Swaminathan, S., Bischofberger, N., Chen, M.S., Mendel, D.B., Tai, C.Y., Laver, W.G., Stevens, R.C. (1997). Influenza neuraminidase inhibitors possessing a novel hydrophobic interaction in the enzyme active site: Design, synthesis, and structural analysis of carbocyclic sialic acid analogues with potent anti-influenza activity. *J. Am. Chem.* Soc. **119**:681–690.

Kim, E.E., Baker, C.T., Owyer, M.D., Murcmu, M.A., Rao, B.G., Tung, R.O., Navia, M.A. (1995). Crystal structure of HIV-1 protease in complex with VX-478, a potent and orally available inhibitor of the enzyme. *J. Am. Chem. Soc.* **117**:1181–1182.

Krupka, H.I., Rupp, B., Segelke, B.W., Lekin, T.P., Wright, D., Wu, H.C., Todd, P., Azarani, A. (2002). The high-speed Hydra-Plus-One system for automated high-throughput protein crystallography. *Acta Crystallogr. D Biol. Crystallogr.* **58**(Pt 10 Pt 1):1523–1526.

Kuhn, P., Wilson, K., Patch, M.G., Stevens, R.C. (2002). The genesis of high-throughput structure-based drug discovery using protein crystallography. *Curr. Opin. Chem. Biol.* **6**(5):704–710.

Lamzin, V.S., and Perrakis, A. (2000). Current state of automated crystallographic data analysis. *Nat. Struct. Biol.* **7**(Suppl):978–981.

Lesley, S.A. (2001). High-throughput proteomics: Protein expression and purification in the postgenomic world. *Protein Expr. Purif.* **22**(2):159–164.

Miyazawa, S., and Jernigan, R.L. (1996). Residue-residue potentials with a favorable contact pair term and an unfavorable high packing density term, for simulation and threading. *J. Mol. Biol.* **256**:623.

Muchmore, S.W., Olson, J., Jones, R., Pan, J., Blum, M., Greer, J., Merrick, S.M., Magdalinos, P., Nienaber, V.L. (2000). Automated crystal mounting and data collection for protein crystallography. *Structure Fold Des.* **8**(12):R243–246.

Murzin, A.G., Brenner, S.E., Hubbard, T., and Chothia, C. (1995). SCOP: A structural classification of proteins database for the investigation of sequences and structures. *J. Mol. Biol.* **247**:536–540.

Najmanovich, R., Kuttner, J., Sobolev, V., and Edelman, M. (2000). Side-chain flexibility in proteins upon ligand binding. *Proteins* **39**(3):261–268.

Ng, J.D., Sauter, C., Lorber, B., Kirkland, N., Amez, J., Giege, R. (2002). Comparative analysis of space-grown and earth-grown crystals of an aminoacyl-tRNA synthetase: Space-grown crystals are more useful for structural determination. *Acta Crystallogr. D Biol. Crystallogr.* **58**(Pt 4):645–652.

Orengo, C.A., Michie, A.D., Jones, S., Jones, D.T., Swindells, M.B., and Thornton, J.M. (1997). CATH—A hierarchic classification of protein domain structures. *Structure* **5**(8):1093–1108.

Page, R., Grzechnik, S.K., Canaves, J.M., Spraggon, G., Kreusch, A., Kuhn, P., Stevens, R.C., and Lesley, S.A. (2003). Shotgun crystallization strategy for structural genomics: An optimized two-tiered crystallization screen against the thermatoga maritima proteome. *Acta Crystallogr. D Biol. Crystallogr.* **59**(Pt 6):1028–1037.

Possee, R.D. (1997). Baculoviruses as expression vectors. *Curr. Opin. Biotechnol.* **8**(5):569–572.

Rupp, B., Segelke, B.W., Krupka, H.I., Lekin, T., Schafer, J., Zemla, A., Toppani, D., Snell, G., Earnest, T. (2002). The TB structural genomics consortium crystallization facility: Towards automation from protein to electron density. *Acta Crystallogr. D Biol. Crystallogr.* **58**(Pt 10 Pt 1):1514–1518.

Russell, R.B., and Barton, G.J. (1993). Protein structure prediction. *Nature* **364**:765.

Santarsiero, B.D., Yegian, D.T., Lee, C.C., Spraggon, C., Gu, J., Scheibe, D., Uber, D.C., Cornell, E.W., Nordmeyer, R.A., Kolbe, W.F., Jin, J., Jones, A.L., Jaklevic, J.M., Schultz, P.G., Stevens, R.C. (2002). An approach to rapid protein crystallization using nanodroplets. *J. Appl. Crystallogr.* **35**:278–281.

Schmid, M.B. (2002). Structural proteomics: The potential of high-throughput structure determination. *Trends Microbiol.* **10**(10 Suppl):S27–31.

Serre, L., Verdon, G., Choinowski, T., Hervouet, N., Risler, J.L., and Zelwer, C.J. (2001). How methionyl-tRNA synthetase creates its amino acid recognition pocket upon L-methionine binding. *Mol. Biol.* **306**(4):863–876.

Snell, G., et al. (2002). Automatic sample mounting and alignment system for macromolecular crystallography at the ALS. Annual meeting of the ACA, San Antonio, TX.

Spraggon, G., Lesley, S.A., Kreusch A., and Priestle, J.P. (2002). Computational analysis of crystallization trials. *Acta Crystallogr. D Biol. Crystallogr.* **58**(Pt 11):1915–1923.

Stevens, R.C. (2003). The cost and value of three-dimensional protein structure. *Drug Discovery World* **4**:35–48.

Stevens, R.C. (2000). High-throughput protein crystallization. *Curr. Opin. Struct. Biol.* **10**(5):558–563.

Stevens, R.C., and Wilson, I.A. (2001). Tech. Sight. Industrializing structural biology. *Science* **293**(5529):519–520.

Stewart, L., Clark, R., and Behnke, C. (2002). High-throughput crystallization and structure determination in drug discovery. *Drug Discov. Today* **7**(3):187–196.

Sugahara, M., and Miyano, M. (2002). [Development of high-throughput automatic protein crystallization and observation system]. *Tanpakushitsu Kakusan Koso* **47**(8 Suppl):1026–1032.

Terwilliger, T.C., and Berendzen, J. (1999). Automated MAD and MIR structure solution. *Acta Crystallogr. D Biol. Crystallogr.* **55**(Pt 4):849–861.

Thorsen, T., Maerkl, S.J., and Quake, S.R. (2002). *Microfluidic large-scale integration.* **298**(5593):580–584.

Tripos, Inc. www.tripus.com

Vekilov, P.G. (1999). Protein crystal growth—microgaravity aspects. *Adv. Space Res.* **24**(10): 1231–1240.

Venclovas, C., Zemla, A., Krzysztof Fidelis, and Moult, J. (2001). Comparison of performance in successive CASP experiments. *Proteins: Structure, Function, and Genetics* Suppl. **5**:163–170.

von Itzstein, M., Wu, W.Y., Kok, G.B., Pegg, M.S., Dyason, J.C., Jin, B., Van Phan, T., Smythe, M.L., White, H.F., Oliser, S.W., Colman, P.M., Varghese, J.N., Ryan, O.M., Woods, J.M., Bethell, R.C., Hotham, V.J., Camerer, J.M., Penn, C.R. (1993). Rational design of potent sialidase-based inhibitors of influenza virus replication. *Nature* **363**(6428):418–423.

Yasutake, Y., Yao, M., and Tanaka, I., (2002). [High-throughput protein crystallography]. *Tanpakushitsu Kakusan Koso* **47**(8 Suppl):1033–1037.

5

HIGH-THROUGHPUT ANALYSIS OF PROTEIN STRUCTURE BY HYDROGEN/DEUTERIUM EXCHANGE MASS SPECTROMETRY

Yoshitomo Hamuro, Patricia C. Weber, and Patrick R. Griffin

ExSAR Corporation, Monmouth Junction, New Jersey

Industrial Proteomics: Applications for Biotechnology and Pharmaceuticals, edited by Daniel Figeys
ISBN 0-471-45714-0

INTRODUCTION

Most currently marketed pharmaceuticals target and alter the function of specific proteins within the body. Many of these therapeutic agents are proteins themselves. An understanding of protein structure and how it relates to function is increasingly critical to the development of safe and efficacious medicines. Chapters 1 through 4 describe technology and methodology applied to the study of protein sequence and protein structure and include methods for protein identification, analysis of protein–protein interactions, the identity and function of posttranslational modifications, and the analysis of protein three-dimensional structure. These technologies provide the foundation for modern research in structural and functional proteomics that offers the opportunity to understand how protein structures regulate biological functions.

With the availability of the sequences of all human proteins and an increasing number of protein three-dimensional structures in public databases, the need for techniques to rapidly evaluate protein structure and protein interactions has increased dramatically. This need has been partially filled by incorporation of advanced instrumentation, automation, and computational approaches into the well-established protein structure determination methods of X-ray crystallography and nuclear magnetic resonance (NMR), described in detail in Chapter 4. Computational analyses of protein sequence is discussed in Chapter 10.

Hydrogen/deuterium exchange (H/D-Ex) detected by mass spectrometry (MS) has emerged as a reliable tool for rapid analysis of protein structure. H/D-Ex methods can quickly provide global structural information detailing the extent of folding within a protein. H/D-Ex coupled with fragmentation methods enables one to pinpoint sites of conformational flexibility with high accuracy. This approach can also be used to map protein interaction sites involving both small molecules and macromolecules. Examples demonstrating the use and application of MS-based methods of H/D-Ex on proteins are reviewed here. The recently published examples include views of global protein conformation by analysis of intact proteins. Others highlight the utility of following H/D-Ex by an enzymatic fragmentation step to provide localized information on protein structure and dynamics. Reflecting the increased activity in this field, several comprehensive reviews have been published (Engen and Smith, 2001; Hernandez and Robinson, 2001; Hoofnagle et al., 2003; Kaltashov and Eyles, 2002). This review focuses on recently published examples of representative research areas.

THEORY OF HYDROGEN/DEUTERIUM EXCHANGE FOR ANALYSIS OF PROTEIN STRUCTURE AND DYNAMICS

Amide Hydrogen/Deuterium Exchange

Because the rates of hydrogen/deuterium exchange of main chain amides reflect the unique environment of individual amino acids in the three-dimensional structure, the measure and analysis of amide H/D-Ex have been active research areas for many years (Englander et al., 1997). The patterns of amide H/D-Ex rates differ dramatically in folded and unfolded proteins. In folded proteins, the backbone amide hydrogens exhibit highly variable H/D-Ex rates that can range over 8 orders of magnitude (Engen and Smith, 2001). In contrast, backbone amide H/D-Ex rates in proteins lacking secondary and tertiary structure vary about 100-fold, and the variations in rate depend primarily on neighboring amino acid side chains (Bai et al., 1993).

The exchange kinetics can be readily followed by stable isotope labeling because the exchange times of many backbone amide hydrogens range from seconds to days. In contrast, hydrogens from amino acid side chains containing –OH, –SH, $–NH_2$, –COOH, and $–CONH_2$ groups and hydrogens from the amino and carboxy termini exhibit exchange rates too fast for real-time measurement. Carbon-bound aliphatic and aromatic hydrogens do not participate in standard exchange reactions and undergo isotope substitution only following activation by chemical treatment, such as reaction with hydroxyl radicals (Goshe and Anderson, 1999).

Protection Factor

Tertiary structural features affect the rate of amide H/D-Ex. These include the protein's structure and dynamical properties and participation in hydrogen bonding (Hilser and

Freire, 1996), amide distance from the protein surface (Resing et al., 1999), and flexibility of the peptide chain (Zhang et al., 1996). The degree of retardation in amide H/D-Ex rate as a result of amide physical environment is termed the *protection factor (pf)*:

$$\text{pf} = k_{\text{ch}}/k_{\text{ex}} \tag{5.1}$$

where k_{ex} is the observed exchange rate and k_{ch} is the "intrinsic" exchange rate calculated by assuming a random coil conformation at a given pH and temperature (Bai et al., 1993).

Backbone Amide Hydrogens as Thermodynamic Sensors

Formalisms to relate the observed rates of amide H/D-Ex to thermodynamic stabilization of proteins have been developed (Englander and Kallenbach, 1984). Amide hydrogens of proteins in the native, folded state are proposed to exchange according to the following equation:

$$\text{closed} \underset{k_{\text{cl}}}{\overset{k_{\text{op}}}{\rightleftarrows}} \text{open} \xrightarrow{k_{\text{ch}}} \text{exchanged} \tag{5.2}$$

$$k_{\text{ex}} = k_{\text{op}} \bullet k_{\text{ch}}/(k_{\text{cl}} + k_{\text{ch}}) \tag{5.3}$$

For most proteins at or below neutral pH, amide H/D-Ex occurs by an Ex2 mechanism (Sivaraman et al., 2001), where $k_{cl} \gg k_{ch}$ and Eq. (5.3) becomes

$$k_{\text{ex}} = k_{\text{op}} \bullet k_{\text{ch}}/k_{\text{cl}} = K_{\text{op}} \bullet k_{\text{ch}} \tag{5.4}$$

The opening equilibrium constant at each amide ($K_{op} = k_{op}/k_{cl}$) is reciprocal of the protection factor (pf) and can be translated into the stabilization free energy (ΔG) by Eq. (5.5):

$$\Delta G = -RT\ln(1/K_{\text{op}}) = -RT\ln(\text{pf}) = -RT\ln(k_{\text{ch}}/k_{\text{ex}}) \tag{5.5}$$

The measured H/D-Ex rates in the folded protein can be compared with the calculated intrinsic rates (k_{ch}) to probe the extent of tertiary structure and resulting dynamics. Frequently, the H/D-Ex rates of two or more physical states of a protein, such as with and without ligand and here represented by k_{ex1} and k_{ex2}, are measured to locate stabilization free-energy changes upon the perturbation ($\Delta\Delta G_{1\rightarrow 2}$):

$$\Delta\Delta G_{1\rightarrow 2} = \Delta G_2 - \Delta G_1 = -RT\ln(k_{\text{ex1}}/k_{\text{ex2}}) \tag{5.6}$$

In this formalism, backbone amide hydrogens serve as thermodynamic sensors of their local environment, given that their exchange rates under Ex2 conditions can be readily quantified using mass spectrometry.

OVERVIEW OF HYDROGEN/DEUTERIUM EXCHANGE TECHNOLOGIES

Several approaches have been developed to assess protein dynamics by H/D-Ex. Measure of amide hydrogen exchange by NMR is widely used, and much of the underlying understanding of protein dynamics has resulted from these studies (Englander and Kallenbach, 1984; Woodward et al., 1982; Bai et al., 1994; Kim et al., 1993). In a typical NMR experiment, exchange rates of individual backbone amides are measured. Deuterium atoms in proteins can also be localized by neutron diffraction methods, and the initial neutron crystallography studies on six proteins demonstrated differences in the overall H/D-Ex properties of α-helixes and β-sheets (Kossiakoff, 1985). Because backbone amides within the protein interior were observed to incorporate deuterium, these experiments provided early evidence of large-scale motions in proteins. Investigations by crystallography and NMR have also demonstrated that certain flexible and solvent-exposed regions are resistant to H/D-Ex. While much H/D-Ex information has been obtained by these techniques, some limits exist. For example, few protein crystals are suitable for neutron crystallography (Niimura, 1999) and comprehensive knowledge of backbone amide hydrogen exchange rates is rarely obtained by NMR.

Mass spectrometry (MS) is the preferred method for rapid analysis of H/D-Ex in proteins. The technique is well-suited to measure the level and rates of deuterium incorporation. Only micrograms of sample are required to follow exchange rates of backbone amide hydrogens throughout the entire sequence. The method can also be applied to very large proteins. Crystallization is not required, and there is no need to produce and purify isotopically enriched proteins. Methods to localize sites of H/D-Ex by proteolysis (Zhang and Smith, 1993) and/or MS fragmentation (Deng et al., 1999; Demmers et al., 2002; Yan et al., 2002) have improved the resolution of the MS-based technique so that information approaching single amide resolution can be obtained.

The experimental protocol for analysis of H/D-Ex by MS coupled with proteolysis is shown in Figure 5.1, and steps in the experimental procedure are described in some detail below. Advances in laboratory instrumentation, mass spectrometry, and the understanding of H/D-Ex phenomenon have resulted in an automated system for high-throughput, high-resolution H/D-Ex analysis (Hamuro et al., 2003a). The system, described at the end of this section, incorporates solid-phase proteolysis, automated liquid handling, and streamlined data reduction software.

On-Exchange

To initiate an H/D-Ex by MS investigation, a protein sample is incubated in a deuterated environment to allow incorporation of deuterium atoms at available hydrogen exchange sites. There are few restrictions on reaction conditions, and H/D-Ex behavior can be studied as a function of protein and buffer concentration and composition, solution pH, or presence of ligands. In a typical H/D-Ex assay, comparative studies are conducted. To follow the deuterium buildup of individual amide hydrogens or sets of hydrogens, several on-exchange time points are sampled for each condition (Tables 5.1 and 5.2).

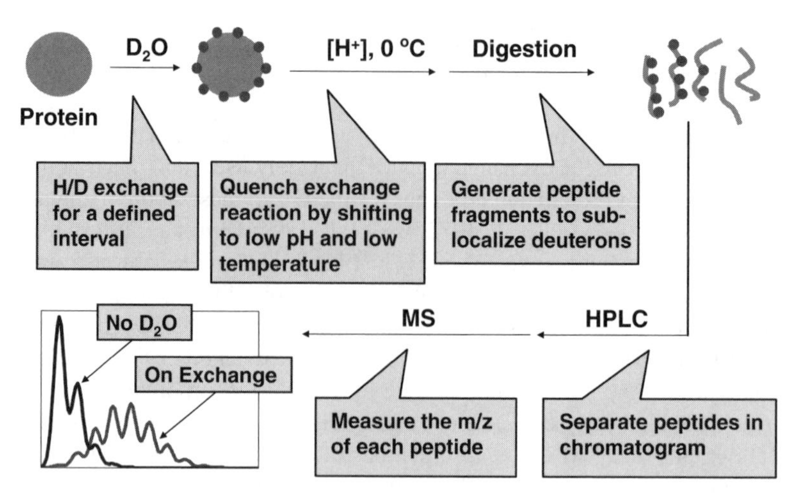

Figure 5.1. Overall H/D exchange experiment. Figure adapted with permission from Hamuro et al., 2003a.

Quench of Exchange Reaction

Following incubation in a deuterated environment for a defined interval, the exchange reaction is quenched by diluting the protein sample with cold, acidic solutions (pH ~ 2.5 and 0°C). The quench conditions significantly slow the amide hydrogen exchange reaction and limit undesirable back exchange to hydrogen. Subsequent procedures are usually conducted near the quench conditions to minimize the loss of incorporated deuterium. After the exchange reaction is quenched, the level of total deuterium incorporation can be obtained by mass analysis of the intact protein.

Protein Fragmentation by Proteolysis

To localize the rate of deuterium buildup to specific amides, the analyte protein is fragmented into a collection of peptides using combinations of endo- and exoproteases. This procedure provides higher resolution information about protein structure and is capable of characterizing any ligand interaction sites on the protein. Due to the low pH of the quench conditions in which the protein and peptide samples are maintained after labeling, the acid-stable protease pepsin is widely used in H/D-Ex studies. Studies with combinations of acid-stable endoproteinases and carboxypeptidases have been employed to achieve greater sequence coverage and higher amide resolution (Woods and Hamuro, 2001; Englander et al., 2003).

Digestion Optimization

The digestion conditions are optimized prior to conducting multiple H/D-Ex experiments, to ensure high sequence coverage and high resolution. To obtain complete sequence coverage, the digestion conditions should result in peptides that can be separated by high-performance liquid chromatography (HPLC), measured by mass

TABLE 5.1. Application of H/D-Ex to Intact Proteins

Protein	Perturbation	Comments	Reference
Cytochrome *c*	temperature	Gas-phase H/D	Valentine and Clemmer, 2002
Myoglobin	± Heme	Gas-phase H/D	Mao et al., 2002
Myoglobin	± Heme	Folding, millisecond exchange	Simmons et al., 2003
Human lysozyme	Mutation D67H	Folding	Canet et al., 2002
Aspartate aminotransferase	Mutation P138A	Folding	Birolo et al., 2002
Pulmonary surfactant protein C	Mutation	Fibril formation	Hosia et al., 2002
Thioredoxin	Oxidation/reduction, C-modification, temperature	Protein stability	Kim et al., 2002
α-Lactabumin β-Lactoglobin	± Ca^{2+}, ± sugar, various protein sources	Protein stability	Alomirah et al., 2003
Cellular retioic acid binding protein I	± Small hydrophobic ligands	Protein–ligand interactions	Xiao et al., 2003
Colicin E9 endonuclease	± Zn^{2+}, ± immunity protein Im9	Protein–metal, protein–protein interactions	van den Bremer et al., 2002
Abelson tyrosine kinase SH3 domain S-protein	± Peptides (GdnHCl)	Protein–peptide interactions	Powell and Fitzgerald, 2003
4 different proteins	± Various ligands (GdnHCl)	Protein–ligand interactions	Powell et al., 2002b
6 different protein	Oligomerizations	Stability of oligomers	Powell et al., 2002a
Influenza A M2 TM fragment	± Lipids	Protein lipid interactions	Hansen et al., 2002
Hydrophobic α-helical peptides	Various sequences in membrane	Effects of tryptophan	de Planque et al., 2003

TABLE 5.2. Applications of H/D-Ex Coupled with Fragmentation

Protein	Perturbation	Number of Residues	Number of Fragments[a]	Exchange (s)	Reference
Macrophage colony stimulating factor-β	n/a	221	22	10–18,000	Yan et al., 2002
Dual-specific A-kinase anchoring protein 2	n/a	375	40	10–3,000	Hamuro et al., 2002a
Amyloid-β peptides	n/a	42	13[d]	1,200	Kraus et al., 2003
Human growth hormone	pH	191	25	10–100,000	Hamuro et al., 2003a
Apomyoglobin	pH and ± hexafluoroisopropanol	154	6	3,600	Sirangelo et al., 2003
Interferon-γ	Aggregation	140	18	180–86,000	Tobler and Fernandez, 2002
HIV capsid protein	Assembly	231	14	30–244,800	Lanman et al., 2003
ε-Subunit of yeast F1-ATPase	Complexation	61	7	600	Nazabal et al., 2003
Factor XIII	Activation	731	22	60–600	Turner and Maurer, 2002
Hemoglobin	± O_2	287[b]	24[b]	on/off[c]	Englander et al., 2003
Endopolygalacturonase II	± Carbohydrate	335	9	86,400, 172,800	King et al., 2002
α-Crystallin	± ATP	357[b]	25[b]	30	Hasan et al., 2002
Protein tyrosine phosphatase 1B	± Inhibitors	321	31	30–1,200	Guo et al., 2002
C-terminal Src kinase	± ADP and ± AMPPNP	462	28	10–72,000	Hamuro et al., 2002b
Regulatory subunit Iα of PKA (94–244)	± cAMP, ± catalytic subunit	151	16	30–600	Anand et al., 2002
Regulatory subunit IIβ of PKA	± cAMP, ± catalytic subunit	416	38	10–3,000	Hamuro et al., 2003b
Thrombin	± Antibody	295	15	600-on, 600-off	Baerga-Ortiz et al., 2002
Interleukin-6	± Antibody	184	15[d]	600-on, 1,200-off	Yamada et al., 2002

[a] The number of peptides used in the analysis of H/D-Ex. Many more peptides are typically identified, e.g., 82 peptides were identified and analyzed, but the H/D-Ex results of only 38 were reported to obtain >99% coverage of the regulatory subunit IIβ of cAMP-dependent protein kinase A (Hamuro et al., 2003).
[b] Total of both α and β subunits.
[c] The protein was on-exchanged (incubated in deuterated environment) in one state and off-exchanged (incubated in nondeuterated environment) in the other state.
[d] Peptides were generated by MS.

spectrometry and account for the entire protein sequence. To achieve near single amide resolution, many overlapping peptides must be generated. Calculation of the difference of deuterium incorporation in overlapping peptides is the preferred method to localize deuterium atoms (Anand et al., 2002; Hamuro et al., 2002a). Digestion parameters optimized include the type and bed volume of the protease columns, the transit time of the protein over the protease column, the type and concentration of denaturant (Hamuro et al., 2002a), and inclusion of reducing reagents such as Tris(2-carboxyethyl)phosphine hydrochloride, TCEP (Yan et al., 2002). For a typical protein of ~50 kDa, pepsin digestion alone can generate more than 100 peptides.

Sublocalization of Deuterium by MS/MS

Several groups have reported success of deuterium sublocalization using tandem mass spectrometry, or MS/MS (Deng et al., 1999; Angeregg et al., 1994; Kim et al., 2001). With these experiments, either native peptide or peptides obtained from proteolysis are further fragmented by collisional-induced dissociation (CID) and the resultant fragments are recorded in the mass spectrometer. Analysis of the CID spectrum allows deuterium atoms within the peptide sequence to be localized to specific amino acids. Although this is a potentially powerful and attractive method, limitations to the technique include the possibility of hydrogen/deuterium rearrangement during the CID process (Kaltashov and Eyles, 2002; Demmers et al., 2002).

High-Performance Liquid Chromatography Separation

The peptides generated from proteolysis are separated using reverse-phase HPLC to minimize mass overlap and ionization suppression caused by ion competition in the electrospray source. The optimized liquid chromatography (LC) gradient parameters efficiently separate peptides while minimizing loss of deuterium through back exchange with solvent. Increased sensitivity can be achieved by using capillary HPLC columns and by precise calibration of LC separation steps (Wang and Smith, 2003).

Mass Analysis by Liquid Chromatography–Mass Spectrometry (LC–MS)

The deuterated sample containing the intact protein or peptides derived from protease fragmentation are separated by reverse-phase HPLC and are introduced directly into the mass spectrometer. The majority of H/D-Ex studies that have been reported employed robust quadrupole ion-trap (qIT) instruments due to their ease of use, excellent sensitivity, ability to perform MS/MS experiments, compact size, and relatively low cost. Other reports discuss the use of instruments with high-mass resolving power such as the hybrid QqTOF mass spectrometry, which has an advantage over ion traps in measuring the accurate average molecular mass (centroid) of peptides (Wang and Smith, 2003). A few groups have utilized Fourier transform ion cyclotron resonance (FT-ICR) mass spectrometry, which offers ultra-high-mass resolving power and improved mass accuracy (Lanman et al., 2003; Akashi and Takio, 2002). In addition,

FT-ICR has been coupled with soft fragmentation methods that may be useful in sub-localization of deuterium within peptides (Yamada et al., 2002; Mao et al., 2002).

Matrix-Assisted Laser Desorption Ionization Methods

The use of matrix-assisted laser desorption ionization (MALDI) time-of-flight (TOF) MS for H/D-Ex studies can offer an advantage in sample throughput compared to LC-based methods since samples can be prepared in parallel and deposited on a single MALDI target to be analyzed without further separation (Mandell et al., 1998). However, MALDI approaches tend to result in poorer sequence coverage for larger molecular weight (MW) proteins than LC-based methods perhaps due to the lack of separation of peptide fragments in the time dimension resulting in suppression of ionization. In addition, MALDI methods have suffered from poor deuterium recovery as compared to LC-based work, although a recent manuscript by Kipping appears to offer a solution to the deuterium loss during the sample crystallization step (Kipping and Schierhorn, 2003). Finally, it has been proposed that H/D-Ex using MALDI MS offers an improvement in sensitivity when compared to the LC-based methods; however, the use of microelectrospray coupled with qIT instruments may offer comparable sensitivity.

Automation of Hydrogen/Deuterium Exchange by Mass Spectrometry

A fully automated system for performing detailed H/D-Ex MS studies has been developed (Fig. 5.2) (Hamuro et al., 2003a). The system requires manual loading of a stock solution of the nondeuterated protein, and the remaining experimental steps as illustrated in Figure 5.1 are automated. Using the system shown in Figure 5.2, more than 100 experiments can be run continuously. This automated system consists of two functional components; the liquid handling device and the protein processing unit. The liquid handling operations (shown at the top of Fig. 5.2) are performed by two robotic arms equipped with low-volume syringes and two temperature controlled chambers, one held near 23°C and the other held near 1°C. To initiate on-exchange, a small amount of protein solution is mixed with a selected deuterated buffer within the syringe. The mixture is then incubated for a programmed period of time in the 23°C chamber. This is immediately followed by mixing the on-exchange sample with a quench solution in the 1°C chamber. The entire sample is then injected onto the protein processing system, which includes injection loops, protease column(s), a trap column, an analytical column, three electronically controlled valves, and isocratic and gradient pumps. The injector, columns, and valves reside in a low-temperature chamber to minimize the loss of deuterium by back exchange (Fig. 5.2). The quenched protein solution is pumped in series through a column containing an immobilized protease and a trap column to capture the peptide fragments. Extra pumping at this step can eliminate any polar additives, such as salts and denaturants. The gradient pump is activated following digestion of the protein on the protease column(s), and the peptides captured on the trap column are eluted and separated over an analytical reverse-phase HPLC column directly into the mass spectrometer.

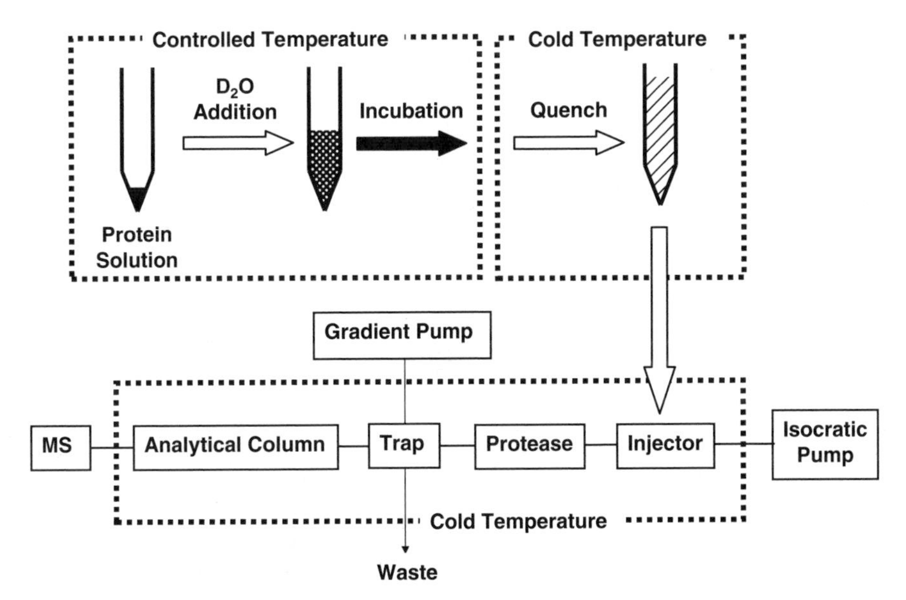

Figure 5.2. Diagram of a fully automated system for acquiring H/D-Ex MS data starting with a stock solution of the nondeuterated protein. In this system (Hamuro et al., 2003), the liquid handler mixes a small amount of concentrated protein solution with a selected deuterated buffer and the mixture is incubated for a programmed period of time. The exchange reaction is conducted in a temperature-controlled chamber held near 23°C. The mixture is then transferred to an acidic quench solution held near 1°C. After quenching the exchange reaction, the entire sample is injected onto an LC-MS system, which includes injection loops, protease column(s), a trap, an analytical column, and isocratic and gradient pumps. The injector, columns, and electronically controlled valves reside in a low-temperature chamber to minimize the loss of deuterium by back exchange. The quenched protein solution is pumped in series over a column containing the immobilized protease and a reverse-phase trap to capture the peptide fragments. The gradient pump is activated following the digestion and the peptides captured on the trap are eluted into the mass spectrometer after separation by the analytical column. Figure adapted with permission from Hamuro et al., 2003a.

Automated Data Analysis

A software system capable of extracting and cataloging the large number of data points obtained during each experiment has been developed (Hamuro et al., 2003a). The automated system streamlines most data handling steps and reduces the potential for errors associated with manual manipulation of large data sets. In the first automated processing step, the centroid mass value is obtained for each peptide ion observed in every LC-MS data file associated with the experiment. This step includes peak detection, selection of retention time window, selection of *m/z* range, and calculation of the average mass for each peptide. In the absence of automated data extracting software, obtaining centroid mass values can be the most time-consuming step of an H/D-Ex experiment. The second automated data processing step involves calibration of deuterium incorporation (Zhang and Smith, 1993) to account for any loss of deuterium atoms during the digestion and

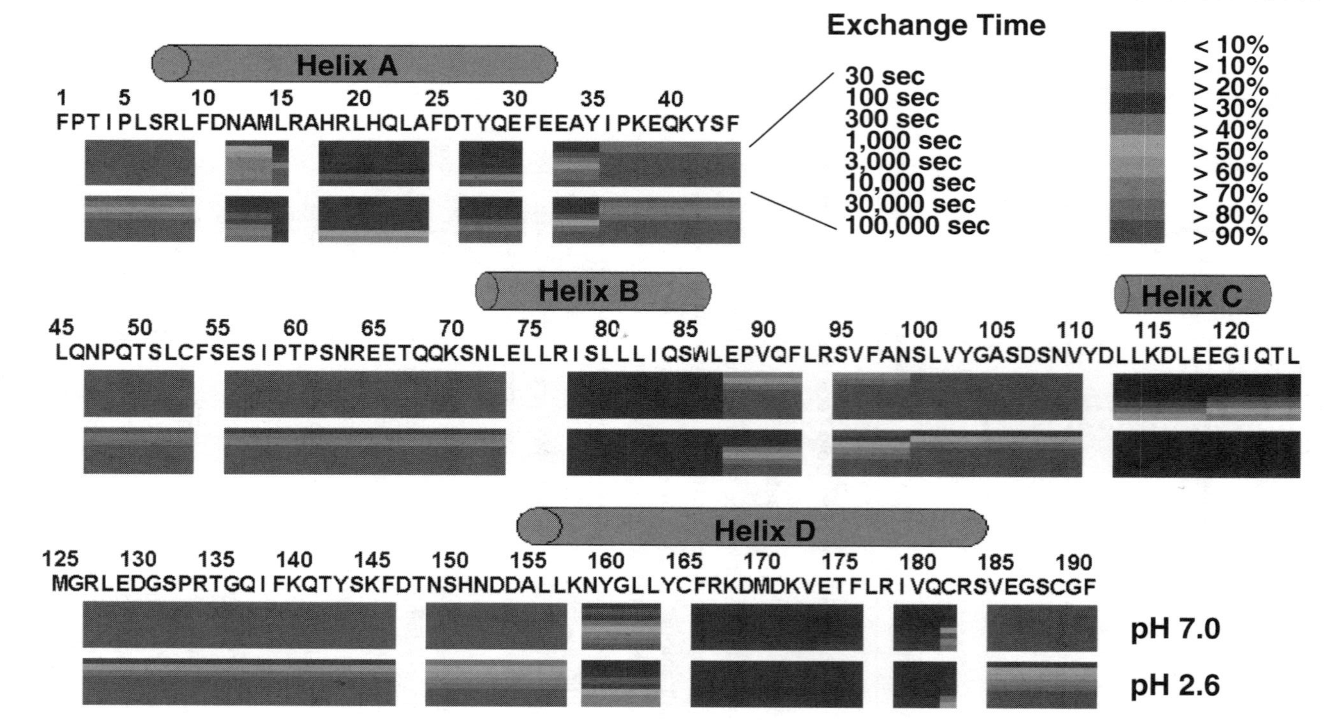

Figure 5.3. (See color insert.) H/D-Ex analysis of hGH at pH 7.0 and 2.6. Each block represents a pepsin-generated peptide. Each block consists of eight rows that represent eight distinct on-exchange time points, shown at the right. The level of deuteration in each peptide at each time point is represented by color according to the diagram displayed at the top right. Blocks representing on-exchange at pH 7.0 are on the top row, while blocks representing on-exchange at pH 2.6 are shown on the bottom. Light blue cylinders above the sequence indicate the helices identified from the X-ray crystal structure of hGH (1HGU). Peptides that contain mostly slow exchanging amide hydrogens are represented by blue bars, while red bars represent peptides that contain mostly rapidly exchanging amide hydrogens. Figure adapted with permission from Hamuro et al., 2003a.

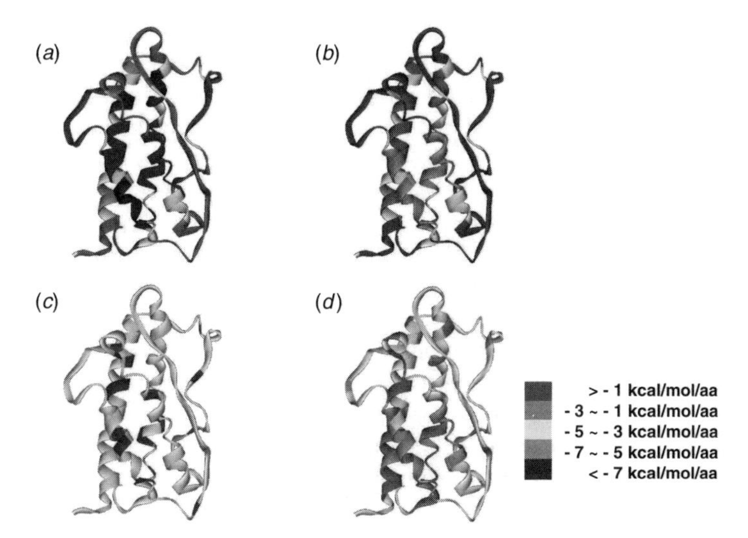

Figure 5.4. (See color insert.) Free-energy Change upon Folding of hGH as determined by MS and NMR: (*a*) pH 7.0 by MS, (*b*) pH 2.6 by MS, (*c*) pH 7.0 by NMR, and (*d*) pH 2.7 by NMR. Folding free energies are mapped on the X-ray structure (1HGU) by colored segments according to the key at the lower right. Gray indicates residues that were not analyzed by the method used. Figure adapted with permission from Hamuro et al., 2003a.

separation steps. After calculating the percent deuterium incorporation for each peptide at each time point, H/D-Ex data is displayed as a stacked bar chart that is aligned with the protein primary sequence (e.g., Fig. 5.3). If the tertiary structure of the protein is known or a homology model is available, the H/D-Ex data can be projected onto a three-dimensional model of the protein structure (e.g., Fig. 5.4).

APPLICATION OF HYDROGEN/DEUTERIUM EXCHANGE MASS SPECTROMETRY WITHOUT FRAGMENTATION

Information about the global conformation of proteins as monomers and in complexes can be obtained by mass analysis of intact proteins. Recent studies of intact proteins by H/D-Ex are summarized in Table 5.1. One advantage of this approach compared with other biophysical methods is its applicability under a wide variety of sample conditions. Highly purified samples may not be required because MS functions as a separation device, as well as a mass detector (Ghaemmaghami et al., 2000). This methodology is also compatible with membrane proteins that are difficult to study by most biophysical methods (Hansen et al., 2002; de Planque et al., 2003). Several examples of H/D-Ex studies without fragmentation are given below.

Glutathione–Thioredoxin Conjugates

Hydrogen/deuterium exchange studies are conducted to probe conformational differences in closely related molecules. An example of this use is provided by studies of the structural consequences of glutathione–thioredoxin conjugation, a possible *in vivo* outcome of exposure to volatile vicinal dihaloethanes (Kim et al., 2002). Time-dependent deuterium buildup in the +8 ion peak was monitored for 4 thioredoxin variants that included oxidized and reduced thioredoxin and the ethyl glutathione and ethyl cysteine adducts at Cys32. Comparison of the charged species distribution (CSD) profiles showed greater compactness for oxidized thioredoxin and the ethylglutathione-modified form. However, both modified molecules and reduced thioredoxin were significantly less stable and exhibited 11° to 14° reductions in melting temperature relative to oxidized thioredoxin. In the H/D-Ex studies, molecules were incubated at 50°C, a temperature below the 67°C melting point of oxidized thioredoxin but near the melting transition temperatures of the other molecules. Only the oxidized form exhibited an EX2 mechanism of exchange, in agreement with the thermal stability measurements in that molecules near a folding transition would be expected to show EX1 mechanism (Sivaraman et al., 2001). Analysis of total deuterium uptake showed less deuterium incorporation for the oxidized and modified molecules. Taken together, the H/D-Ex studies showed that the chemically modified forms retained features of the compact, H/D exchange-resistant structure of oxidized thioredoxin without increasing protein stability.

Retinoic Acid Binding Protein–Ligand Interactions

Xiao et al. (2003) developed a method to assess relative binding affinities between a protein and hydrophobic ligands. In this method, the H/D-Ex profiles of cellular retinoic acid binding protein in the presence of several hydrophobic ligands in the same molar excess were measured. The ligand-induced protection from H/D-Ex decreased in the following order: trans-retinoic acid, 9-cis retinoic acid, and 13-cis retinoic acid with retinol showing no protection. Increasing magnitude of protection against H/D-Ex correlated with higher binding constants as determined by a fluorometric assay.

Metal and Protein Complexes of Colicin E9

The complex between colicin E9, its cognate immunity protein Im9, and Zn^{2+} demonstrates the use of H/D-Ex to determine the effects of complex formation on global protein structure (van den Bremer et al., 2002). Colicin proteins function as bacterial antibiotics by transporting a deoxyribonuclease (DNase) domain across cell membranes to induce cell death by cleavage of chromosomal deoxyribonucleic acid (DNA). High-affinity complexation between colicin DNase domains and immunity proteins prevent bacterial self-destruction. H/D-Ex on the intact E9 DNase domain and Im9, determined by both electrospray ionization (ESI)–MS and CID, showed that the relatively large changes in deuterium levels on binding of Zn^{2+} to the DNase–Im9 complex could be attributed solely to structural changes in the DNase domain because the deuteration levels in Im9 were independent of Zn^{2+} binding. As Zn^{2+} binding is not required for DNase activity and Zn^{2+} release may accompany DNase domain transmembrane transport, this finding supports the hypothesis that for preservation of Im9 function, its interactions with colicin must be independent of metal-induced conformational changes in the DNase domain.

Hydrogen/Deuterium Exchange of Whey Proteins

Similar to protein drugs, food proteins must be stored under conditions that preserve protein tertiary structure for extended periods. Recently, H/D-Ex studies were used to assess the stability of β-lactoglobulin and α-lactalbumin under various storage conditions and as isolated from several whey protein sources (Alomirah et al., 2003). After sampling H/D-Ex kinetics from 30 s to 10 days, the number of exchangeable hydrogens were partitioned into slow, intermediate, and rapid exchange categories. A greater number of slowly exchanging hydrogens were found in proteins isolated from liquid preparations, and the authors concluded that production methods for whey protein liquid concentrates were more effective in preserving protein tertiary structure than those for production of whey protein solids. In these studies, opposite effects of glycosylation on protein stability were found. Addition of four or more hexose residues stabilized α-lactalbumin, whereas increased H/D-Ex in two variants of β-lactoglobulin having only two hexose residues provided evidence of protein destabilization by glycosylation.

Probing the Mechanism of Hydrogen/Deuterium Exchange

The H/D-Ex studies of gas-phase protein ions offer the opportunity to examine the conformations of anhydrous proteins, such as those found in lyophilized powders. In a recent study of the temperature-dependent H/D-Ex of cytochrome *c*, unique features of the gas-phase H/D exchange reaction were also revealed (Valentine and Clemmer, 2002). Structurally stable, gas-phase ions of cytochrome *c* include a +5 species that retains a compact structure well modeled by the solution NMR structure and an elongated +9 ion with very little tertiary structure. Temperature-dependent H/D-Ex studies showed that near 385 K more deuterium atoms were incorporated into the +5 ion, despite its more compact structure that would be expected to afford H/D exchange protection as observed in solution H/D-Ex studies. However, H/D exchange in the gas phase is thought to involve formation of bridging hydrogen bonds between D_2O and two proton sites on the protein, one of which will undergo exchange. Molecular dynamics simulations of both ions indicated that while more potential deuteration sites were accessible to solvent D_2O in the elongated +9 ion, far fewer secondary proton sites were within the hydrogen bonding distance required for gas-phase H/D exchange.

Protein Stability Titration Curves by Hydrogen/Deuterium Exchange Mass Spectrometry

The extent of deuterium incorporation as a function of denaturant concentration can be determined by MS, and the data used to prepare titration curves assessing protein stability (Ghaemmaghami et al., 2000). The deuteration levels of an intact protein with various concentrations of guanidine hydrochloride (GdnHCl) were measured by MALDI-MS. An advantage of this method over conventional GdnHCl titrations is the ability to use partially purified proteins or even defined protein mixtures. Recently, the technology has been expanded to measure the binding constants of protein–ligand interactions (Powell and Fitzgerald, 2003; Powell et al., 2002b) as well as to determine the stability of multimeric proteins (Powell et al., 2002a).

Hydrogen/Deuterium Exchange Mass Spectrometry to Study Membrane Proteins

The H/D-Ex by MS has been extended to the study the structure of membrane proteins (Akashi and Takio, 2001; Demmers et al., 2000), a large and medically important class of proteins where analyses by crystallography and NMR have been limited to a few membrane proteins that can be solubilized or crystallized. Hansen et al. studied the H/D-Ex patterns of the transmembrane fragment of the influenza A M2 protein (M2-TM) (Hansen et al., 2002). Variable H/D-Ex rates of M2-TM were found in methanol, Triton X-100 micelles, and DMPC vesicle preparations, indicating that the H/D-Ex rates of membrane protein backbone amides are affected by the protein structure and the properties of the lipids. Similar results were found by de Planque et al. in study of the effects of tryptophan residues in transmembrane protein–lipid interactions (de Planque et al., 2003).

APPLICATIONS OF HYDROGEN/DEUTERIUM EXCHANGE COUPLED WITH FRAGMENTATION

The H/D-Ex MS analysis coupled with proteolytic fragmentation has become a powerful approach for localization of deuterium atoms within a protein. This approach affords information on protein structure and dynamics over segments several amino acids in length. The H/D-Ex MS is more widely applicable than the NMR and crystallographic methods. There are few limitations on sample composition, and it is possible to analyze protein structure and dynamics over a wide range of protein concentrations, in many buffer systems, and in the presence of most types of binding partner. In addition, automated systems offer the possibility of high-throughput analysis. H/D-Ex coupled with proteolytic fragmentation is especially useful to compare the structure and dynamics of a protein in several states. For this reason, the majority of the recent publications using this technology have focused on the analysis of alterations in protein structure as a result of ligand binding or environmental changes.

Analysis of Therapeutic Proteins

Many new protein drug candidates are anticipated from knowledge of the human genome and differential analyses of the proteomes from diseased and healthy individuals. Currently, MS-based methods are widely employed in research and development of protein therapeutics. Addition of H/D-Ex MS methods, such as those provided by the examples summarized in Tables 5.1 and 5.2, significantly enhance the scope and utility of MS methods for protein drug analysis. Automated systems capable of high-throughput H/D-Ex experiments, such as recently described by Hamuro et al. (2003a), promote the widespread application of H/D-Ex into proteomics research. Recent examples demonstrating an increased understanding of protein conformation afforded by H/D-Ex studies include the characterizations of the acid-induced transitions in human growth hormone (Hamuro et al., 2003a) and storage-induced aggregation of human interferon-γ (Tobler and Fernandez, 2002).

pH-Dependent Conformations of Human Growth Hormone

At acidic and neutral pH, human growth hormone (hGH) exhibits distinct conformations that share virtually identical secondary structures, but differ in global fold (DeFelippis et al., 1995). The acidic conformation is less stable and has been implicated as the intermediate for undesirable aggregation (DeFelippis et al., 1995). H/D-Ex results of hGH at pH 2.6 and 7.0 are summarized in Figure 5.3 (Hamuro et al., 2003a). At both pH values, the slow exchange regions correspond to helical regions identified by X-ray crystallography. In these studies, the sampling of deuterium buildup at time points ranging from 30 to 100,000 s allowed estimation of the free-energy change upon folding. Because the analysis according to Eq. (5.5) eliminates effects arising from differences in intrinsic exchange rates, the approach can be applicable for studies of pH-dependent structural changes. Figures 5.4*a* and 5.4*b* illustrate the changes in localized free energy for hGH as determined by H/D-Ex analysis at pH 7.0 and 2.6, respectively. These data demonstrate that the overall hGH structure is significantly more stable at neutral pH. Regions of greatest stabilization are found within the helix bundle, but the stability of these regions markedly shifts as a function of pH (Fig. 5.4). At neutral pH, regions of high stability are located at the central portion of the helix bundle. At pH 2.6 the region of highest stability in the structure is located near the end of the bundle containing the N- and C-termini.

Comparison of Hydrogen/Deuterium Exchange by Mass Spectrometry and Nuclear Magnetic Resonance

When the results of H/D-Ex studies on hGH carried out by NMR (Kasimova et al., 2002) and by MS (Hamuro et al., 2003a) are compared, the stabilization free energies are in good agreement (Fig. 5.4). The two studies also illustrate differences in coverage and resolution obtained by NMR and MS methods. In the NMR studies, individual amides were analyzed, whereas the MS study averaged approximately 6 amide resolution (149 amino acid residues monitored by 25 peptides). The MS study offered much higher amino acid sequence coverage. Using H/D-Ex MS, 149 amide hydrogens of hGH were studied while only a fraction of the amide hydrogens (22 and 69 amide hydrogens, at neutral and acidic pH conditions, respectively) were followed by NMR.

Aggregation of Interferon-γ

Hydrogen/deuterium exchange followed by protease digestion was used to determine the structure of interferon-γ within aggregates and to study the mechanism of aggregation (Tobler and Fernandez, 2002). Helix C of interferon-γ remained intact in aggregates formed on addition of the chaotropic salts, GdnHCl, or KSCN. Although at equilibrium similar structures of interferon-γ were found in the aggregates, the protein aggregation mechanism differed. GdnHCl acted in solution to unfold most of the protein tertiary and secondary structures. In contrast, the H/D-Ex patterns of interferon-γ in KSCN-containing solutions resembled those of the native protein. Thus KSCN decreased the solubility of the native protein and facilitated unfolding of the protein within the precipitate. Such detailed understanding of protein aggregation mechanism is important for developing strategies that prevent unwanted aggregation side reactions during handling and storage of proteins and protein therapeutics.

Application of Hydrogen/Deuterium Exchange to Hemoglobin

H/D-Ex has been used to study the allosteric motion of hemoglobin, (Englander et al., 2003). In a typical experiment, the oxy form of hemoglobin (Hb) was on-exchanged in a deuterated environment for short periods of time, then off-exchanged in a non-deuterated environment after the protein was converted to the deoxy form. In this manner, the amide hydrogens that did not change their exchange rates upon deoxygenation lose most of deuteriums and only allosterically sensitive amino acid residues retain significant portions of deuteriums. This type of on-/off-exchange experiments should be applicable not only for protein conformational changes but also protein–ligand or protein–protein interactions. After the identification of allosterically sensitive regions, the regions were sub-localized by generating even smaller peptide fragments using multiple protease columns. This suggests a general method to obtain higher resolution by H/D-Ex.

Application of Hydrogen/Deuterium Exchange to Kinases and Phosphatases

Understanding the role and regulation of protein phosphorylation in human disease represents a major task for modern proteomic research. Nearly 1000 kinases and phosphatases, enzymes that add and remove protein phosphate groups, have been identified in the human genome. Because many are involved in cell cycle regulation, kinases and phosphatases constitute new targets for treatment of cancer, inflammation, and other diseases. The clinical success of a few kinase inhibitors clearly demonstrates the need for detailed structural information on these drug target proteins (Buchman, 2003). Several recent H/D-Ex studies have extended the understanding of conformational changes accompanying the interactions of kinases with substrates (Hamuro et al., 2002b), kinases and kinase regulatory proteins (Anand et al., 2002; Hamuro et al., 2003b), kinase localization proteins (Hamuro et al., 2002a) and phosphatases with inhibitors (Guo et al., 2002). Taken together, these studies clearly demonstrate the utility of H/D-Ex to study complex protein systems involving multiple small molecule and protein ligands.

Hydrogen/Deuterium Exchange of cAMP or Cyclic Adenosine Monophosphate Dependent Kinase Regulatory Subunit Activation

For the cyclic-adenosine-monophosphate (cAMP) dependent kinase, activity of the catalytic subunit is controlled by the regulatory subunit. Release of the active catalytic domain follows binding of two molecules of cAMP to distinct subdomains on the regulatory subunit. H/D-Ex studies of the regulatory subunit in the full-length molecule of isotype IIβ (Hamuro et al., 2003b) and a truncated form of isotype Iα (Anand et al.,

2002) showed that the effects of cAMP binding were relatively localized (Fig. 5.5). Figure 5.5*a* shows the H/D-Ex of RIIβ alone, RIIβ with cAMP bound, and the average difference of deuteration level of the two states. Only two regions corresponding to the cAMP binding sites identified by crystallography show significantly lower deuteration levels upon cAMP binding, indicating lack of widespread conformational changes upon cAMP binding. In contrast, binding of the catalytic subunit altered H/D exchange patterns at several regions of the regulatory subunit (Fig. 5.5*b*). Several regions mostly in cAMP binding domain B exhibit faster exchange upon catalytic subunit binding, while three regions were protected from exchange. The regions with altered exchange rates upon complex formation are overlaid onto the crystal structure of cAMP-bound RIIβ in Figure 5.6*a* and 5.6*b*. These results show the dynamics of the cAMP-dependent kinase activation at the submolecular level and exemplify the utility of H/D-Ex to investigate the dynamic nature of a protein.

COOH-Terminal Src Kinase with Nucleotides

Reduced exchange rate for residues at nucleotide binding sites has been observed on protein binding of cAMP (Anand et al., 2002; Hamuro et al., 2003), adenosine triphosphate (ATP) (Hasan et al., 2002), and adenosine diphosphate (ADP) and ATP analogs (Hamuro et al., 2002b). For some kinases, catalytic cycling is limited by relatively slow conformational changes that accompany nucleotide binding and release. The COOH-terminal Src kinase (CSK) is representative of the conformationally regulated kinases, and nucleotide-dependent alterations in H/D-Ex were observed at the nucleotide binding site and at regions removed from the nucleotide binding site (Hamuro et al., 2002b). As observed for other kinases, the H/D-Ex results showed the protection of CSK upon ADP and ATP analog binding; however, in this case the binding of structurally related nucleotides produced very different protection patterns. ADP showed more protection near the glycine-rich loop, while the ATP analog protected more distal regions than ADP. The data implied major conformational changes during the catalytic cycle.

Protein–Tyrosine Phosphatase1B with Ligands

H/D-Ex studies (Guo et al., 2002) aided understanding of the structural basis for selectivity of a potent inhibitor of protein–tyrosine phosphatase1B (PT1B), a molecule involved in negative regulation of the insulin-stimulated signal transduction pathway. Inhibitor-induced conformational changes were localized to three polypeptide segments, Ala17-Leu71, His175-Phe191, and Leu251-Leu267. Given previously determined crystal structures of PT1B and other phosphatases, the regions of decreased H/D-Ex were segregated into those that interacted with the phosphotyrosine mimic of the inhibitor and those responsible for inhibitor specificity. Site-directed mutagenesis, along with activity and inhibition assays, further defined the structural basis for selectivity and identified key interactions with residue 50, which is serine in PT1B and leucine in many other phosphatases. The results show that H/D-Ex can be used to define detailed features of drug candidate binding sites, in addition to estimating binding affinities (Powell et al., 2002b).

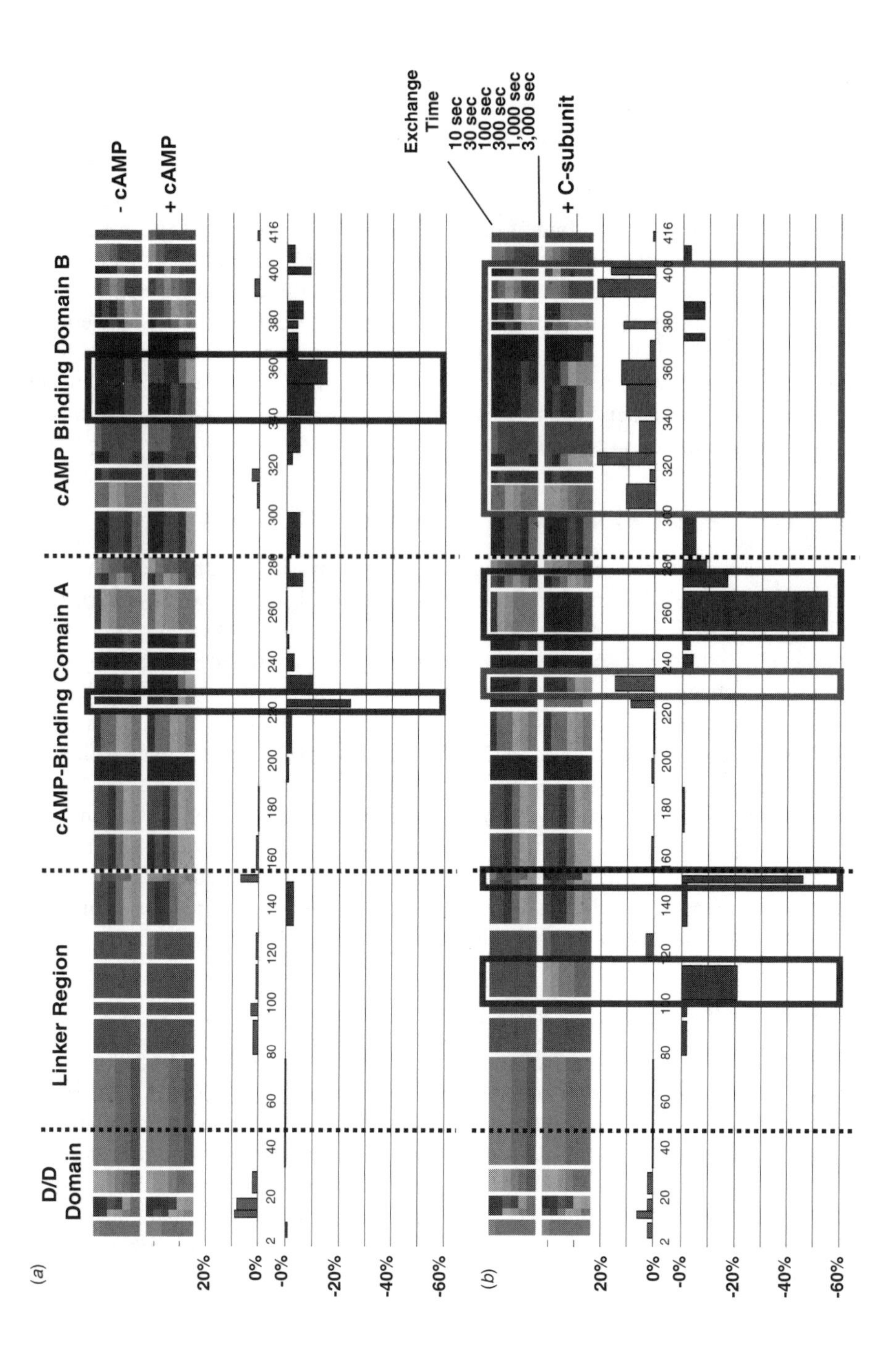
(a)
D/D Domain
Linker Region
cAMP-Binding Comain A
cAMP Binding Domain B
- cAMP
+ cAMP
20%
0%
-0%
-20%
-40%
-60%
(b)
Exchange Time
10 sec
30 sec
100 sec
300 sec
1,000 sec
3,000 sec
+ C-subunit

Figure 5.6. (See color insert.) Average differences in deuteration of RIIβ upon ligand binding overlaid on the crystal structure of an RIIβ fragment (1CX4): (*a*) with/without cAMP and (*b*) with/without catalytic subunit. Blue indicates protected regions, and red indicates regions exhibiting increased exchange upon ligand binding.

Protein–Carbohydrate Interactions

Interactions between carbohydrates and proteins can also be studied by H/D-Ex MS methods. For example, an overall decrease in H/D-Ex was observed for lysozyme complexed with its carbohydrate substrate (King et al., 2002). When a plant cell wall degrading enzyme from *Aspergillus niger* was studied by H/D-Ex, low overall levels of H/D-Ex of the endopolygalacturonase were thought to reflect its fold. Typically proteins with mostly β sheets exhibit less H/D-Ex and the endopolygalacturonase fold includes an extended parallel β helix of about 10 turns (van Santen et al., 1999). Binding of the substrate oligosaccharide increased H/D-Ex in one β-helix turn near the center of the molecule. No H/D-Ex was observed in this region in either the unliganded or inhibited forms of endopolygalacturonase.

Figure 5.5. (*a*) H/D-Ex results of apo RIIβ (*top*), H/D-Ex results of cAMP-bound RIIβ (*middle*), and the average difference in deuteration levels with and without cAMP (*bottom*). Each block represents a pepsin-generated peptide and has six time points (10, 30, 100, 300, 1000, and 3000 s). Levels of deuterium incorporation are indicated according to the color scheme used in Figure 5.3. The average difference in deuteration level is the average of the deuteration level with ligand subtracted by that without ligand. A negative value (blue) indicates the region is protected upon ligand binding. A positive value (red) indicates the region exchanges faster on ligand binding. Solid blue rectangles indicate regions significantly protected upon ligand binding. (*b*) H/D-Ex results of apo RIIβ (*top*), H/D-Ex results of catalytic-subunit-bound RIIβ (*middle*), and the average difference in deuteration levels with and without the catalytic subunit (*bottom*). Red rectangles in the bottom part indicate that the region exchanges faster upon catalytic-subunit binding and reveal conformational changes induced by complexation.

Application of Hydrogen/Deuterium Exchange to Antibodies

Defining the antibody binding site on a protein antigen assists understanding of the antibody action, whether the antibody is used in a clinical or laboratory setting. Because antibodies frequently interact with multiple sites arising from distinct regions of the protein antigen sequence, the most reliable methods of epitope mapping utilize intact protein antigens, rather than antigen-derived peptides. The ability of H/D-Ex to localize protein regions involved in protein antigen–antibody contacts makes the technology ideally suited for epitope mapping.

Epitope Mapping of Antibody against Thrombin

In a recent study, the binding sites of a monoclonal antibody to thrombin were mapped using H/D-Ex coupled with MALDI MS (Baerga-Ortiz et al., 2002). Analogous to a hemoglobin study by Englander et al. (2003), the epitope mapping studies used on-off-exchange to map the thrombin–antibody interaction sites. Both proteins were first deuterated separately (on-exchange): The monoclonal antibody attached to beads was incubated in deuterated buffers, and lyophilized thrombin was resuspended in D_2O. After allowing the deuterated proteins to react, the complex was resuspended in water (off-exchange), so that only interprotein contact regions remained deuterated. The solution used to quench the exchange reaction also dissociated the thrombin–antibody complex and facilitated removal of the antibody and analysis of protease-derived thrombin peptides. The antibody was found to recognize a discontinuous epitope located near the thrombomodulin binding site on thrombin.

Epitope Mapping of Antibody against Interleukin-6

In a similar study, the binding sites of a monoclonal antibody to interleukin-6 (IL-6) were mapped using infrared multiphoton dissociation (IRMPD) (Yamada et al., 2002). IL-6 was first on-exchanged in deuterated buffer and then off-exchanged in normal buffer after complexation with the immobilized monoclonal antibody. Instead of proteolysis, the intact IL-6 was introduced directly into the Fourier transform ion cyclotron resonance mass spectrometer (FT-ICR-MS) and fragmented by IRMPD to localize the antibody binding sites. Although this is an attractive approach, there will likely be an upper limit to the size of the protein that can be directly analyzed without further proteolytic fragmentation. In addition, it is possible that deuteriums scramble during the activation of the protein.

CONCLUSIONS

With the increased number of new protein sequences determined by genomic and proteomic efforts, a high-throughput and widely applicable protein structure analysis technology is desirable. H/D-Ex coupled with MS is widely applicable to the investigation of protein structure, protein dynamics, and protein–ligand interactions. The recent improvements of MS instrumentation, fluidics, automation, fragmentation chemistry, and data analysis software have made this technology very sensitive, robust, high-throughput, and high resolution. Examples cited here demonstrate that the emerging MS-based methods have exceptional capabilities of measuring and localizing H/D-Ex events in proteins and that the H/D-Ex data can provide invaluable insights into the protein dynamics.

Abbreviations

ADP	adenosine diphosphate
ATP	adenosine triphosphate
cAMP	cyclic adenosine monophosphate
CSD	charged species distribution
CSK	COOH-terminal Src kinase
ESI	electrospray ionization
FT-ICR-MS	Fourier transform ion cyclotron resonance mass spectrometer
GdnHCl	guanidine hydrochloride
H/D-Ex	hydrogen/deuterium exchange
hGH	human growth hormone
HPLC	high-performance liquid chromatography
IL-6	interleukin-6
IRMPD	infrared multiphoton dissociation
KSCN	potassium thiocyanate.
MS	mass spectrometry
MALDI	matrix-assisted laser desorption ionization
NMR	nuclear magnetic resonance
PT1B	protein–tyrosine phosphatase 1B
RIIβ	cAMP-dependent kinase regulatory subunit isotype IIβ

REFERENCES

Akashi, S., and Takio, K. (2002). Melittin-diacylphosphatidylcholine interaction examined by electrospray ionization Fourier transform ion cyclotron resonance mass spectrometry. *J. Mass Spectrom. Soc. Jpn.* **50**:67–71.

Akashi, S., and Takio, K. (2001). Structure of melittin bound to phospholipid micelles studied using hydrogen-deuterium exchange and electrospray ionization Fourier transform ion cyclotron resonance mass spectrometry. *J. Am. Soc. Mass Spectrom.* **12**:1247–1253.

Alomirah, H., Alli, I., and Konishi, Y. (2003). Charge state distribution and hydrogen/deuterium exchange of α-lactalbumin and β-lactoglobulin preparations by electrospray ionization mass spectrometry. *J. Agricul. Food Chem.* **51**:2049–2057.

Anand, G.S., Hughes, C.A., Jones, J.M., Taylor, S.S., and Komives, E.A. (2002). Amide H/2H exchange reveals communication between the cAMP and catalytic subunit-binding sites in the RIα subunit of protein kinase A. *J. Mol. Biol.* **323**:377–386.

Angeregg, R.J., Wagner, D.S., and Stevenson, C.L. (1994). The mass spectrometry of helical unfolding in peptides. *J. Am. Soc. Mass Spectrom.* **5**:425–433.

Baerga-Ortiz, A., Hughes, C.A., Mandell, J.G., and Komives, E.A. (2002). Epitope mapping of a monoclonal antibody against human thrombin by H/D-exchange mass spectrometry reveals selection of a diverse sequence in a highly conserved protein. *Protein Sci.* **11**:1300–1308.

Bai, Y., Milne, J., Mayne, L., and Englander, S. (1994). Protein stability parameters measured by hydrogen exchange. *Proteins: Struct. Funct. Genet.* **20**:4–14.

Bai, Y., Milne, J.S., Mayne, L.C., and Englander, S.W. (1993). Primary structure effects on peptide group hydrogen exchange. *Proteins: Struct. Funct. Genet.* **17**:75–86.

Birolo, L., Dal Piaz, F., Pucci, P., and Marino, G. (2002). Structural characterization of the M* partly folded intermediate of wild type and P138A aspartate aminotransferase from *Escherichia coli*. *J. Biol. Chem.* **277**:17428–17437.

Buchman, S. (2003). Protein structure: Discovering selective protein kinase inhibitors. *TARGETS* **2**:101–108.

Canet, D., Last, A.M., Tito, P., Sunde, M., Spencer, A., Archer, D.B., Redfield, C., Robinson, C.V., and Dobson, C.M. (2002). Local cooperativity in the unfolding of an amyloidogenic variant of human lysozyme. *Nature Struct. Biol.* **9**:308–315.

DeFelippis, M.R., Kilcomons, M.A., Lents, M.P., Youngman, K., and Havel, H.A. (1995). Acid stabilization of human growth hormone equilibrium folding intermediates. *Biochim. Biophys. Acta* **1247**:35–45.

Demmers, J.A.A., Rijkers, D.T.S., Haverkamp, J., Killian, J.A., and Heck, A.J.R. (2002). Factors affecting gas-phase deuterium scrambling in peptide ions and their implications for protein structure determination. *J. Am. Chem. Soc.* **124**:11191–11198.

Demmers, J., Haverkamp, J., Heck, A., Koeppe II, R., and Killian, J. (2000). Electrospray ionization mass spectrometry as a tool to analyze hydrogen/deuterium exchange kinetics of transmembrane peptides in lipid bilayers. *Proc. Natl. Acad. Sci. USA* **97**:3189–3194.

Deng, Y., Pan, H., and Smith, D.L. (1999). Selective isotope labeling demonstrates that hydrogen exchange at individual peptide amide linkages can be determined by collision-induced dissociation mass spectrometry. *J. Am. Chem. Soc.* **121**:1966–1967.

de Planque, M.R.R., Bonev, B.B., Demmers, J.A.A., Greathouse, D.V., Koeppe, R.E., II, Separovic, F., Watts, A., and Killian, J.A. (2003). Interfacial anchor properties of tryptophan residues in transmembrane peptides can dominate over hydrophobic matching effects in peptide-lipid interactions. *Biochemistry* **42**:5341–5348.

Engen, J.R., and Smith, D.L. (2001). Investigating protein structure and dynamics by hydrogen exchange MS. *Anal. Chem.* **73**:256A–265A.

Englander, S.W., and Kallenbach, N.R. (1984). Hydrogen exchange and structural dynamics of proteins and nucleic acids. *Quart. Rev. Biophys.* **16**:521–655.

Englander, J., Del Mar, C., Li, W., Englander, S., Kim, J., Stranz, D., Hamuro, Y., and Woods Jr., V. (2003). Protein structure change studied by hydrogen-deuterium exchange, functional labeling, and mass spectrometry. *Proc. Natl. Acad. Sci. USA* **100**:7057–7062.

Englander, S.W., Mayne, L., Bai, Y., and Sosnick, T.R. (1997). Hydrogen exchange: The modern legacy of Linderstrom-Lang. *Protein Sci.* **6**:1101–1109.

Ghaemmaghami, S., Fitzgerald, M., and Oas, T. (2000). A quantitative, high-throughput screen for protein stability. *Proc. Natl. Acad. Sci. USA* **97**:8296–8301.

Goshe, M.B., and Anderson, V.E. (1999). Hydroxyl radical-induced hydrogen/deuterium exchange in amino acid carbon-hydrogen bonds. *Radiat. Res.* **151**:50–58.

Guo, X., Shen, K., Wang, F., Lawrence, D., and Zhang, Z. (2002). Probing the molecular basis for potent and selective protein-tyrosine phosphatase 1B inhibition. *J. Biol. Chem.* **277**: 41014–41022.

Hamuro, Y., Coales, S.J., Southern, M.R., Nemeth-Cawley, J.F., Stranz, D.D., and Griffin, P.R. (2003a). Rapid analysis of protein structure and dynamics by hydrogen/deuterium exchange (H/D-Ex) mass spectrometry. *J. Biomolec. Tech.* **14**(3), 171–182.

Hamuro, Y., Zawadzki, K.M., Kim, J.S., Stranz, D.D., Taylor, S.S., and Woods, V.L. (2003b). Dynamics of cAPK type IIβ activation revealed by enhanced amide H/2H exchange mass spectrometry (DXMS). *J. Mol. Biol.* **327**:1065–1076.

Hamuro, Y., Burns, L.L., Canaves, J.M., Hoffman, R.C., Taylor, S.S., and Woods Jr., V.L. (2002a). Domain organization of D-AKAP2 revealed by enhanced deuterium exchange–mass spectrometry (DXMS). *J. Mol. Biol.* **321**:703–714.

Hamuro, Y., Wong, L., Shaffer, J., Kim, J.S., Stranz, D.D., Jennings, P.A., Woods, V.L., Jr., and Adams, J.A. (2002b). Phosphorylation driven motions in the COOH-terminal Src kinase, Csk, revealed through enhanced hydrogen-deuterium exchange and mass spectrometry (DXMS). *J. Mol. Biol.* **323**:871–881.

Hansen, R.K., Broadhurst, R.W., Skelton, P.C., and Arkin, I.T. (2002). Hydrogen/deuterium exchange of hydrophobic peptides in model membranes by electrospray ionization mass spectrometry. *J. Am. Soc. Mass Spectrom.* **13**:1376–1387.

Hasan, A., Smith, D.L., and Smith, J.B. (2002). α-Crystallin regions affected by adenosine 5′-triphosphate identified by hydrogen-deuterium exchange. *Biochemistry* **41**:15876–15882.

Hernandez, H., and Robinson, C.V. (2001). Dynamic protein complexes: Insights from mass spectrometry. *J. Biol. Chem.* **276**:46685–46688.

Hilser, V.J., and Freire, E. (1996). Structure-based calculation of the equilibrium folding pathway of proteins. Correlation with hydrogen exchange protection factors. *J. Mol. Biol.* **262**: 756–772.

Hoofnagle, A.N., Resing, K.A., and Ahn, N.G. (2003). Protein analysis by hydrogen exchange mass spectrometry. *Annu. Rev. Biophys. Biomol. Struct.* **32**:1–25.

Hosia, W., Johansson, J., and Griffiths, W.J. (2002). Hydrogen/deuterium exchange and aggregation of a polyvaline and a polyleucine α-helix investigated by matrix-assisted laser desorption ionization mass spectrometry. *Mol. Cellular Proteomics* **1**:592–597.

Kaltashov, I.A., and Eyles, S.J. (2002). Crossing the phase boundary to study protein dynamics and function: Combination of amide hydrogen exchange in solution and ion fragmentation in the gas phase. *J. Mass Spectrom.* **37**:557–565.

Kasimova, M.R., Kristensen, S.M., Howe, P.W.A., Christensen, T., Matthiesen, F., and Petersen, J. (2002). NMR studies of the backbone flexibility and structure of human growth hormone: A comparison of high and low pH conformations. *J. Mol. Biol.* **318**:679–695.

Kim, M.-Y., Maier, C.S., Reed, D.J., and Deinzer, M.L. (2002). Conformational changes in chemically modified *Escherichia coli* thioredoxin monitored by H/D exchange and electrospray ionization mass spectrometry. *Protein Sci.* **11**:1320–1329.

Kim, M.-Y., Maier, C.S., Reed, D.J., and Deinzer, M.L. (2001). Site-specific amide hydrogen/deuterium exchange in *E. coli* thioredoxins measured by electrospray ionization mass spectrometry. *J. Am. Chem. Soc.* **123**:9860–9866.

Kim, K., Fuchs, J., and Woodward, C. (1993). Hydrogen exchange identifies native-state motional domains important in protein folding. *Biochemistry* **32**:9600–9608.

King, D., Lumpkin, M., Bergmann, C., and Orlando, R. (2002). Studying protein–carbohydrate interactions by amide hydrogen/deuterium exchange mass spectrometry. *Rapid Commun. Mass Spectrom.* **16**:1569–1574.

Kipping, M., and Schierhorn, A. (2003). Improving hydrogen/deuterium exchange mass spectrometry by reduction of the back-exchange effect. *J. Mass Spectrom.* **38**:271–276.

Kossiakoff, A. (1985). The application of neutron crystallography to the study of dynamic and hydration properties of proteins. *Annu. Rev. Biochem.* **54**:1195–1227.

Kraus, M., Bienert, M., and Krause, E. (2003). Hydrogen exchange studies on Alzheimer's amyloid-β peptides by mass spectrometry using matrix-assisted laser desorption/ionization and electrospray ionization. *Rapid Commun. Mass Spectrom.* **17**:222–228.

Lanman, J., Lam, T.T., Barnes, S., Sakalian, M., Emmett, M.R., Marshall, A.G., and Prevelige, P.E. (2003). Identification of novel interactions in HIV-1 capsid protein assembly by high-resolution mass spectrometry. *J. Mol. Biol.* **325**:759–772.

Mandell, J.G., Falick, A.M., and Komives, E.A. (1998). Identification of protein–protein interfaces by decreased amide proton solvent accessibility. *Proc. Natl. Acad. Sci. U.S.A.* **95**:14705–14710.

Mao, D., Ding, C., and Douglas, D.J. (2002). Hydrogen/deuterium exchange of myoglobin ions in a linear quadrupole ion trap. *Rapid Commun. Mass Spectrom.* **16**:1941–1945.

Nazabal, A., Laguerre, M., Schmitter, J.-M., Vaillier, J., Chaignepain, S.T., and Velours, J. (2003). Hydrogen/deuterium exchange on yeast ATPase supramolecular protein complex analyzed at high sensitivity by MALDI mass spectrometry. *J. Am. Soc. Mass Spectrom.* **14**:471–481.

Niimura, N. (1999). Neutrons expand the field of structural biology. *Curr. Opin. Struct. Biol.* **9**:602–608.

Powell, K., and Fitzgerald, M. (2003). Accuracy and precision of a new H/D exchange- and mass spectrometry-based technique for measuring the thermodynamic properties of protein-peptide complexes. *Biochemistry* **42**:4962–4970.

Powell, K., Wales, T., and Fitzgerald, M. (2002a). Thermodynamic stability measurements on multimeric proteins using a new H/D exchange- and matrix-assisted laser desorption/ionization (MALDI) mass spectrometry-based method. *Protein Sci.* **11**:841–851.

Powell, K.D., Ghaemmaghami, S., Wang, M.Z., Ma, L., Oas, T.G., and Fitzgerald, M.C. (2002). A general mass spectrometry-based assay for the quantitation of protein–ligand binding interactions in solution. *J. Am. Chem. Soc.* **124**:10256–10257.

Resing, K.A., Hoofnagle, A.N., and Ahn, N.G. (1999). Modeling deuterium exchange behavior of ERK2 using pepsin mapping to probe secondary structure. *J. Am. Soc. Mass Spectrom.* **10**:685–702.

Simmons, D.A., Dunn, S.D., and Konermann, L. (2003). Conformational dynamics of partially denatured myoglobin studied by time-resolved electrospray mass spectrometry with online hydrogen-deuterium exchange. *Biochemistry* **42**:5896–5905.

Sirangelo, I., Dal Piaz, F., Malmo, C., Casillo, M., Birolo, L., Pucci, P., Marino, G., and Irace, G. (2003). Hexafluoroisopropanol and acid destabilized forms of apomyoglobin exhibit structural differences. *Biochemistry* **42**:312–319.

Sivaraman, T., Arrington, C.B., and Robertson, A.D. (2001). Kinetics of unfolding and folding from amide hydrogen exchange in native ubiqutin. *Nature Struct. Biol.* **8**:331–333.

Tobler, S.A., and Fernandez, E.J. (2002). Structural features of interferon-γ aggregation revealed by hydrogen exchange. *Protein Sci.* **11**:1340–1352.

Turner, B.T., Jr., and Maurer, M.C. (2002). Evaluating the roles of thrombin and calcium in the activation of coagulation factor XIII using H/D exchange and MALDI-TOF MS. *Biochemistry* **41**:7947–7954.

Valentine, S., and Clemmer, D. (2002). Temperature-dependent H/D exchange of compact and elongated cytochrome *c* ions in the gas phase. *J. Am. Soc. Mass Spectrom.* **13**:506–517.

van den Bremer, E.T.J., Jiskoot, W., James, R., Moore, G.R., Kleanthous, C., Heck, A.J.R., and Maier, C.S. (2002). Probing metal ion binding and conformational properties of the colicin E9 endonuclease by electrospray ionization time-of-flight mass spectrometry. *Protein Sci.* **11**:1738–1752.

van Santen, Y., Benen, J., Schroter, K., Kalk, K., Armand, S., Visser, J., and Dijkstra, B. (1999). 1.68-A crystal structure of endopolygalacturonase II from *Aspergillus niger* and identification of active site residues by site-directed mutagenesis. *J. Biol. Chem.* **274**:30474–30480.

Wang, L., and Smith, D.L. (2003). Downsizing improves sensitivity 100-fold for hydrogen exchange-mass spectrometry. *Anal. Biochem.* **314**:46–53.

Woods, V.L., and Hamuro, Y. (2001). High resolution, high-throughput amide deuterium exchange-mass spectrometry (DXMS) determination of protein binding site structure and dynamics: Utility in Pharmaceutical design. *J. Cell. Biochem.* **S37**:89–98.

Woodward, C., Simon, I., and Tuchsen, E. (1982). Hydrogen exchange and the dynamic structure of protein. *Mol. Cell. Biochem.* **48**:135–160.

Xiao, H., Kaltashov, I.A., and Eyles, S.J. (2003). Indirect assessment of small hydrophobic ligand binding to a model protein using a combination of ESI MS and HDX/ESI MS. *J. Am. Soc. Mass Spectrom.* **14**:506–515.

Yamada, N., Suzuki, E.-I., and Hirayama, K. (2002). Identification of the interface of a large protein–protein complex using H/D exchange and Fourier transform ion cyclotron resonance mass spectrometry. *Rapid Commun. Mass Spectrom.* **16**:293–299.

Yan, X., Zhang, H., Watson, J., Schimerlik, M.I., and Deinzer, M.L. (2002). Hydrogen/deuterium exchange and mass spectrometric analysis of a protein containing multiple disulfide bonds: Solution structure of recombinant macrophage colony stimulating factor-beta (rhM-CSFb). *Protein Sci.* **11**:2113–2124.

Zhang, Z., and Smith, D.L. (1993). Determination of amide hydrogen exchange by mass spectrometry: A new tool for protein structure elucidation. *Protein Sci.* **2**:522–531.

Zhang, Z., Post, C.B., and Smith, D.L. (1996). Amide hydrogen exchange determined by mass spectrometry: application to rabbit muscle aldolase. *Biochemistry* **35**:779–791.

6

PROTEOMICS TECHNOLOGIES FOR IDENTIFICATION AND VALIDATION OF PROTEIN TARGETS

John E. Hale, Weijia Ou, Pavel Shiyanov, Michael D. Knierman, and James R. Ludwig

Enabling Biology Department, Lilly Research Labs, Indianapolis, Indiana

INTRODUCTION

The utilization of protein targets in drug development is a fundamental process in the pharmaceutical industry today. Simply defined, a protein target is a molecule, the func-

Industrial Proteomics: Applications for Biotechnology and Pharmaceuticals, edited by Daniel Figeys
ISBN 0-471-45714-0

tional inhibition or stimulation of which modulates a disease process or state. Screening compounds against targets has been the backbone of drug discovery for many years, and industries have sprung up around automation and increased throughput of screening operations. Yet it has been estimated that the majority of this effort has been focused on only 500 different targets (Drew, 1996). Genomics efforts have demonstrated that the number and diversity of potential protein targets is much larger than this. Therefore, much anticipation has surrounded projects promising large numbers of new targets. Through genomics we are able to identify potential target proteins and, where information is available about the functions of these proteins, infer disease associations. Genomics alone cannot determine protein function, interactions, modifications, intracellular locations, and expression levels. Of these, expression level and function have been approached using a combination of genomics database mining and transcript profiling (Shimkets et al., 1996).

Transcript profiling (Lockhart et al., 1996) determines basal messenger ribonucleic acid (mRNA) levels in tissues or cells of interest and changes in these levels in response to various perturbations. These include onset of disease, disease progression, treatment, disease regression in response to treatment, and toxicities associated with treatment among others. It is possible to extrapolate changes in protein level based on mRNA level, although the kinetic relationships between the two vary significantly for individual proteins (Ideker et al., 2001). Factors such as mRNA half-life (Even et al., 2002) and protein processing time (Garlisi et al., 2003) impact this, but, in general, a rise in mRNA level is followed by a rise in protein synthesis, and a fall in mRNA level is reflected in a fall in protein synthesis. However, the specific relationship, in terms of magnitude, duration, and temporal displacement of change, varies from protein to protein (Seilhamer, 1997; Gygi et al., 1999a). The existence of gene families has allowed a combination of genomics and transcriptional profiling to suggest logical hypotheses regarding the possible function of some novel proteins discovered. However, this is balanced by the tendency of biological systems to reuse certain domains in molecules that have evolved to very different functions. Therefore these hypotheses need to be tested in valid biological systems to confirm function.

As for modifications, intracellular locations, and protein interactions, genomics and transcript profiling provide less valuable information. While the sequence of the nascent polypeptide chain is determined by the sequence of mRNA, only potential sites for modification can be deduced from this. This can, of course, be confused by the existence of multiple splice variants for certain genes. While a signal sequence can be found in many cases on protein molecules that are transported, the final intracellular location of the protein is not yet directly predictable from the amino acid sequence alone (Blobel, 2000). While potential binding sites can be suggested from structural analysis of related proteins, the total number of other proteins, substrates, or cofactors associating with a specific protein is not predictable from structural data alone, not to mention relative affinities of different potential binders for the same site. Finally, a large number of proteins are not synthesized in their active form, and their local chemical environment, the presence of appropriate modifying enzymes, and the level of activation of those modifying enzymes affect the ultimate structure the processed proteins take in a cell.

Protein analytical technologies have experienced an explosion in development and utilization in response to these needs. The issues and challenges surrounding this field of research, now known as proteomics, arise from the great complexity that exists in the proteome and the wide range of expressional levels. Estimates of the size of the human genome suggest a relatively small number of genes, on the order of 40,000. The number of proteins derived from the genome is much larger with estimates ranging from hundreds of thousands to millions of discrete chemical entities. In many cases, one gene may result in the production of multiple different proteins. Reasons for this include expression of splice variants and posttranslational modifications, as mentioned above. Frequently, posttranslational modifications will alter the functional activity of a protein; thus, simple knowledge of gene expression levels may not indicate the levels of functional protein that is present. In addition, proteins rarely function individually but rather work in concert with networks of other proteins. Understanding these networks as well as points of intersection of multiple networks will allow more informed selection of target proteins while minimizing unwanted, off-target effects of compounds developed in screens. Just as the proteome is complex, the tools and technologies are numerous and varied. Perhaps no single review could do justice to the multitude of strategies that have been used to address this field. In this review, we will focus on the protein analytical tools and technologies that are used to identify proteins, study their posttranslational modifications and their interactions with other proteins. We will also summarize biological strategies used to study the function of proteins and infer disease associations as part of the validation process. We hope this will give the reader a sense of the current state and use of the technologies and provide some examples of real-world applications.

TOOLS FOR TARGET DISCOVERY

Separation/Identification

One-Dimensional Gel Electrophoresis. Perhaps the simplest differential protein display technology is the one-dimensional (1D) polyacrylamide gel. As a tool for target discovery, this technology is rapid (requiring as little as 1 h), relatively sensitive (silver staining and fluorescence detection visualizing low nanogram quantities of protein), and represents an integral component of the biochemists tool kit. This technology began as a descriptive technique (Laemmli-Laemmli, 1970) used to globally visualize protein components. It began its transformation to an analytical technique with the advent of the Western blot (Towbin et al., 1979). Application of blotting technology to automated protein sequencing expanded the analytical capability of the 1D gel by providing the ability to identify unknown proteins directly (Matsudaira, 1987). The progression of the analytical utility continued to grow with the development of techniques to proteolytically digest proteins in situ in the gel (Rosenfeld et al., 1992). At the time that this technique was being developed, ionization techniques were being refined for the analysis of peptide mixtures. Early applications of protein identification by peptide fingerprinting utilized electrospray ionization or matrix-assisted laser desorption ionization (MALDI) time-of-flight (TOF) mass spectrometry coupled with database searching algorithms (Rasmussen et al., 1994; Ji et al., 1994; Cottrell, 1994). The advantages this technology provides over protein sequencing includes speed of analysis and increased sensitivity. While undeniably powerful, this approach is somewhat limited in that identifications are still difficult when multiple proteins migrate together in a single band on the gel. In this situation, liquid chromatography tandem mass spectrometry (LC-MS/MS) technology is applicable. Combination of reversed-phase separations with electrospray ionization technology (Fenn et al., 1989) makes it possible to introduce peptides into mass spectrometers. In addition to a mass fingerprint, triple quadrupole and ion trap instruments offer the added capability of peptide fragmentation, which can yield peptide sequence information and fragmentation patterns that can be utilized for database searching (see below). Increased sensitivity of LC-MS/MS has been achieved through the development of nanospray (Shevchenko et al., 1996) and capillary LC methods (Battersby et al., 1994).

With all of these developments in the back-end analytical capability for protein identification, one-dimensional gels have been applied to a number of front-end biological strategies for target identification. Some of these strategies are summarized below.

Two-Dimensional Gel Electrophoresis. Two dimensional (2D) gel electrophoresis is the technology most frequently associated with proteomics. Since its inception in the mid-1970s (O'Farrell, 1975), 2D gel electrophoresis has provided the highest resolving power for the study of complex protein mixtures. This is due to the orthogonal separations in the first and second dimensions with proteins being separated by charge first and then by size. A series of improvements in the first dimensional separation has simplified and increased the robustness of the technique (Righetti, 1990). More recently, differential fluorescent staining has been used in order to determine relative differences in expression levels between control and perturbed states (Valdes et al., 2000). Many studies have utilized 2D gels to visualize differential expression of proteins and identify candidate target proteins. Some recent examples of these include the use of 2D gels to identify differentially expressed proteins in cancerous tissues (Adam et al., 2003), cardiovascular diseases (Macri and Rapundalo, 2001), and neurological disorders (Tsuji et al., 2002) among many, many others. As with one-dimensional gels, 2D gels experienced a dramatic increase in utility with the advent of the microanalytical techniques described below. The major drawback to 2D gel electrophoresis is the time-consuming nature of the technique. Examples of the use of 1D and 2D gels in target identification are outlined below.

Differential Display. Identification of protein targets for drug development requires the demonstration of the involvement of the protein in a disease state or in a pathway involved in that disease state. Direct analysis of perturbed versus normal tissues or cells is better achieved with 2D gel electrophoresis (see below) since many proteins may co-migrate on a 1D gel. However 1D gels are still quite useful in identifying pathway components. Prefractionation strategies have been utilized to simplify protein mixtures prior to gel separation. This approach has been used to identify components of signaling pathways. Stimulation of a signaling pathway is achieved through treatment of cells with the ligand, immunoprecipitation with antireceptor antibodies, and separation/visualization on a 1D gel. Comparison of the immunoprecipitated proteins from stimulated and unstimulated cells provides a differential display of pathway associations. Novel members of the epidermal growth factor (EGF) pathway have been identified through their physical association with the receptor (Blagoev et al., 2003). Immunoprecipitation with class-specific antibodies has been utilized to map pathways. For instance, after stimulation of cells with a ligand such as EGF or platelet derived growth factor (PDGF) (Pandey et al., 2000), extracted proteins were immunoprecipitated with antiphosphotyrosine antibodies. By separating the precipitated proteins on 1D gels and comparing the protein profile to that of unstimulated cells, individual members of the signaling pathway were visualized and subsequently identified by mass spectrometric techniques. Proteins whose phosphorylation state are altered by PDGF treatment have also been identified by 2D gel electrophoresis. In one example, proteins from lysates of cells stimulated and unstimulated by PDGF were separated by 2D electrophoresis and phosphotyrosine-containing proteins determined by Western blotting with antiphosphotyrosine antibodies. Proteins whose phosphorylation states were modulated were identified by mass spectral techniques (Soskic et al., 1999).

Pathway Mapping. Identification of physical interactions in a signaling pathway provides additional target proteins and offers the possibility of increased specificity in drug development. For instance, a number of signaling pathways may converge in a downstream messenger such as mitogen activated protein (MAP) kinase. Thus inhibition of a particular pathway at this point may have unwanted consequences through inhibition of all of the other pathways signaling through this downstream process. By understanding the upstream transmission of signal, different points for intervention may be selected that will inhibit only a desired pathway. Immunoprecipitation of a messenger protein and its associated partners under physiological conditions can allow mapping of all interactions of all components of signaling pathways. An example of such an interaction map is illustrated in Figure 6.1, which shows the results of an immunoprecipitation of the protein tyrosine phosphatase src homology 2 phosphotyrosine phosphatase (SHP2) from a cell lysate. SHP2 is known to mediate a number of signaling pathways through interaction with receptor tyrosine kinases including cell proliferation and differentiation (Okuda et al., 1999). Because of its plieotropic nature, the inhibition of SHP2 could yield unwanted effects. Mapping the interactions of other proteins with SHP2 could provide additional targets with increased specificity. The immunoprecipitation of SHP2 was performed and proteins separated and visualized by 1D gel electrophoresis. A number of proteins were seen in addition to SHP2 and the antibody used to precipitate it. Some of these proteins are known to interact with SHP2 (for instance Gab 2 and CD-31) in the transduction of signal and may provide different points of intervention in potential drug development. The changes in these interactions may be studied in response to different stimuli (i.e., different growth factor stimulation of cells) in order to more clearly understand the contributions of individ-

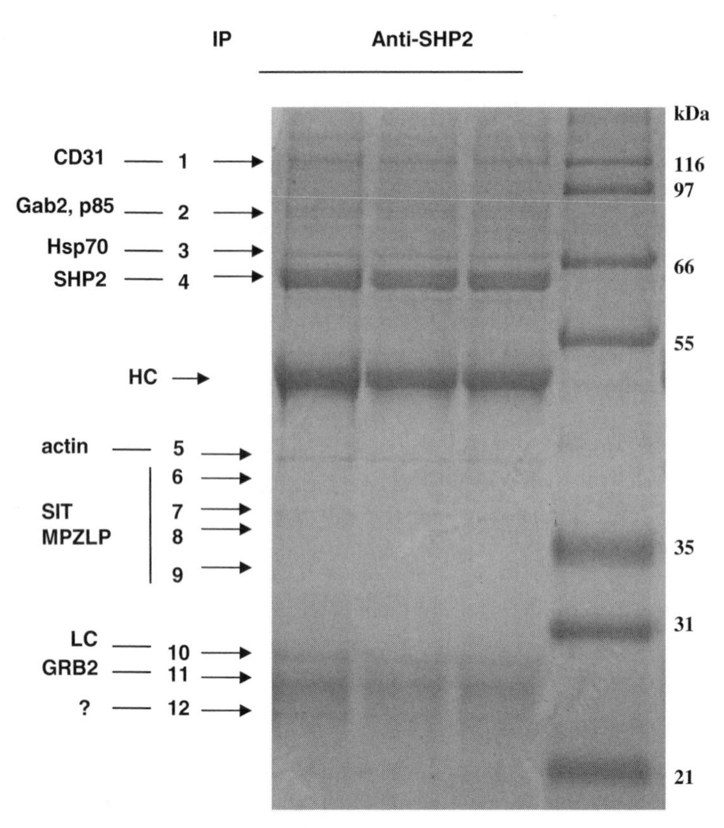

Figure 6.1. Identification of proteins co-immunoprecipitated with SHP-2. Cell lysates were incubated with antibody to SHP-2. Antibody was precipitated with immobilized protein A. Bound proteins were released with SDS-polyacrylamide gel electrophoresis (PAGE) sample buffer and proteins separated under reducing conditions. Protein bands were excised, digested with trypsin, and peptides were analyzed on an ion trap mass spectrometer equipped with a 75-μm C-18 capillary column. Peptides were fragmented by collisional induced dissociation (CID) and MS/MS spectra were used to search protein databases with the program Sequest. Identified proteins are labeled.

ual protein–protein interactions and select targets of the highest selectivity. Interaction maps have also been constructed by expressing proteins tagged with purification handles in cells followed by isolation of the protein under conditions of relatively low stringency so that protein–protein interactions are not disrupted. These complexes are separated on 1D gels and members of the complex then identified (Ho et al., 2002; Gavin et al., 2002).

Gel-Free Methods. While still very useful, gel-based separations have drawbacks as components of protein analytical schemes. Identification of proteins from gel spots or slices involves many time-consuming processing steps. Additionally, losses occur in the course of running the gel and in subsequent extraction of peptides. For this and other reasons, alternative methods were developed for the identification of proteins in complex mixtures. With the development of LC/MS methods it became apparent that pure protein was not a requisite for successful identification (Link et al., 1999). An example of a one-dimensional LC/MS/MS separation is shown in Figure 6.2*a*. In this example, proteins that were tyrosine phosphorylated were immunoprecipitated from lysates of cells with and without ligand stimulation (see below). The sample was reduced, alkylated, digested with trypsin, and finally injected onto a capillary, reversed-phase column. The effluent of the column was sprayed into the orifice of an ion trap mass spectrometer, programmed to isolate and perform collisional induced dissociation (CID) on ions above a certain ion intensity threshold (Figs. 6.2*b* and 6.2*c*). The data was acquired in a continuous format and each spectrum required roughly 1 s to acquire. In a complex sample like this a thousand or more MS/MS spectra may be acquired in a 1-h run. Spectra were subsequently fed into searching programs, proteins identified, and differences in phosphorylated protein component were noted.

Multidimensional Liquid Chromatography Tandem Mass Spectrometry. More complex samples (such as serum) will require additional fractionation steps in order to identify proteins of lower abundance. In practice, LC-MS/MS methods have a dynamic range of roughly 3 orders of magnitude. While the detection limit of the mass spectrometer may be femtomolar or less, the presence of peptides at high concentrations will mask the presence of lower abundance ions. In order to extend the dynamic range of measurement, additional chromatographic or affinity steps have been developed to simplify mixtures as a part of the identification strategy. An online procedure was recently developed and named MudPit, for multidimensional protein identification (Washburn et al., 2001). In this format, a strong cation exchange column is plumbed in front of a reversed-phase column. Classes of peptides are eluted from the cation exchange column with pH or salt steps, and peptides are separated in the second dimension with reversed-phase gradients. This spreads out the analytical window and effectively increases the dynamic range of the analysis.

Posttranslational Modifications. The functionality of proteins is frequently modulated by posttranslational modifications (PTMs). While the variety of PTMs is vast, a few of these are of particular interest in the identification of target proteins. Some of the more thoroughly studied are summarized below.

PHOSPHORYLATION. Addition or removal of a phosphate group on a protein is a modification known to alter the biological activity of many proteins. Indeed, kinases (the enzymes that add phosphate groups to proteins) are one of the most popular classes of targets for drug development and are often modified in this way themselves. Proteomic technologies have been developed specifically aimed at the study of this modification and have been used both to identify potential targets and to identify natural substrates for targets. The utility of mass spectrometry (MS) to study phosphorylation

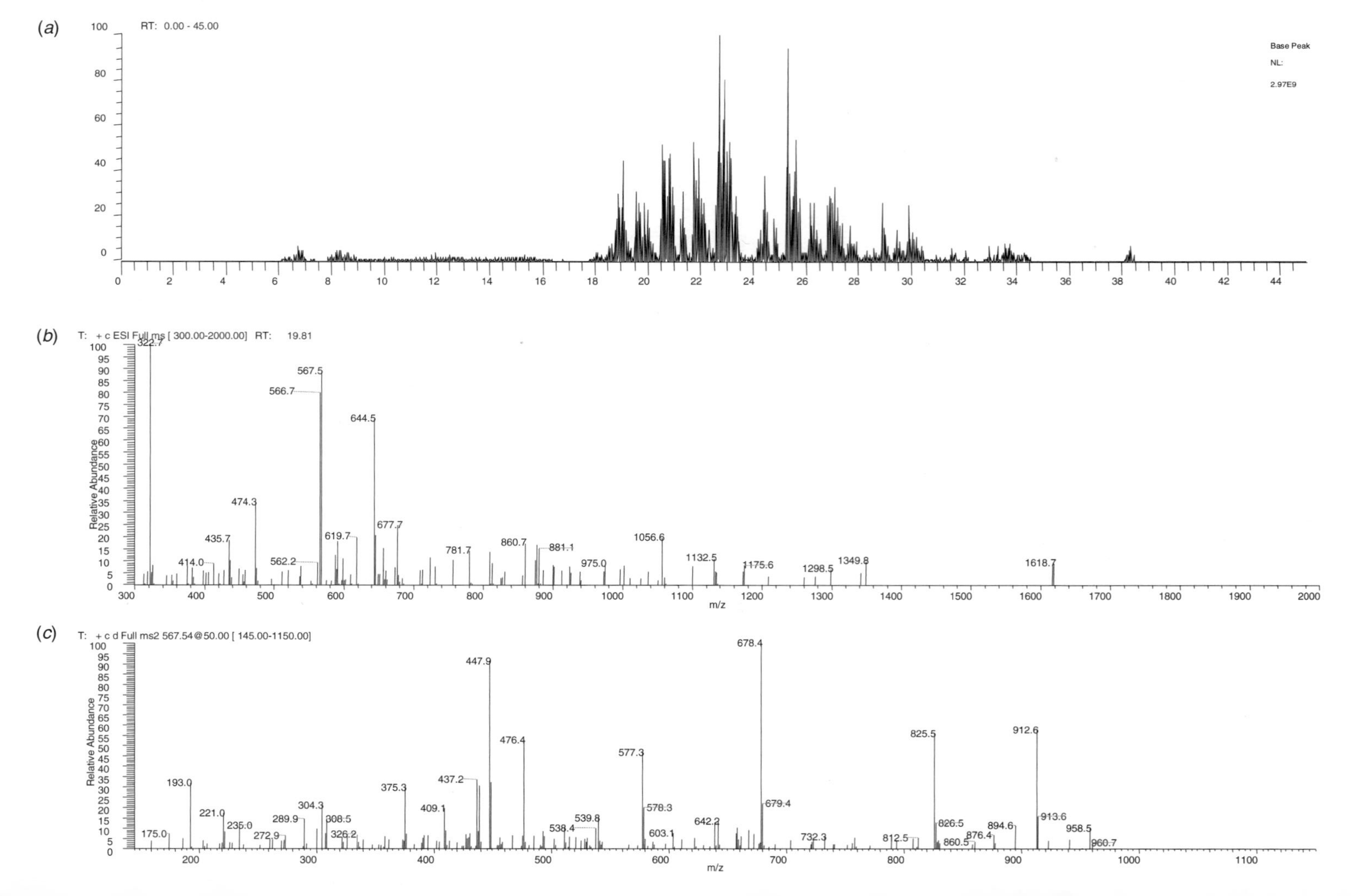
(a)
RT: 0.00 - 45.00
Base Peak
NL:
2.97E9
(b)
T: + c ESI Full ms [300.00-2000.00] RT: 19.81
Relative Abundance
322.7
567.5
566.7
644.5
474.3
677.7
435.7
619.7
781.7
860.7
881.1
1056.6
414.0
562.2
975.0
1132.5
1175.6
1298.5
1349.8
1618.7
m/z
(c)
T: + c d Full ms2 567.54@50.00 [145.00-1150.00]
Relative Abundance
678.4
447.9
476.4
577.3
825.5
912.6
193.0
375.3
437.2
221.0
304.3
409.1
578.3
679.4
175.0
235.0
272.9
289.9
308.5
326.2
538.4
539.8
603.1
642.2
732.3
812.5
826.5
860.5
876.4
894.6
913.6
958.5
960.7
m/z

COLOR PLATES

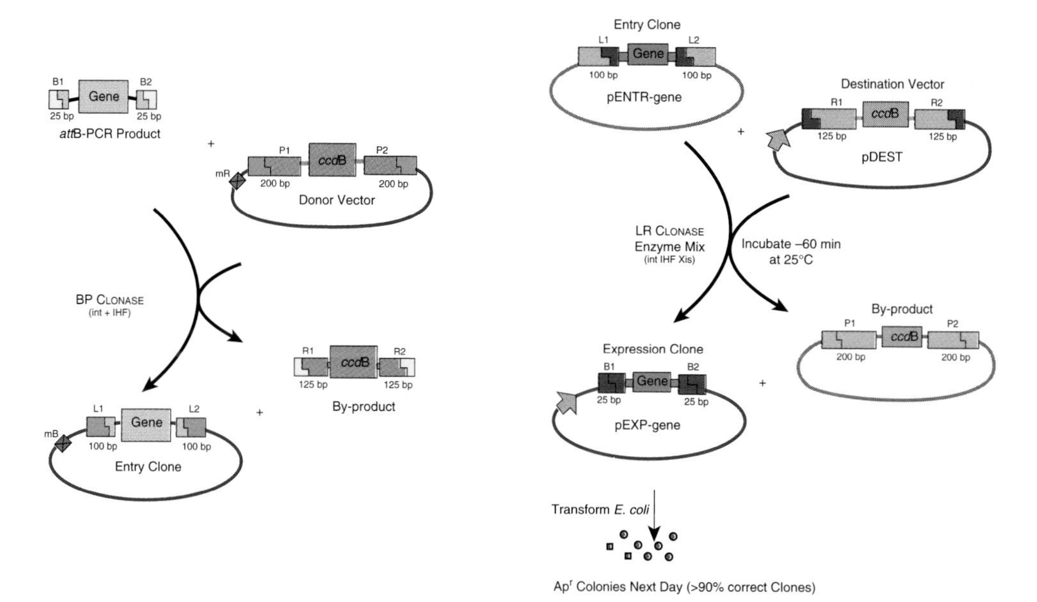

Figure 2.2. Cartoon view of the Gateway cloning system. Primers are first designed to PCR the coding region of interest and flank it with *att*ß sequences that are used with the BP clonase to introduce the gene of interest into the entry clone. The entry clone is then shuffled into the destination vectors using the LR clonase.

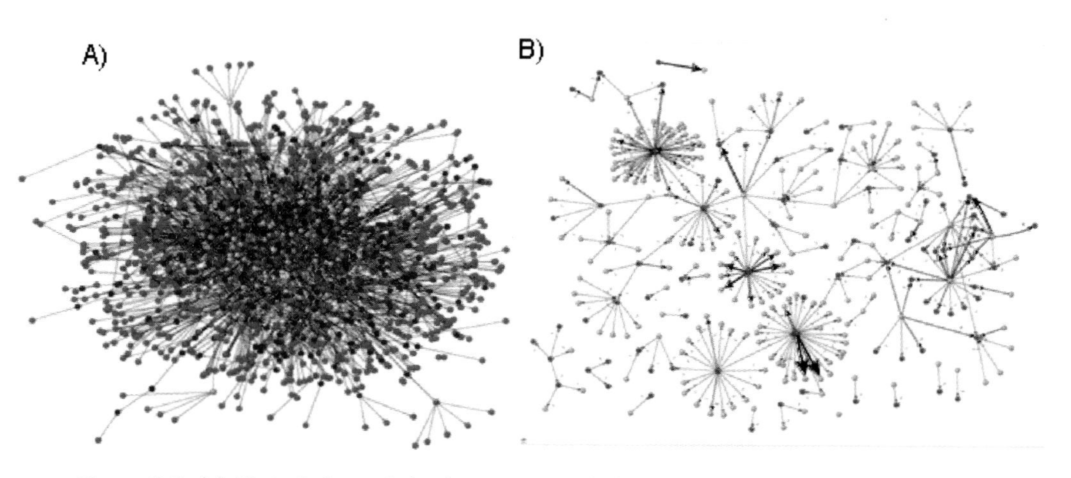

Figure 2.8. (*a*) Global view of the human protein interactions generated using 500 distinct human bait proteins. (*b*) Interaction view for some of the kinases that are part of the network. (The kinases are in red.)

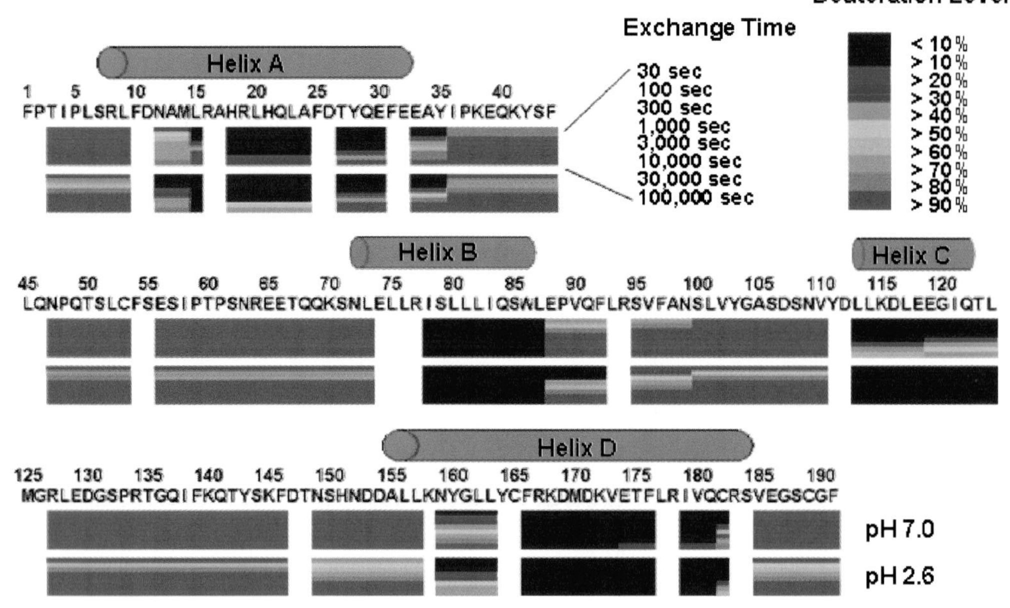

Figure 5.3. H/D-Ex analysis of hGH at pH 7.0 and 2.6. Each block represents a pepsin-generated peptide. Each block consists of eight rows that represent eight distinct on-exchange time points, shown at the right. The level of deuteration in each peptide at each time point is represented by color according to the diagram displayed at the top right. Blocks representing on-exchange at pH 7.0 are on the top row, while blocks representing on-exchange at pH 2.6 are shown at the bottom. Light blue cylinders above the sequence indicate the helices identified from the X-ray crystal structure of hGH (1HGU). Peptides that contain mostly slow exchanging amide hydrogens are represented by blue bars, while red bars represent peptides that contain mostly slow exchanging amide hydrogens are represented by blue bars, while red bars represent peptides that contain mostly rapidly exchanging amide hydrogens. Figure adapted with permission from Hamuro et al., 2003a.

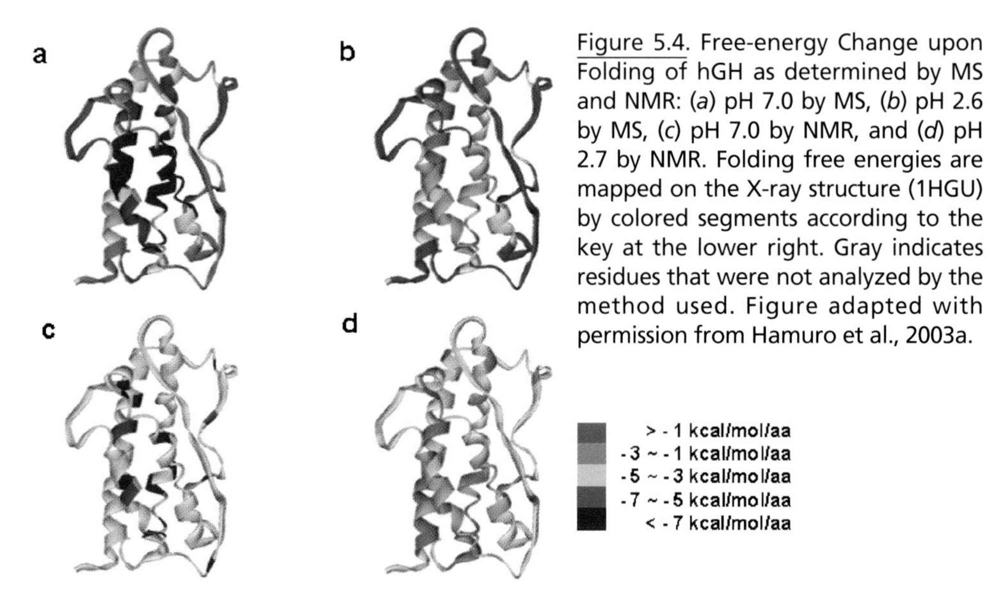

Figure 5.4. Free-energy Change upon Folding of hGH as determined by MS and NMR: (*a*) pH 7.0 by MS, (*b*) pH 2.6 by MS, (*c*) pH 7.0 by NMR, and (*d*) pH 2.7 by NMR. Folding free energies are mapped on the X-ray structure (1HGU) by colored segments according to the key at the lower right. Gray indicates residues that were not analyzed by the method used. Figure adapted with permission from Hamuro et al., 2003a.

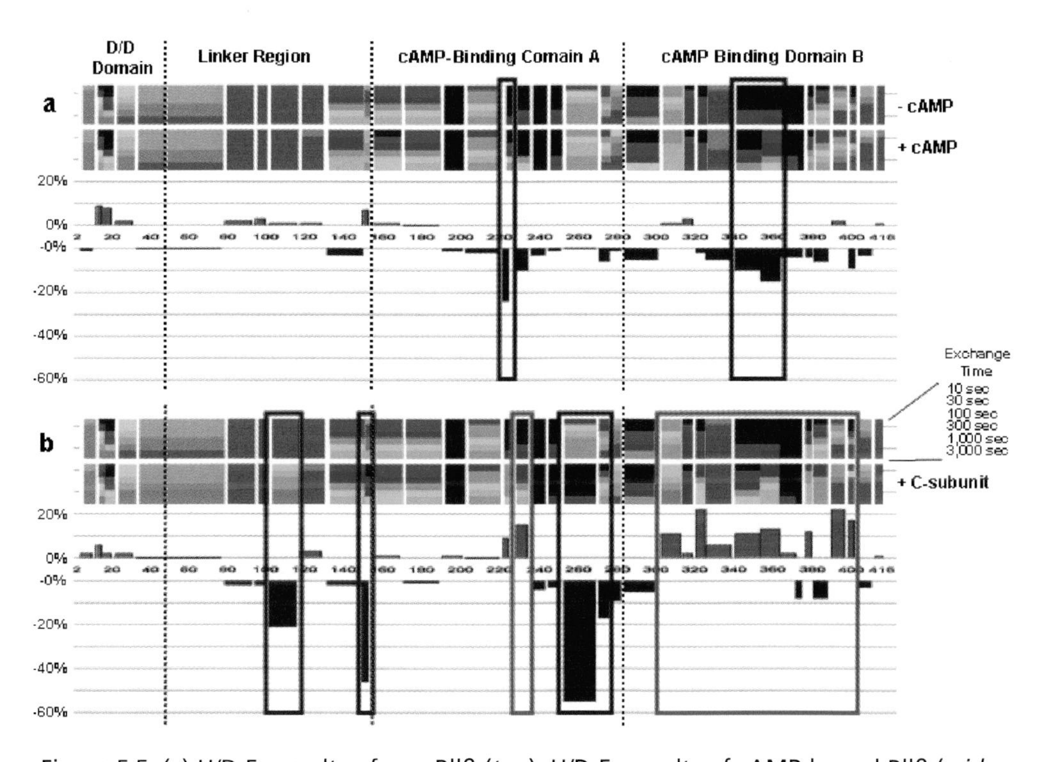

Figure 5.5. (*a*) H/D-Ex results of apo RIIß (*top*), H/D-Ex results of cAMP-bound RIIß (*middle*), and the average difference in deuteration levels with and without cAMP (*bottom*). Each block represents a pepsin-generated peptide and has six time points (10, 30, 100, 300, 1000, and 3000s). Levels of deuterium incorporation are indicated according to the color scheme used in Figure 5.3. The average difference in deuteration level is the average of the deuteration level with ligand subtracted by that without ligand. A negative value (blue) indicates the region is protected upon ligand binding. A positive value (red) indicates the region exchanges faster on ligand binding. Solid blue rectangles indicate regions significantly protected upon ligand binding. (*b*) H/D-Ex results of apo RIIß (*top*) H/D-Ex results of catalytic-subunit-bound RIIß (*middle*), and the average difference in deuteration levels with and without the catalytic subunit (*bottom*). Red rectangles in the bottom part indicate that the region exchanges faster upon catalytic-subunit binding and reveal conformational changes induced by complexation.

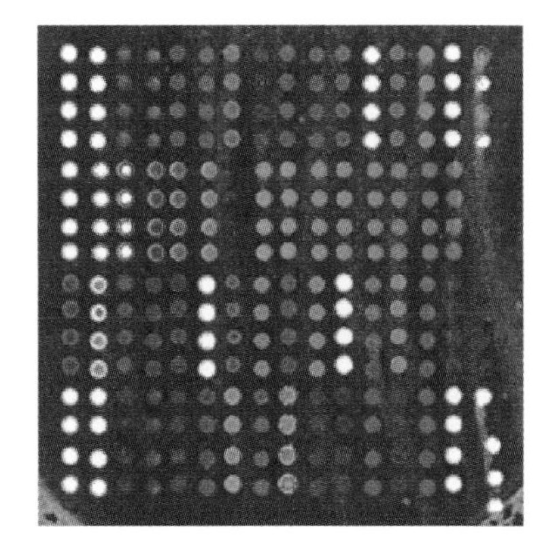

Figure 11.3. Fluorescence image of a multiplexed cytokine sandwich assay using rolling circle amplification of the signal. Signal intensity is represented in pseudocolor. (With kind permission from Molecular Staging, Inc.)

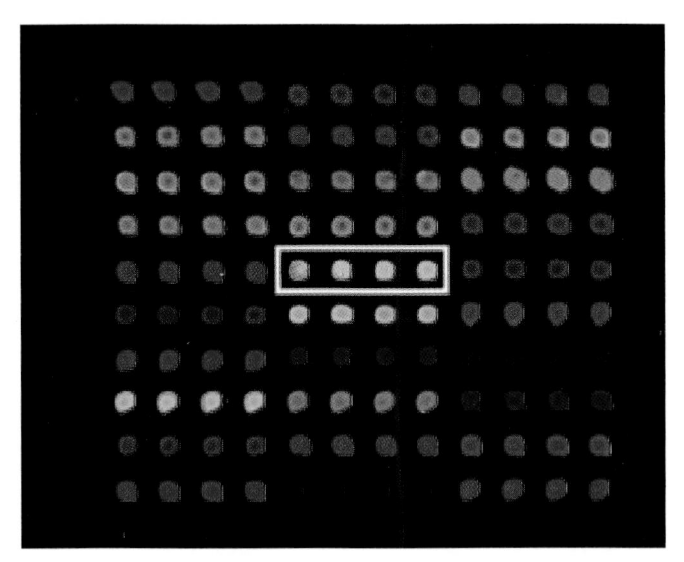

Figure 11.4. Fluorescence image of a photoaptamer microarray. Human serum, diluted fourfold into assay buffer, was incubated on an array of 24 unique aptamers, spotted in quadruplicate, for 60 min. The array was washed, photo-cross-linked, and washed again to remove unbound proteins. The boxed spots are the von Willebrand factor protein captured on the cognate aptamer. (With kind permission from Somalogic, Inc.)

Figure 11.5. Pseudocolor fluorescence image of a 512-member antibody array incubated with a Cy5-labeled extract of RANTES-treated macrophages and a Cy3-labeled extract of RANTES treated lymphocytes. (With kind permission from BD-Biosciences Clontech.)

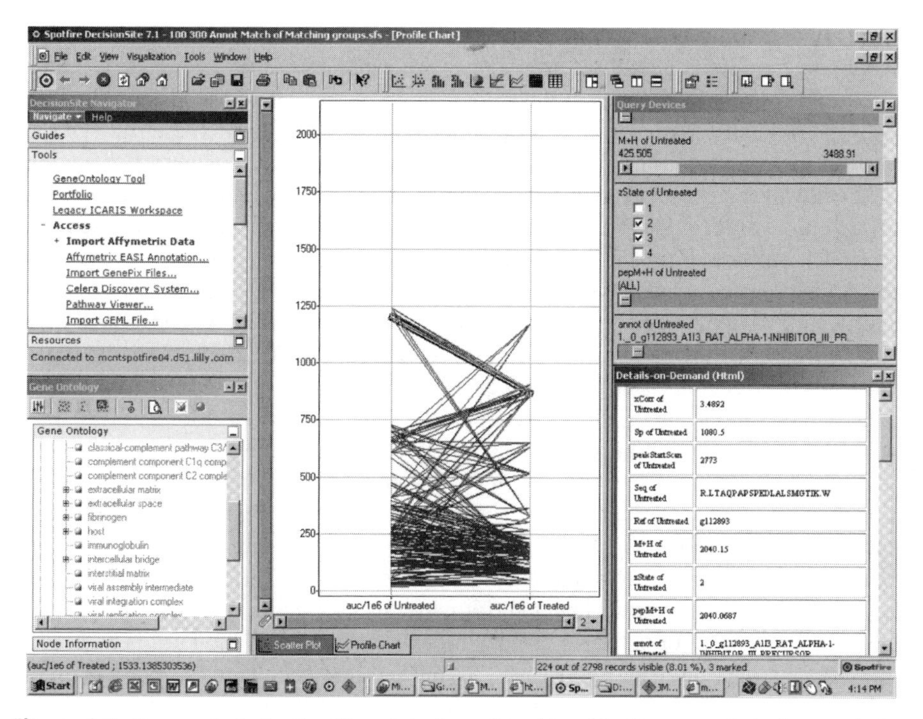

Figure 6.3. Screen shot of a Spotfire plot of peptides identified in a sample before and after perturbation. The right-hand axis represents levels of peptides in a perturbed sample; the left-hand axis represents peptide levels in an untreated sample. Individual peptides and classes of peptides may be selected using the slider bars and buttons. Information on the protein from which the peptide arose may be displayed in the details window, and additional information is available with additional tools.

of proteins has been known for years. Many studies have combined peptide mapping with LC/MS technologies to identify sites of phosphorylation in proteins [reviewed in (McLachlin and Chait (2001)]. Advances in MS technology such as the triple quadrupole allowed for identification of phosphorylated peptides in mixtures using strategies such as precursor ion scanning (Carr et al., 1996) followed by MS/MS identification.

Figure 6.2. LC-MS/MS analysis of a protein mixture. Cellular lysates were incubated with antibody to phosphotyrosine. Antibody was precipitated with immobilized protein A. The protein mixture was reduced, alkylated, and digested with trypsin. The resultant peptides were analyzed on an ion trap mass spectrometer equipped with a 75-μm C-18 capillary column. Peptides were fragmented by collisional induced dissociation (CID) and MS/MS spectra were used to search protein databases with the program Sequest. (*a*) Base peak display of the total ion chromatogram of the LC separation of the peptides. (*b*) Mass spectrum acquired at one time point (19.8 min). (*c*) MS/MS spectrum of ion 567.5 from panel (*b*).

By combining MS identification technologies with fractionation technologies such as immunoprecipitation and gel electrophoresis, information can be assimilated in much more rapid fashion. Many recent developments have been aimed at specific isolation of phosphorylated peptides from digests of very complex mixtures. These strategies are aimed at increasing the throughput as well as the information content of proteomic-based pathway analysis and are based on unique chemical properties of phosphorylated amino acids. This provides a basis for the affinity isolation of phosphorylated peptides. A complex mixture of proteins from cellular lysates, for example, may be digested, en masse, yielding thousands of individual peptides. By isolation of just the phosphorylated peptides, a much smaller number of identifications need to be performed. From these peptide identifications, the proteins from which they are derived can be identified. Comparing peptides from control and perturbed states can implicate proteins as potential target molecules, pathway members, or natural substrates for kinases of interest. Immobilized metal affinity chromatography (IMAC) chromatography can be used for this purpose, as metals such as iron or gallium have an affinity for acidic amino acids, and phosphorylated amino acids are among the most acidic. Chromatographic conditions have been designed that enrich extracts for phosphopeptides (Neville et al., 1997). In addition, some groups have studied the selectivity of different metals for phosphopeptides and have reported better performance using gallium (Posewitz and Tempst, 1999). Neutralization of carboxyl groups in peptides by methylation has been used to reduce the binding of nonphosphorylated residues to IMAC columns, as phosphoamino acids are not neutralized by this chemistry. Coupling this strategy with online capillary chromatography has allowed automation of phosphorylated peptide identification from cellular lysates (Ficarro et al., 2002). Elimination chemistries have also been used to isolate phosphorylated peptides. Phosphate groups may be eliminated from serine and threonine residues leaving reactive groups that may be captured selectively and separated from nonphosphrorylated residues (Zhou et al., 2001). These technologies must be quantitative or compatible with quantification technologies to be of the greatest value (see below).

Glycosylation. Alteration in the glycosylation pattern of proteins is a well-known result of cellular transformation, and altered glycoproteins have been proposed as targets for cancer therapies (Chekenya et al., 2002; Ceriani et al., 1993; Wick and Groner, 1997), including those employing such targets to develop therapeutic antibodies. One of the more common technologies for visualization of changes in glycosylation patterns is two-dimensional gel electrophoresis (see below). Changes in the charge-train pattern are indicative of glycosylation changes. Mass spectrometric methods are widely used in glycoprotein analysis (Harvey, 2001). Increasingly sensitive techniques are being developed that allow for the identification of carbohydrate structures from very small quantities of protein (Mechref and Novotny, 2002) both in solution and from gel spots (Charlwood et al., 2001). The study of glycoproteins is complex as dozens of different carbohydrate structures may exist at any individual glycosylation site and variations in glycoproteins may involve differences in the relative ratios of these structures within the protein. Different strategies must be utilized for the release and analysis of N-linked and O-linked carbohydrates. Databases of carbohydrate structures have begun to appear recently such as the database at *www.glycosuite.com*, but essential informatics tools are only beginning to be developed for this field.

Quantification

Identification of a protein as a target candidate requires involvement of that protein in a disease state or a pathway associated with a disease state. Changes in the level of a protein in response to a perturbation may be used to implicate the protein as having a function in the system being perturbed. In gel-based separations, differential protein expression may be determined by increases or decreases in individual band or spot intensities between perturbed and unperturbed states. Many strategies have been developed recently to determine relative levels of proteins in gel-free separations of protein mixtures. Most of the newer strategies involve isotopic labeling of proteins in normal and perturbed states and subsequent analysis of the ratios of these proteins utilizing mass spectrometry.

Isotope-Coded Affinity Tags. Isotope-coded affinity tags (ICAT) is perhaps the best-known strategy for relative quantification. The reagent, a modified iodoacetamide with a biotin group attached, may be enriched with deuterium to produce a mass shift of eight mass units between peptides labeled with deuterated and nondeuterated reagent (Gygi et al., 1999b). An additional feature of this reagent is the ability to dramatically reduce the number of peptides from a complex mixture. In a typical application, proteins in a mixture are reduced and then alkylated with the ICAT reagent. Proteins from an unperturbed state are alkylated with nondeuterated ICAT; those from a perturbed state are labeled with deuterated ICAT. The protein mixtures are combined and then digested with trypsin. Tryptic peptides labeled with ICAT are isolated with an avidin column and analyzed by LC-MS/MS. This yields pairs of peptide ions separated by eight mass units. The ratio of the heights of the peaks seen for these peptide ions is proportional to the ratio of the abundance in the mixture of the proteins from which they are derived.

Global Internal Standard Technology. Global internal standard technology (GIST) is another isotopic labeling strategy that was developed around the same time as ICAT (Chakraborty and Regnier, 2002). This strategy differs in some respects in that it labels primary amine groups causing shifts of six mass units per amino group between the deuterated and nondeuterated states. In addition, peptides are labeled after tryptic digestion and perturbed and nonperturbed peptides are then mixed. Enrichment of classes of peptides has been accomplished (e.g., isolation of histidine-containing peptides) chromatographically (Wang et al., 2002). Much work has been done to optimize the labeling chemistries as well as use of ^{13}C instead of deuterium to introduce the mass shift. This has been done in order to minimize differences in elution of peptides that has been observed between deuterated and nondeuterated species (Zhang et al., 2002).

Methylation. Methylation of acidic groups is another strategy for introduction of an isotopic tag into a peptide mixture. Deuterated and nondeuterated methanol has been used to esterify the acidic amino acids, aspartate and glutamate, as well as the C-terminus of the peptide in a procedure similar to GIST. This strategy has been applied to the quantification of phosphorylated peptides (Goodlett et al., 2001) that can be affinity isolated.

Mass-Coded Abundance Tags (MCAT). The reagent O-methyl isourea may be used to specifically guanidinate the epsilon amine group of lysine residues. This reagent has been used to label tryptic peptides from a perturbed state, which were then combined with unlabeled peptides from an unperturbed state. It was reported that these peptides eluted closely enough in the reversed-phase separations used in an LC/MS run to quantify peptides using the mass shift of 42 Da (Cagney and Emili, 2002).

Direct Quantification. Recent publications have demonstrated that individual ion intensities extracted from total ion chromatograms may be used to determine relative abundances of peptides between LC/MS runs (Bondarenko et al., 2002). In this strategy, multiple peptides from the same protein may be quantified to give independent values for the parent protein. These may then be utilized to calculate a mean relative abundance of the protein between two states. It is important to normalize the data from each run to proteins that are known not to change.

VALIDATION

As with most tools, proteomics technology is most valuable when used in concert with other tools employed in systems biology. Proteomics findings and transcript profiling findings can certainly be used together with information known about specific proteins to prioritize specific putative targets for validation. However, for both methods, once a protein target has been suggested, there are a number of tools that can be used to validate the relevance of this target.

The ultimate validation of a target occurs in the clinic when a specific drug interacts with a specific target in a manner that reverses or arrests a disease process. However, in the drug discovery process there are a number of intermediate levels of validation (Bumol and Watanabe, 2001) that occur between target identification and clinical validation. Each of these improves the probability that a putative target will lead to a valuable therapy. These intermediate validation states include: (1) association of the target with the disease state, (2) mechanistic association of the target with a particular pathway or structure relevant to the disease state, (3) demonstration that stimulation or inhibition of the target affects the pathway or structure in vitro, (4) demonstration that stimulation or inhibition of the target affects critical disease parameters in a relevant animal model, and (5) demonstration that the stimulation or inhibition of the target by a clinically acceptable therapeutic affects critical disease parameters in humans.

Genomic analysis, transcript profiling, proteomics, immunohistochemistry, in situ hybridization, and other approaches can be used individually or, more effectively, in concert to demonstrate the association of putative target proteins with a disease state. This is often the starting point for a validation campaign but is never sufficient to establish a protein as a drugable target. Disease association does not imply causation and, in fact, the majority of proteins associated with a particular disease are a result of the disease state and do not have a causative role. However, it is essential at this point to define what one means by disease state. In this context, it is critical to be certain that the disease state is that condition which you intend to reverse with therapy. Clinicians are extremely familiar with agents that resolve one or more symptoms but do not have any significant effect on the underlying disease process.

In Vitro Techniques

There are a number of components to mechanistic association of a putative target protein to a particular disease. In the simplest case genomic analysis can show that all patients with a particular mutation in their gene for hemoglobin causes their red blood cells to assume an aberrant morphology, sickle cell, which results in a disease defined as sickle cell anemia. In this case a great deal of analysis of the disease state occurred before the specific genetic mutation was sought and found. In today's era of total genomic databases for an increasing number of organisms, we often find the gene associated with the disease before we know anything about the functional gene product, except a potential sequence. We say potential sequence for reasons mentioned above relating to splice variants, posttranslation modification, cycles of activation and inactivation, and intracellular localization. The quest for mechanistic association takes us from the analysis of deoxyribonucleic acid (DNA) in the nucleus or RNA in the cytoplasm to the analysis of specific proteins, structures, and pathways in every intracellular compartment and on to molecular events that occur at the cell surface or in extracellular milieu. Proteomics tools such as those focused on protein–protein, protein–nucleic acid, or protein–small molecule binding (Ho et al., 2002; Greenbaum et al., 2002; Steiner et al., 2002) can also be used to see what a protein is capable of binding in vitro. However, it is important to remember that this does not necessarily prove that the binding actually occurs in the cell or, if it does occur, that it plays any role in the disease.

Advanced microscopy and imaging reagents, often incorporating antibodies or other selectively binding molecules, allow us to identify components of functional complexes within specific intracellular compartments and extracellularly (Wouterlood et al., 2002). Cell-based assays coupled with neutralizing antibodies (Fishwild et al., 1996), antisense oligonucleotides (Weston et al., 1999), RNA interference (RNAi) (Elbashir et al., 2001; Nasir, 2001), or "validating" therapeutic compounds (Smith, 2003) can help us to implicate proteins as critical components of pathways or functional structures by examining in detail the result of the protein's absence, inhibition, or diminution in the cell. Finally, even if we are able to show that a particular protein has a particular effect in a cultured cell system, or can be inhibited by bathing that culture in a particular therapeutic agent, we have established very little about drugability in an organism.

In Vivo Techniques

The biggest challenge to the validation of a protein target as a valid drug target in an intact organism is the selection and validation of a relevant preclinical animal model. These models may include transgenic or gene knockout animals (Abuin et al., 2002) or in vivo treatment with antisense oligonucleotides (Friedman and Veliskova, 1999; Uhlmann and Peyman, 1990) or RNAi (McCafferty et al., 2002), which demonstrate the effect of lowered or raised levels of target in the organism. It is important in these cases to determine whether the molecule that is increased or decreased is in a functional form and is present in the intracellular or extracellular compartment where it exerts its effects. Many excellent molecules have failed as drugs in the clinic because the animal model selected was not predictive of the disease state in humans. However, even when a relevant model is selected, it is critical to design the preclinical protocol in a manner that generates statistically significant information. It is also critical to measure every relevant parameter, including those at a point distant from the putative location of the drug target. At this point the importance of "off-target" effects of therapeutics are as important as their effects on the desired target in the tissue of interest. It is also important to recognize that many proteins are promiscuous, which can lead to on-target effects at the protein level that are off-target effects at the tissue level. This is extremely critical in cancer where most putative targets are tumor associated and not tumor specific. This is an area where comprehensive proteomic analysis can be extremely helpful. In transcript profiling, we often are able to determine relative expression of specific genes in specific tissues. However, low or undetectable expression is not equal to no expression. Further, many proteins are able to travel from their site of synthesis to other sites of action. Using sensitive tools of proteomic analysis, we can look for specific proteins in specific compartments and even look for the presence or absence of specific posttranslationally modified forms in different compartments. This may be extremely relevant given recent reports suggesting that the same target protein with different glycosylation may be preferentially or uniquely associated with specific tumors (Taylor-Papadimitriou and Epenetos, 1994).

Excellent performance of a specific therapeutic acting on a specific target in a valid animal model justifies the move to clinical validation, but it is not always predictive of what will occur in a patient population. Patient to patient variability has led to the failure of a number of drugs. This has prompted a significant interest in pharmacogenetics and pharmacogenomics as tools to understand patient-specific events and ultimately to predict these prior to administration of a drug. In this way we can limit adverse events and more accurately tailor the dosage to a particular patient's system. The term *pharmacoproteomics* has been coined to look at this at the protein level as well, but in this case the reference is primarily to analysis from readily obtainable patient specimens such as serum, plasma, sputum, urine, cerebrospinal fluid, or biopsy materials. This could limit the utility of pharmacoproteomics relative to pharmacogenetics, as every cell in the body contains the entire genetic complement of the patient. Pharmacogenomics, which follows the patient's response to a particular drug, shares a limitation with pharmacoproteomics as to samples that can be taken for analysis, but its relatively higher sensitivity at present may make this less of a restriction.

INFORMATICS

Primary Data Analysis

As outlined in the previous sections, protein identification is accomplished through the application of database searching algorithms to primary mass spectral data. There are many software packages for the identification of proteins from mass analysis of peptides produced by enzymatic digestion of the parent proteins. Some of the more popular packages for mass fingerprints are Mascot (Perkins et al., 1999), ProFound (Zhang and Chait, 2000), and Protein Prospector (Clauser et al., 1995). Popular packages for MS/MS-based identification are Sequest (Eng et al., 1994), Mascot (Perkins et al., 1999), and Protein Prospector (Clauser et al., 1995). All of these packages have some scoring associated with them as to the correctness of the identification assignment; however, none of these tools has to date replaced manual inspection of the identification for absolute confidence in the assignment. The necessity for manual review has been cumbersome but attainable on simple samples and few runs. The last year or two has seen a number of reports attempting to design better scoring to eliminate the need for manual review of the data (MacCoss et al., 2002; Keller et al., 2002; Moore et al., 2002; Gatlin et al., 2000). The flood of data from automated mass spectrometry solutions makes manual data review impossible. The lack of a clear delineation between good and bad calls can erode confidence in the identification. A list of identifications is only the beginning of the process for target discovery and validation. The identified proteins then need to be harmonized with alternative nomenclature of the proteins and placed in an appropriate biological context.

Tools for Data Manipulation and Data Reduction

Due to the sheer volume of data from proteomic-based experiments, the data must be presented to researchers in a way in which they can understand the data and draw conclusions. It is impossible for a person to look at 60,000 possible identifications and reduce the data to comprehension without appropriate software tools.

Table Manipulation. The first category of tools includes those for table manipulation. Programs such as Excel or JMP are useful for getting data into a table format, performing statistical analysis and extracting additional information missing from the identification software output such as quantitative comparisons.

Table Storage and Joining. The second category of tools includes those for table storage and joining. Through these, expanded tables of primary data can be stored directly into a database such as Access. Placing the tables into a database allows one to add value to the primary information by merging it with additional information about proteins such as gene ontology information, pathway involvement, experimental details, additional analyses, and the like. This additional information is the beginning of establishing the biological context essential for meaningful data interpretation. The data can then be parsed in multiple ways with queries and stored in the database. However, looking at big tables is not a convenient way for researchers to interact with the data and draw conclusions, thus the need for the third category of tools, those for data reduction and visualization.

Data Reduction. Data reduction is best achieved via graphics. The adage "a picture is worth a thousand words" is especially true in this context. A software package well suited for this is Spotfire. It can take a single or multiple tables of data and enable an interactive graphic representation of the table or tables' contents. Sliders and buttons that update the graphics in real time can create multilevel queries simply (see Figure 6.3). There are additional tools in Spotfire that allow viewers for biological pathways and gene ontology to interact with the data in real time. This creates a simple environment through which the end user of complex proteomics output can interact with the data and draw conclusions. The database architecture must enable one database to create multiple views of the same data for different people or views incorporating different data sets for different applications. The mass spectrometrist will want to look at different data than the biologist, and they both may want to look at similar data in different ways.

Interaction Maps. The ultimate fusion of protein data involves construction of protein interaction maps. Databases such as MINT provide information on networks of protein interactions that graphically illustrate protein interactions both upstream and downstream of any particular protein component. These databases are the cornerstones of systems biology databases that will ultimately enable the storage and retrieval of information at all levels of biological interactions.

CONCLUSION

The challenges of target identification and validation are larger now than ever due to the flood of genomic data. The renaissance of protein chemistry coupled with the continuing advances in microanalytical technology provides a formidable arsenal to meet these challenges and advances in these technologies appear in the literature almost daily. We have summarized a number of different applications in the area of target identification/validation. However, we realize that this is by no means an exhaustive review of all of the possible strategies available. Limitations and gaps still exist in the field that will impact these efforts. As more targets are validated the issues surrounding lack of reagent availability will become more and more acute. Additionally, current analytical technologies may still exhibit limitations when applied to protein variants that are very similar in structure. The field of glycoprotein analysis is not nearly as advanced as conventional proteomics. Ultimately, protein analytical methods will provide information for the construction of multiplexed assays, which are being developed in a multitude of formats including protein chip technologies. The entire field of protein bioinfomatics is at a very early stage of development and the exponential increases in data generation, analysis, and archival will need to be addressed. These are a few of the gaps and challenges that will need to be addressed in the coming years as the demand for innovative drug discovery continues its explosive growth.

REFERENCES

Abuin, A., Holt, K.H., Platt, K.A., Sands, A.T., and Zambrowicz, B.P. (2002). Full-speed mammalian genetics: In vivo target validation in the drug discovery process. *Trends Biotechnol.* **20**(1):36–42.

Adam, P.J., Boyd, R., Tyson, K.L., Fletcher, G.C., Stamps, A., Hudson, L., Poyser, H.R., Redpath, N., Griffiths, M., Steers, G., Harris, A.L., Patel, S., Berry, J., Loader, J.A., Townsend, R.R., Daviet, L., Legrain, P., Parekh, R., and Terrett, J.A. (2003). Comprehensive proteomic analysis of breast cancer cell membranes reveals unique proteins with potential roles in clinical cancer. *J. Biol. Chem.* **278**(8):6482–6489.

Battersby, J.E., Guzzetta, A.W., and Hancock, W.S. (1994). Application of capillary high-performance liquid chromatography to biotechnology, with reference to the analysis of recombinant DNA-derived human growth hormone. *J. Chromatogr. B: Biomed. Applic.* **662**(2):335–342.

Blagoev, B., Kratchmarova, I., Ong, S.E., Nielsen, M., Foster, L.J., and Mann, M. (2003). A proteomics strategy to elucidate functional protein–protein interactions applied to EGF signaling. *Nature Biotechnol.* **21**(3):315–318.

Blobel, G. (2000). Protein targeting (Nobel lecture). *Chembiochem.* **1**(2):86–102.

Bondarenko, P.V., Chelius, D., and Shaler, T.A. (2002). Identification and relative quantitation of protein mixtures by enzymatic digestion followed by capillary reversed-phase liquid chromatography–tandem mass spectrometry. *Anal Chem.* **74**(18):4741–4749.

Bumol, T.F., and Watanabe, A.M. (2001). Genetic information, genomic technologies, and the future of drug discovery. *JAMA* **285**(5):551–555.

Cagney, G., and Emili, A. (2002). De novo peptide sequencing and quantitative profiling of complex protein mixtures using mass-coded abundance tagging. *Nature Biotechnol.* **20**(2):163–170.

Carr, S.A., Huddleston, M.J., and Annan, R.S. (1996). Selective detection and sequencing of phosphopeptides at the femtomole level by mass spectrometry. *Anal. Biochem.* **239**(2):180–192.

Ceriani, R.L., Peterson, J.A., Blank, E.W., and Lamport, D.T. (1993). Epitope expression on the breast epithelial mucin. *Breast Cancer Res. Treat.* **24**(2):103–113.

Chakraborty, A., and Regnier, F.E. (2002). Global internal standard technology for comparative proteomics. *J. Chromatogr. A* **949**(1–2):173–184.

Charlwood, J., Bryant, D., Skehel, J.M., and Camilleri, P. (2001). Analysis of N-linked oligosaccharides: Progress towards the characterisation of glycoprotein-linked carbohydrates. *Biomol. Eng.* **18**(5):229–240.

Chekenya, M., Enger, P.O., Thorsen, F., Tysnes, B.B., Al-Sarraj, S., Read, T.A., Furmanek, T., Mahesparan, R., Levine, J.M., Butt, A.M., Pilkington, G.J., and Bjerkvig, R. (2002). The glial precursor proteoglycan, NG2, is expressed on tumour neovasculature by vascular pericytes in human malignant brain tumours. *Neuropathol. Appl. Neurobiol.* **28**(5):367–380.

Clauser, K.R., Hall, S.C., Smith, D.M., Webb, J.W., Andrews, L.E., Tran, H.M., Epstein, L.B., and Burlingame, A.L. (1995). Rapid mass spectrometric peptide sequencing and mass matching for characterization of human melanoma proteins isolated by two-dimensional PAGE. *Proc. Natl. Acad. Sci. USA* **92**:5072–5076.

Cottrell, J.S. (1994). Protein identification by peptide mass fingerprinting. *Peptide Res.* **7**(3):115–124.

Drews, J. (1996). Genomic sciences and the medicine of tomorrow. *Nature Biotechnol.* **14**:1516–1518.

Elbashir, S.M., Harborth, J., Lendeckel, W., Yalcin, A., Weber, K., and Tuschl, T. (2001). Duplexes of nucleotide RNAs mediate RNA interference in cultured mammalian cells. *Nature* **411**(6836):494–498.

Eng, J.K., McCormack, A.L., and Yates, J.R. (1994). An approach to correlate tandem mass spectral data of peptides with amino acid sequences in aprotein database. *J. Am. Soc. Mass Spec.* **5**:976.images; *Anal. Chem.* 1999, **71**:4981–4988.

Even, S., Lindley, N.D., Loubiere, P., and Cocaign-Bousquet, M. (2002). Dynamic response of catabolic pathways to autoacidification in *Lactococcus lactis*: Transcript profiling and stability in relation to metabolic and energetic constraints. *Mol. Microbiol.* **45**(4):1143–1152.

Fenn, J.B., Mann, M., Meng, C.K., Wong, S.F., and Whitehouse, C.M. (1989). Electrospray ionization for mass spectrometry of large biomolecules. *Science* **246**(4926):64–71.

Ficarro, S.B., McCleland, M.L., Stukenberg, P.T., Burke, D.J., Ross, M.M., Shabanowitz, J., Hunt, D.F., and White, F.M. (2002). Phosphoproteome analysis by mass spectrometry and its application to *Saccharomyces cerevisiae*. *Nature Biotechnol.* **20**(3):301–305.

Fishwild, D.M., O'Donnell, S.L., Bengoechea, T., Hudson, D.V., Harding, F., Bernhard, S.L., Jones, D., Kay, R.M., Higgins, K.M., Schramm, S.R., and Lonberg, N. (1996). High-avidity human IgG kappa monoclonal antibodies from a novel strain of minilocus transgenic mice. *Nature Biotechnol.* **14**(7):845–851.

Friedman, L.K., and Veliskova, J. (1999). GluR2 antisense knockdown produces seizure behavior and hippocampal neurodegeneration during a critical window. *Ann. NY Acad. Sci.* **868**:541–545.

Garlisi, C.G., Zou, J., Devito, K.E., Tian Fang Zhu Feng X Liu Jianjun Shah Himanshu Wan Yuntao Billah M. Motasim Egan Robert W Umland Shelby P Human (2003). ADAM33: Protein maturation and localization. *Biochem. Biophys. Res. Commun.* **301**(1):35–43.

Gatlin, C.L., Eng, J.K., Cross, S.T., Detter, J.C., and Yates, J.R. 3rd. (2000). Automated identification of amino acid sequence variations in proteins by HPLC/microspray tandem mass spectrometry. *Anal. Chem.* **72**(4):757–763.

Gavin, A.C., Bosche, M., Krause, R., Grandi, P., Marzioch, M., Bauer, A., Schultz, J., Rick, J.M., Michon, A.M., Cruciat, C.M., Remor, M., Hofert, C., Schelder, M., Brajenovic, M., Ruffner, H., Merino, A., Klein, K., Hudak, M., Dickson, D., Rudi, T., Gnau, V., Bauch, A., Bastuck, S., Huhse, B., Leutwein, C., Heurtier, M.A., Copley, R.R., Edelmann, A., Querfurth, E., Rybin, V., Drewes, G., Raida, M., Bouwmeester, T., Bork, P., Seraphin, B., Kuster, B., Neubauer, G., and Superti-Furga, G. (2002). Functional organization of the yeast proteome by systematic analysis of protein complexes. *Nature* **415**(6868):141–147.

Goodlett, D.R., Keller, A., Watts, J.D., Newitt, R., Yi, E.C., Purvine, S., Eng, J.K., von Haller, P., Aebersold, R., and Kolker, E. (2001). Differential stable isotope labeling of peptides for quantitation and de novo sequence derivation. *Rapid Commun. Mass Spectrom.* **15**(14):1214–1221.

Greenbaum, D., Baruch, A., Hayrapetian, L., Darula, Z., Burlingame, A., Medzihradszky, K.F., and Bogyo, M. (2002). Chemical approaches for functionally probing the proteome. *Mol. Cell. Proteomics* **1**(1):60–68.

Gygi, S.P., Rochon, Y., Franza, B.R., and Aebersold, R. (1999). Correlation between protein and mRNA abundance in yeast. *Mol. Cell. Biol.* **19**(3):1720–1730.

Gygi, S.P., Rist, B., Gerber, S.A., Turecek, F., Gelb, M.H., and Aebersold, R. (1999). Quantitative analysis of complex protein mixtures using isotope-coded affinity tags. *Nature Biotechnol.* **17**(10):994–999.

Harvey, D.J. (2001). Identification of protein-bound carbohydrates by mass spectrometry. *Proteomics* **1**(2):311–328.

Ho, Y., Gruhler, A., Heilbut, A., Bader, G.D., Moore, L., Adams, S.L., Millar, A., Taylor, P., Bennett, K., Boutilier, K., Yang, L., Wolting, C., Donaldson, I., Schandorff, S., Shewnarane, J., Vo, M., Taggart, J., Goudreault, M., Muskat, B., Alfarano, C., Dewar, D., Lin, Z., Michalickova, K., Willems, A.R., Sassi, H., Nielsen, P.A., Rasmussen, K.J., Andersen, J.R., Johansen, L.E., Hansen, L.H., Jespersen, H., Podtelejnikov, A., Nielsen, E., Crawford, J., Poulsen, V., Sorensen, B.D., Matthiesen, J., Hendrickson, R.C., Gleeson, F., Pawson, T., Moran, M.F., Durocher, D., Mann, M., Hogue, C.W., Figeys, D., and Tyers, M. (2002). Systematic identification of protein complexes in *Saccharomyces cerevisiae* by mass spectrometry. *Nature* **415**(6868):180–183.

Ideker, T., Thorsson, V., Ranish, J.A., Christmas, R., Buhler, J., Eng, J.K., Bumgarner, R., Goodlett, D.R., Aebersold, R., and Hood, L. (2001). Integrated genomic and proteomic analyses of a systematically perturbed metabolic network. *Science.* **292**(5518):929–934.

Ji, H., Whitehead, R.H., Reid, G.E., Moritz, R.L., Ward, L.D., and Simpson, R.J. (1994). Two-dimensional electrophoretic analysis of proteins expressed by normal and cancerous human crypts: Application of mass spectrometry to peptide-mass fingerprinting. *Electrophoresis* **15**(3–4):391–405.

Keller, A., Nesvizhskii, A.I., Kolker, E., and Aebersold, R. (2002). Empirical statistical model to estimate the accuracy of peptide identifications made by MS/MS and database search. *Anal. Chem.* **74**(20):5383–5392.

Laemmli, U.K. (1970). Cleavage of structural proteins during the assembly of the head of bacteriophage T4. *Nature* **227**:680–685.

Link, A.J., Eng, J., Schieltz, D.M., Carmack, E., Mize, G.J., Morris, D.R., Garvik, B.M., and Yates, J.R. 3rd. (1999). Direct analysis of protein complexes using mass spectrometry. *Nature Biotechnol.* **17**(7):676–682.

Lockhart, D.J., Dong, H.L., Byrne, M.C., Follettie, M.T., Gallo, M.V., Chee, M.S., Mittmann, M., Wang, C.W., Kobayashi, M., Horton, H., and Brown, E.L. (1996). Expression monitoring by hybridization to high-density oligonucleotide arrays. *Nat. Biotechnol.* **14**(13):1675–1680.

MacCoss, M.J., Wu, C.C., and Yates, J.R. 3rd. (2002). Probability-based validation of protein identifications using a modified SEQUEST algorithm. *Anal. Chem.* **74**(21):5593–5599.

Macri, J., and Rapundalo, S.T. (2001). Application of proteomics to the study of cardiovascular biology. *Trends Cardiovasc. Med.* **11**(2):66–75.

Matsudaira, P. (1987). Sequence from picomole quantities of proteins electroblotted onto polyvinylidene difluoride membranes. *J. Biolog. Chem.* **262**(21):10035–10038.

McCafferty, A.P., Meuse, L., Pham, T., Conklin, D.S., Hannon, G.J., and Kay, M.A. (2002). Gene expression: RNA interference in adult mice. *Nature* **418**(6893):38–39.

McLachlin, D.T., and Chait, B.T. (2001). Analysis of phosphorylated proteins and peptides by mass spectrometry. *Curr. Opin. Chem. Biol.* **5**(5):591–602.

Mechref, Y., and Novotny, M.V. (2002). Structural investigations of glycoconjugates at high sensitivity. *Chem. Rev.* **102**(2):321–369.

Moore, R.E., Young, M.K., and Lee, T.D. (2002). Qscore: An algorithm for evaluating SEQUEST database search results. *J. Am. Soc. Mass Spectrom.* **13**(4):378–386.

Nasir, J. (2001). Shutting off mammalian gene expression the easy way. *Clin. Genet.* **60**(5):332–333.

Neville, D.C., Rozanas, C.R., Price, E.M., Gruis, D.B., Verkman, A.S., and Townsend, R.R. (1997). Evidence for phosphorylation of serine 753 in CFTR using a novel metal-ion affinity resin and matrix-assisted laser desorption mass spectrometry. *Protein Sci.* **6**(11):2436–2445.

O'Farrell, P.H. (1975). High resolution two-dimensional electrophoresis of proteins. *J. Biol. Chem.* **250**:4007–4021.

Okuda, K., Foster, R., and Griffin, J.D. (1999). Signaling domains of the beta c chain of the GM-CSF/IL-3/IL-5 receptor. *Ann. NY Acad. Sci.* **872**:305–312.

Pandey, A., Podtalejnikov, A.V., Blagoy, B., Bustelo, X.R., Mann, M., and Lodish, H.F. (2000). Analysis of receptor signaling pathways by mass spectrometry: Identification of vav-2 as a substrate of the epidermal and platelet-derived growth factor receptors. *Proc. Natl. Acad. Sci. USA* **97**(1):179–184.

Perkins, D.N., Pappin, D.J., Creasy, D.M., and Cottrell, J.S. (1999). Probability-based protein identification by searching sequence databases using mass spectrometry data. *Electrophoresis* **20**(18):3551–3567.

Posewitz, M.C., and Tempst, P. (1999). Immobilized gallium(III) affinity chromatography of phosphopeptides. *Anal. Chem.* **71**(14):2883–2892.

Rasmussen, H.H., Mortz, E., Mann, M., Roepstorff, P., and Celis, J.E. (1994). Identification of transformation sensitive proteins recorded in human two-dimensional gel protein databases by mass spectrometric peptide mapping alone and in combination with microsequencing. *Electrophoresis* **15**(3–4):406–416.

Righetti, P.G. (1990). Isoelectric focusing in immobilized pH gradients of phosphoglucomutase and esterases from the spiny lobster. *Electrophoresis* **11**(10):810–812.

Rosenfeld, J., Capdevielle, J., Guillemot, J.C., and Ferrara, P. (1992). In-gel digestion of proteins for internal sequence analysis after one- or two-dimensional gel electrophoresis. *Anal. Biochem.* **203**(1):173–179.

Anderson, L., and Seilhamer, J. (1997). A comparison of selected mRNA and protein abundances in human liver. *Electrophoresis* **18**(3–4):533–537.

Shimkets, R.A., Lowe, D.G., Tai, J.T., Sehl, P., Jin, H., Yang, R., Predki, P.F., Rothberg, B.E., Murtha, M.T., Ruth, M.E., Shenoy, S.G., Windermath, A., Simpson, J.W., Simon, J.F., Daley, M.P., Gold, S.A., McMenna, M.P., Aillan, K., Went, G.T., and Rothberg, J.M. (1996). Gene expression analysis by transcript profiling coupled to a gene database query. *Nat Biotechnol.* **17**:798–803.

Smith, C. (2003). Hitting the target. *Nature* **422**(6929):341, 343, 345, 347, 349, 351.

Wilm, M., Shevchenko, A., Houthaeve, T., Breit, S., Schweigerer, L., Fotsis, T., and Mann, M. (1996). Femtomole sequencing of proteins from polyacrylamide gels by nano-electrospray mass spectrometry. *Nature* **379**(6564):466–469.

Soskic, V., Gorlach, M., Poznanovic, S., Boehmer, F.D., and Godovac-Zimmermann, J. (1999). Functional proteomics analysis of signal transduction pathways of the platelet-derived growth factor beta receptor. *Biochemistry* **38**(6):1757–1764.

Steiner, T., Kaiser, J.T., Marinkovic, S., Huber, R., and Wahl, M.C. (2002). Crystal structures of transcription factor NusG in light of its nucleic acid- and protein-binding activities. *EMBO J.* **21**(17):4641–4653.

Taylor-Papadimitriou, J., and Epenetos, A.A. (1994). Exploiting altered glycosylation patterns in cancer: Progress and challenges in diagnosis and therapy. *Trends Biotechnol.* **12**:227–233.

Towbin, H., Staehelin, T., and Gordon, J. (1979). Electrophoretic transfer of proteins from polyacrylamide gels to nitrocellulose sheets: Procedure and some applications. *Proc. Natl. Acad. Sci. USA* **76**(9):4350–4354.

Tsuji, T., Shiozaki, A., Kohno, R., Yoshizato, K., and Shimohama, S. (2002). Proteomic profiling and neurodegeneration in Alzheimer's disease. *Neurochem. Res.* **27**(10):1245–1253.

Uhlmann, E., and Peyman, A. (1990). Antisense oligonucleotides: A new therapeutic principle. *Chem. Rev.* **90**:544–584.

Valdes, I., Pitarch, A., Gil, C., Bermudez, A., Llorente, M., Nombela, C., and Mendez, E. (2000). Novel procedure for the identification of proteins by mass fingerprinting combining two-dimensional electrophoresis with fluorescent SYPRO red staining. *J. Mass Spectrom.* **35**(6):672–682.

Wang, S., Zhang, X., and Regnier, F.E. (2002). Quantitative proteomics strategy involving the selection of peptides containing both cysteine and histidine from tryptic digests of cell lysates. *J. Chromatogr. A* **949**(1–2):153–162.

Washburn, M.P., Wolters, D., and Yates, J.R. 3rd. (2001). Large-scale analysis of the yeast proteome by multidimensional protein identification technology. *Nat Biotechnol.* **19**(3):242–247.

Weston, B.W., Hiller, K.M., Mayben, J.P., Manousos, G.A., Bendt, K.M., Liu, R., and Cusack, J.C. Jr. (1999). Expression of human alpha(1,3)fucosyltransferase antisense sequences inhibits selectin-mediated adhesion and liver metastasis of colon carcinoma cells. *Cancer Res.* **59**(9):2127–2135.

Wick, B., and Groner, B. (1997). Evaluation of cell surface antigens as potential targets for recombinant tumor toxins. *Cancer Lett.* **118**(2):161–172.

Wouterlood, F.G., van Haeften, T., Blijleven, N., Perez-Templado, P., and Perez-Templado, H. (2002). Double-label confocal laser-scanning microscopy, image restoration, and real-time three-dimensional reconstruction to study axons in the central nervous system and their contacts with target neurons. *Appl. Immunohistochem.* **10**(1):85–95.

Zhang, R., Sioma, C.S., Thompson, R.A., Xiong, L., and Regnier, F.E. (2002). Controlling deuterium isotope effects in comparative proteomics. *Anal. Chem.* **74**(15):3662–3669.

Zhang, W., and Chait, B.T. (2000). ProFound—an expert system for protein identification using mass spectrometric peptide mapping information. *Anal Chem.* **72**:2482–2489.

Zhou, H., Watts, J.D., and Aebersold, R. (2001). A systematic approach to the analysis of protein phosphorylation. *Nat Biotechnol.* **19**(4):375–378.

7

PROTEOMICS DISCOVERY OF BIOMARKERS

Robert Massé and Bernard F. Gibbs

Applied R&D Department, Early Clinical Research and Bioanalysis, MDS Pharma Services, Montreal, Quebec, Canada

INTRODUCTION

The recent completion of the draft sequencing of the human genome (*http://www.ncbi.nlm.nih.gov/gemone/seq/*) and that of a variety of simple and complex biological organisms (*http://www.ncbi.nlm.nhi.gov/entrez/genome/main_genomes.html*) and subsequent proteomics initiatives (*http://www.hupo.org*) have changed the

Industrial Proteomics: Applications for Biotechnology and Pharmaceuticals, edited by Daniel Figeys
ISBN 0-471-45714-0

classical paradigm of biological and medical research. Albeit genome-sequencing projects are genuine scientific tour de force, they are actually the starting points of an intellectually challenging scientific quest. They aim at understanding how the thousands of encoded genes, transcribed ribonucleic acid (RNA), and the hundreds of thousands of translated proteins and posttranslational products work together to orchestrate the most complex biological phenomena. Ongoing industrial-scale projects to decipher the functions and regulation of these genes, and the proteins they encode, are generating data in great volume and increasing complexity. Turning the terabytes of data into knowledge about life's complex processes and applying it to a variety of practical and widely applicable goals, such as the development of new therapeutics or diagnostics, is an immense and intellectually challenging undertaking.

It is noteworthy that the term *biomarker* has appeared in the scientific literature long before modern genomics and proteomics initiatives were initiated. Actually, Taniguchi (1965) first used the expression "biomarker" almost four decades ago in a report on the clinical application of urinary metanephrine and normetanephrine. Since then, an increasing interest in biomarker research emanated from all spheres of life and health sciences (Fig. 7.1). Interestingly, both the concept and scope of applications of biomarker-related research have paralleled advances in science and technology. The advent of the Human Genome Project in the mid-1980s and major scientific and technological advances have driven subsequent proteomics endeavors in the 1990s. Development of enabling technologies such as the deoxyribonucleic acid (DNA) sequencer (Smith et al., 1986), polymerase chain reaction (PCR) (Mullis et al., 1986), DNA synthesizer (Horvath et al., 1987), and oligonucleotide arrays (Blanchard et al., 1996) have played a significant role in the rapid completion of the Human Genome Project. Similarly, by allowing the soft ionization of complex biopolymers, matrix-assisted laser desorption ionization (MALDI) (Karas and Hillenkamp, 1988) and electrospray ionization (ESI) (Fenn et al., 1989) provided mass spectrometry (MS) with unique and powerful tools in the burgeoning field of proteomics. These and other key technologies played a significant role in the recent expansion of research initiatives for the discovery of novel biomarkers and their subsequent applications in a variety of sectors of the life and health sciences. This trend is reflected in the number of biomarker-related studies published since 1996, which exceeded, on a yearly basis, the total number of studies published over the entire 1965 to 1985 period (Fig. 7.1).

Although particular genes and gene products such as DNA, RNA, and variants in the human genome such as single nucleotide polymorphisms (SNPs) are biomarkers often associated with genetic disorders (Collins and McKusick, 2001; King et al., 2002; Kwok, 2001), they only provide partial molecular clues to deciphering the functions of cells. In addition, one gene is not necessarily translated only into one protein. In contrast to the genome, which is relatively static and where all information can be obtained from a single cell, the proteome is by essence very dynamic and dependent, not only on the type of cell but also on the physiological state of the cell, which can be altered by endogenous and/or exogenous factors. The observation of Hoogland et al. (1999) of 22 different forms of human α-1-antitrypsin in human plasma depicts the intrinsic biochemical dynamism of the proteome. The issue becomes more involved as it was demonstrated that protein networks can be significantly dissimilar in differentiated cells

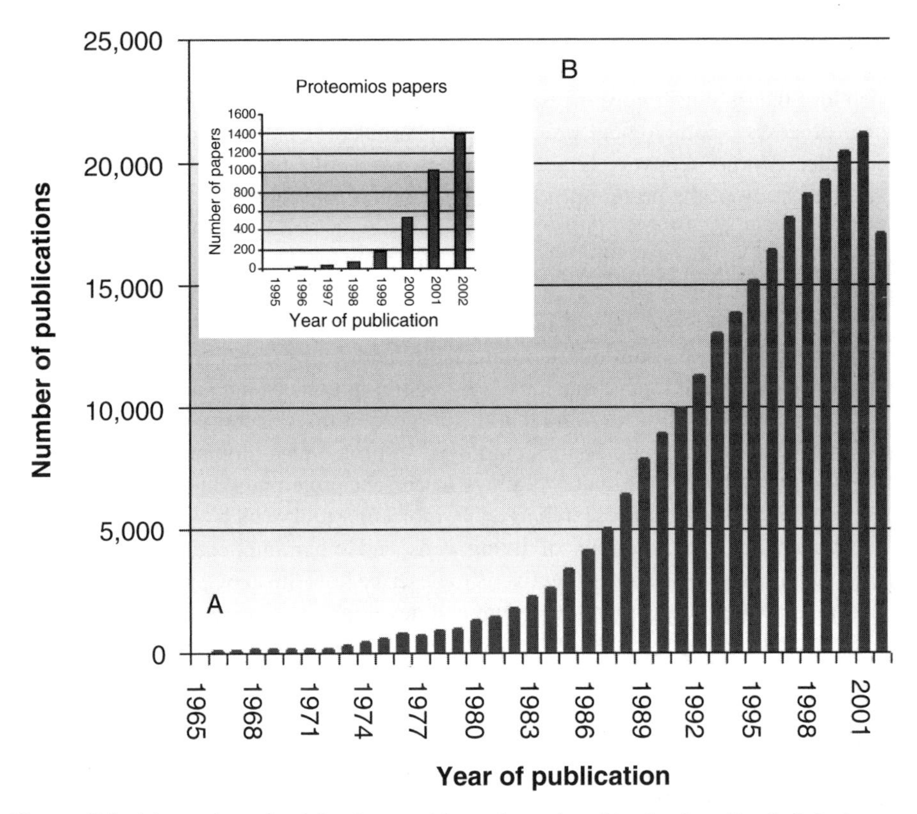

Figure 7.1. (*a*) Number of publications on biomarker-related topics since the first study mentioning the word *biomarker* was published in 1965 and (*b*) number of proteomics research articles published since 1995. The PubMed database was searched for the works that comprise of the term *biomarkers* (Fig. 7.1*a*) or *proteome/proteomics* (Fig. 7.1*b*).

from the same organism. A typical example being insulinlike growth factor-1 (IGF-1) and IGF binding protein-3 (IGFBP-3), both of which use diverging signaling pathways to express dissimilar biological activities in different cell types or cell states (Baxter, 2001; Valentinis and Baserga, 2001).

In this nascent biomarker area, the challenges are to obtain an integrated view of the functioning of living cells at the genomic and proteomic levels, so as to enhance our understanding of the complex and interconnected biological pathways that modulate and relate protein expression and their role in health and disease (Pandey and Mann, 2000). It is therefore quite evident that novel analytical strategies and technology platforms capable of simultaneous or parallel qualitative and quantitative analyses of complex proteins or peptide mixtures are required to discover and identify biologically relevant biomarkers, so as to monitor their levels, and temporal differential expression and locations in tissues and whole organisms, and finally map out their interaction pathways in health and disease.

To aid the reader of this review, we use the term *biomarker* as recently defined by the Biomarkers Definitions Working Group (2001) as "a characteristic (e.g., a protein or peptide) that is objectively measured and evaluated as an indicator of normal biological processes, pathological processes or pharmacological response(s) to a therapeutic intervention." Conversely, proteomics, as a scientific field, can be defined as the large-scale study of the protein products of a genome, and their interactions and functions (Wilkins et al., 1996). Similarly, the proteins expressed in a biological system or living organism at a given time, in a given environment, and under specific physiological conditions constitute a proteome. Hence, proteomes are highly active networks of complex and sometime elusive molecules, whose functions and fate are influenced, among other things, by subtle structural modifications in protein frameworks or via interactions with small ligand molecules or large biopolymers such as DNA or RNA.

The reader should bear in mind that the purpose of this chapter is to briefly review the current experimental approaches employed in proteomics biomarker research and to highlight the pivotal role that MS plays in current proteomics initiatives. We consider proteomics biomarker research as a key molecular tool, which is becoming essential to understand the functions of living cells and other intellectually challenging biological processes. Also, our objective is to describe how this new proteomics knowledge can be applied to assist in the design and development of new therapeutic agents and the creation of highly sensitive and specific diagnostic tests.

MASS-SPECTROMETRY-BASED PROTEOMICS APPROACHES

Over the past decade, MS has become a core enabling technology for performing proteomics investigations. So far MS has been employed for structural elucidation and patterning of individual proteins or protein mixtures to determine the composition of protein complexes, to detect and characterize posttranslational modifications (PTM) of proteins, conduct sequence-based protein identification, and to characterize biopolymers such as recombinant proteins (Chapman, 2000). Mass analyzers with different designs and performances, namely Fourier transform ion cyclotron MS (FT-ICR-MS), quadrupole time-of-flight (qTOF), and ion-trap (IT), are currently used in proteomics. The successful implementation of MS-based proteomics strategies also requires seamless integration of various analytical and biological front-end techniques and methods such as two-dimensional gel electrophoresis (2DE), multidimensional liquid chromatography (MDLC), antibody affinity capture, proteinchips, and micro- and nano-LC. These procedures are used to isolate, purify, and separate the proteins or peptides of interest from complex biological samples. Of equal importance to MS advances is the development of comprehensive proteomic strategies to investigate and solve biologically relevant problems. Expression proteomics investigates the up- or down-regulation of protein levels, and functional proteomics aims at the study of signaling pathways, protein interactions, protein complexes, and cellular compartments (Mann et al., 2001).

After its introduction, 2DE rapidly became the method of choice for the separation of complex mixtures of proteins and the technical basis of classical proteomics investigations (O'Farrell, 1975; Scheele, 1975). The reader is referred to Chapter 1 of this book in which Figeys provides a basic overview of proteomics. However, the separation of proteins by their isoelectric point and molecular weight is often insufficient for unambiguous identification. Also, the lack of reproducibility of 2DE, its failure to resolve most proteins with molecular weight greater than 100 kDa, and its inability to separate the majority of membrane and glycoproteins make it very difficult to achieve comprehensive analysis of complex proteomes. The protein load can be distributed among several thousand proteins with relative abundances covering large dynamic ranges. In spite of recent advances in 2DE (Görg, 2000) and its usage as a front-end separation technique in MS studies (Figeys, 2001), alternative MS-based technology platforms have been recently developed and proven effective for performing complex proteomics analyses. For example, several research groups used various MS-based strategies to answer different proteomics questions. Using the yeast *Saccharomyces cerevisiae* as a model organism, Washburn et al. (2001) and Wolters et al. (2001) performed large-scale protein identification studies using 2D LC/MS, Ficarro et al. (2002) explored its phosphoproteome, while Ho et al. (2002) characterized novel protein interactions within the organism.

Previous to these studies, Link et al. (1999) and Washburn and Yates (2000) reported a substitute approach to 2DE based on multidimensional liquid chromatography separations with stepwise salt gradient elution of the peptides combined with online tandem MS analysis to generate peptide ion fragment data that is subsequently mined by standard database searching procedures to determine the identity of the related proteins. In this shotgun methodology, named multidimensional protein identification technology (MudPIT), an enzymatic digest of a complex protein mixture was infused on a capillary biphasic column. Peptides were sequentially separated, first by electrostatic charge on a strong cation exchange material (SCX) followed by hydrophobicity on a revere-phase (RP) resin. Illustrations of one-dimensional (1D) LC/MS, 2-phase MudPIT, and 3-phase MudPIT setups are presented in Figure 7.2. In a typical experiment, *S. cerevisiae* ribosomal proteins were isolated and digested. The resulting protein digests were desalted (offline) and the peptides bound to a resin at low pH. The peptides were subsequently eluted using multiple salt pulses of increasing concentration followed by reverse-phase separation of each subfraction of the original peptide load. Seventy-five of the 78 predicted proteins in the ribosomal complex were identified. Two noteworthy biological features were observed in this research. First, the MS method used allowed the identification of peptides arising from 59 proteins that are encoded by duplicate genes. Second, a protein bel1, which had not been previously identified as a component of the yeast ribosome, was found to have strong homology to human rack1, a protein associated to human ribosome 40S subunit, which had been identified as an intracellular receptor for activated protein kinase C; this interaction is thought to target specific protein substrates for phosphorylation by protein kinase C. These findings indicated that ribosomal proteins are likely involved in cellular processes, for example, signaling pathways, other than protein synthesis.

In an effort to detect a larger number of proteins in a single run, the 2D-LC/MS/MS method was subsequently modified (Wolters et al., 2001) to improve the LC separation process and ionization efficiency by substituting potassium chloride and acetic acid for

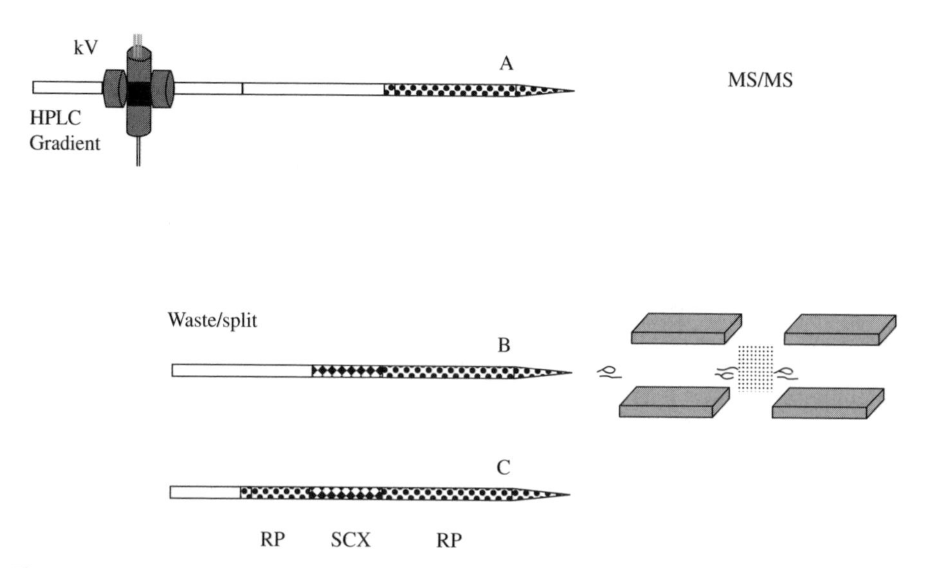

Figure 7.2. General diagrams of three-phase MudPIT, two-phase MudPIT, and 1D LC-MS/MS adapted with permission from McDonald et al. (2002). All columns have the same configuration. (*a*) An HPLC gradient is delivered to a junction where the voltage is applied and a portion of the flow is split off to reduce the flow rate through column to approximately 300 nL/min; 6.5 cm of 5 μm C-18 material (RP) are packed into a 100-μm ID fused-silica capillary whose tip had been pulled down to approximately 5 μm. (*b*) Two-phase MudPIT column. Three centimeters of strong cation exchange (SCX) was added upstream of the RP phase. (*c*) Three-phase MudPIT column identical to the two-phase MudPIT column except that an additional 3.0 cm of RP is placed upstream the SCX phase for the purpose of online desalting.

volatile ammonium acetate and heptafluoro butyric acid, respectively, in the elution buffer. By performing a 15-cycle MudPIT analysis of the insoluble and soluble fractions of an *S. cerevisiae* whole-cell lysate, a total of 5540 unique peptides were detected. This lead to the identification of nearly 1500 unique proteins, which represents 24 percent of the complete *S. cerevisiae* proteome comprising, to date, approximately 6200 proteins (*http://ebi.ac.uk/proteome/YEAST/interpro/stat.html*). The number of peptides per fraction was fairly consistent, with an average of 383 peptides per cycle (range: 218 to 505 peptides). This compares advantageously with previous labor-intensive 2DE gel studies where about 1000 to 1400 protein spots have been observed with an isoelectric point (pI) of 4 to 8 and molecular weight (MW) in the 15- to 150-kDa range (Gygi et al., 1999b; Perrot et al., 1999). These results indicate the high analytical value of each chromatography cycle for detecting and identifying a relatively large number of peptides in complex biological samples including several low abundance proteins not detected by 2DE.

A report from the same laboratory (McDonald et al., 2002) recently reported an improved MudPIT method where an additional reverse-phase bed was added to the original biphasic microcolumn permitting online desalting during the first cycle. The relative efficacy of this three-phase MudPIT method was compared with 1D LC/MS/MS and two-phase MudPIT methods using approximately 15 μg of a digested mixture of proteins from bovine brain microtubules. The corresponding peptide ion chromatograms of three of the six-cycle MudPIT analyses are presented in Figure 7.3. Significant differences were observed between the 1D LC/MS/MS separation and the six-cycle MudPIT analyses, particularly the total number of peptides detected and proteins identified (Table 7.1).

Although the three-phase MudPIT column was found to be superior both in number of proteins and peptides that were sampled, the authors stressed the fact that the selection of appropriate proteomics strategies must be dictated by the relative complexity of the sample, the amount of the starting peptide load, and the level and extent of analysis needed. Hence, the determination of the most abundant proteins should be preferentially performed by 1D LC/MS/MS and three-phase MudPIT chosen for the analysis of low abundant proteins or very small amounts of a protein mixture. The usefulness of this general approach to proteome analysis was further demonstrated by Lipton et al. (2002) who, by combining 1D-LC separation with high-resolution FT-ICR-MS were capable of identifying more than 1900 protein products of distinct open reading frames (ORF) in the bacterium *Deinococcus radiodurans*.

These methods can be ingeniously applied in biomarker discovery studies owing to their ability to quantitate proteome expression differences between samples and to detect various PTMs as well as low abundant proteins (MacDonald and Yates, 2002). They can be used (a) to decipher the etiology of disease and underlying biochemical mechanisms, (b) to discover new drug targets and design proteome-based investigations, and (c) to provide biologists and clinicians with novel molecular tools to assess drug toxicity and efficacy, thereby improving the design of clinical trials. These features of the MudPIT approach are of particular importance since many potential biomarkers and drug targets such as membrane receptors, transcription and growth factors, and protein kinases are low abundance proteins.

PHOSPHOPROTEOME ANALYSIS

Proteome analysis is further complicated not only by the sheer number of different proteins but also by the extensive posttranslational reactions such as phosphorylation and proteolytic processing. Annan and Zappacosta in Chapter 3 of this book present a detailed overview of the MS-based techniques currently used to map PTMs. Phosphorylation cascades are one of the most ubiquitous features of cell-signaling pathways that involve the phosphorylation and dephosphorylation of proteins, from transcription factors to membrane receptors and other proteins engaged in downstream responses such as cytoskeletal proteins. The fact that approximately 30 percent of cellular proteins are subject to phosphorylation indicates that this process is a major cellular regulation mechanism. The phosphorylation state of specific proteins can be clinically and pharmacologically relevant. These findings can be used as valuable biomarkers of disease and other physiological conditions. Although MS can be the technique of choice to investigate the phosphoproteome, there are several analytical issues to be addressed.

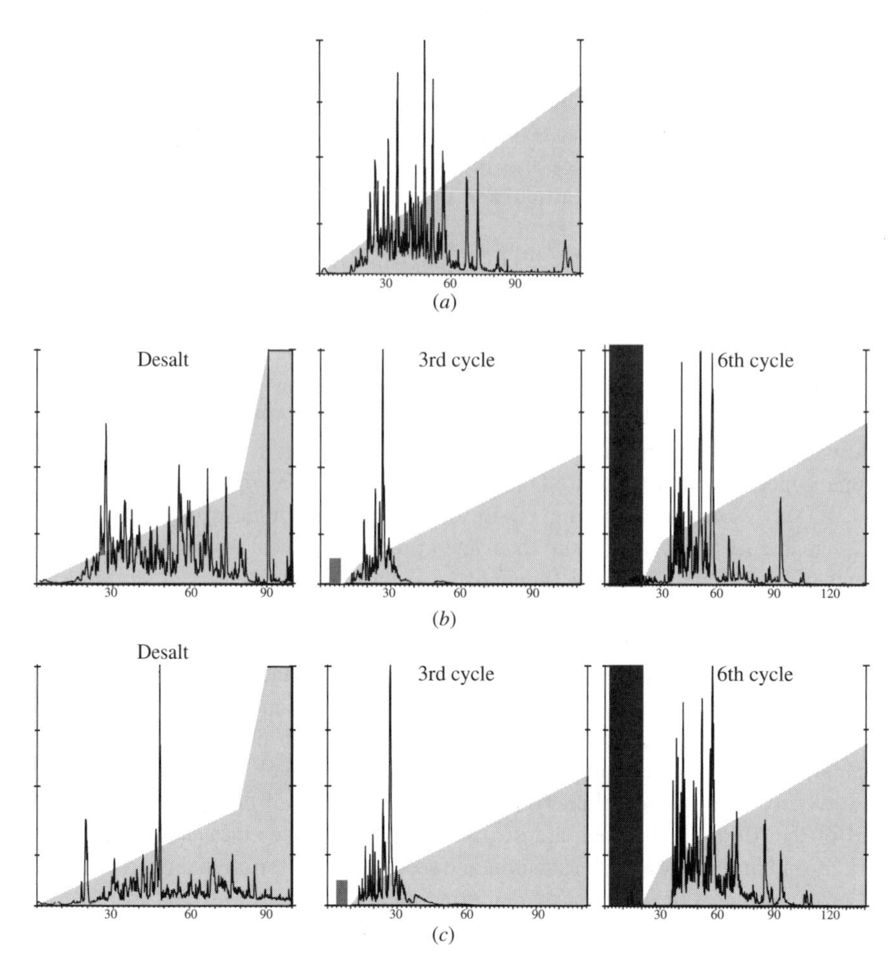

Figure 7.3. Peptide chromatograms and diagrams of chromatographic conditions. (*a*) 1D LC-MS/MS. Shows a base-peak peptide chromatogram and the linear gradient profile with the percentage of buffer B (80% ACN, 0.05% HFBA) indicated by gray shading. (*b*) Two-phase MudPIT. As in (*a*) except that profile for three out of six steps are included. The darker gray box indicated percentage of buffer C (250 mM ammonium acetate, 5% ACN, 0.05% HFBA, third cycle—section 2) and black shaded box indicated percentage of buffer D (500 mM ammonium acetate, 5% ACN, 0.5% HFBA, sixth cycle—section 2). Elution for second, fourth, and fifth steps were as for the third cycle except with differing percentage of buffer C. (*c*) Three-phase MudPIT. As in (*b*) except for three-phase MudPIT analysis. [Reproduced from McDonald et al. (2002) with permission from authors.]

TABLE 7.1. Total Proteins, Peptides, and Total Spectra Returned after Filtering 1D, Two-phase MudPIT and Three-phase MudPIT MS Data (see Fig. 7.2)

Analysis Mode	Number of Proteins	Number of Peptides	Total Spectra Matched
One dimensional	26	147 (82%,[a] 64%[b])	186
Two-phase MudPIT	55	341 (91%,[a] 64%[b])	634
Three-phase MudPIT	62	431 (85%,[a] 57%[b])	996

[a] Percent of "at least half tryptic" peptides (one or both of the cleavages specific for trypsin).
[b] Percent of "true tryptic" peptides (both cleavages specific for trypsin).
Source: Reproduced from McDonald et al. (2002) with permission from authors.

Phosphorylated proteins need to be concentrated prior to analysis due to the fact that identification of the phosphorylation sites requires more extensive sample preparation and analytical work than classical sequencing by database mining using a limited number of peptides. Also, it is preferable to separate phosphopeptides from nonphosphorylated peptides in protein digests to improve the selectivity of the overall identification process. Dephosphorylation of the peptide of interest with phosphatases with subsequent measurement of the *m/z* difference of the peptides before and after treatment can assist in ascertaining the identity of phosphorylated peptides and conversely the sites of phosphorylation.

Ficarro et al. (2002) developed an elegant MS-based phosphoproteome analysis method that they applied to *S. cerevisiae*. Proteins isolated from whole-cell lysate were digested with trypsin and the resulting peptides converted to methyl esters, which were subsequently enriched for phosphopeptides by immobilized metal-affinity chromatography (IMAC), and analysed by nanoflow high-performance LC/ESI-MS. The analytical system was operated in the data-dependent mode throughout the high-performance liquid chromatography (HPLC) gradient, where every 12 to 15 s, one full-scan spectrum and five collisionally induced decomposition (CID) spectra were recorded sequentially on the five most abundant peptides present. Data was analyzed using an in-house computer algorithm, the Neutral Loss Tool, and SEQUEST (Eng et al., 1994). As a result, a total of 216 peptides sequences defining 383 sites of phosphorylation were determined, including 60, 145, and 11, which were singly, doubly, and triply phosphorylated, respectively. Of these, 18 had been previously identified, including the doubly phosphorylated motif pTXpY derived from the activation loop of two mitogen-activated proteins (MAP) kinases. By including an esterification step, this method virtually eliminates nonspecific binding to the IMAC column, thus allowing for the systematic identification of phosphorylation sites on proteins isolated from whole-cell lysate in a single analysis. Also, the sensitivity of this methodology was demonstrated by three major features: (a) the detection of 5 fmol of a phosphorylated peptide in a standard mixture, (b) the mapping of phosphorylation sites on proteins with very low codon bias, and (c) by the identification of sites of tyrosine phosphorylation on proteins in a whole-cell lysate. Another advantage of this method is the ability to detect phosphotyrosine-containing peptides, which is not possible using the protocol proposed by Oda et al. (2001) where the phosphate groups of serine and threonine are replaced by ethanediol, and the resulting derivatives attached to a biotin affinity tag, separated from nonphosphorylated tryptic peptides by affinity chromatography on NeutrAvidin beads prior to MS analysis. Finally, this methodology can be adapted to allow quantitation and/or differential display of phosphoproteins expressed in two different cell systems or cell states. In such experiments, peptides are converted to methyl esters in one sample using methanol and deuterated methanol for the second sample. The two samples are then mixed and fractionated on an IMAC column prior to nanoflow HPLC-ESI FT/MS. Signals for peptides detected in the mixture appear as doublets separated by n (3 Da)/z, where n is the number of carboxylic acid groups in the peptide. The ratio of the two signals in the doublet changes proportionally with the expression level of the corresponding proteins in each sample, thus allowing for phosphoprotein profiling as a measure of cellular states. Given the importance of protein phosphorylation/dephosphorylation in signal transduction pathways, this approach offers the possibility of identifying new drug targets and biomarkers involved in disease-related metabolic cascades.

PROTEIN COMPLEXES AND SIGNALING PATHWAYS

Proteomics now faces the same problem that current genomics investigations are confronted with, that is, the identification of the functional roles of genes and proteins. It is recognized that the identification and mapping of all the proteins present in cells represents the first step in a proteomics journey which aims to characterize their interactions with other proteins and other biopolymers and decipher their various functions in health and disease states along life's developmental stages. The biology of proteins is further complicated by the fact that many of their biological functions involve protein–protein complexes (Jones and Thornton, 1996). MS-based approaches have been developed to characterize such complexes. However, this task is not trivial given that such complexes involve multiple components such as proteins, nucleic acids, and carbohydrates with diverse life spans such as the transient complexes involved in signaling pathways and the stable complexes of the ribosome. In addition, there is no clear definition of what is a complex or whether two complexes are of different types (Sali et al., 2003). Thus far, generation of large-scale protein–protein interaction maps has relied on the yeast two-hybrid system, which detects binary interactions through activation of reporter gene expression (Ito et al., 2001). Using 725 of the predicted yeast proteins as baits, including protein kinases, phosphatases, regulatory subunits, and proteins involved in the DNA damage response, in combination with a high-throughput MS protein complex identification (HMS-PCI) method, Ho et al. (2002) detected 3617 associated proteins covering 25 percent of the yeast proteome. The HMS-PCI method has identified known complexes from a variety of subcellular compartments including the cytoplasm, cytoskeleton, nucleus, nucleolus, plasma membrane, mitochondrion, and vacuole. Analysis of a complex including the mitogen-activated protein kinase (MAPK) Kss1 as bait identified interactions of potential biological significance, including Bem3, which is a guanosine triphosphate hydrolase (GTPase)-activating protein that may

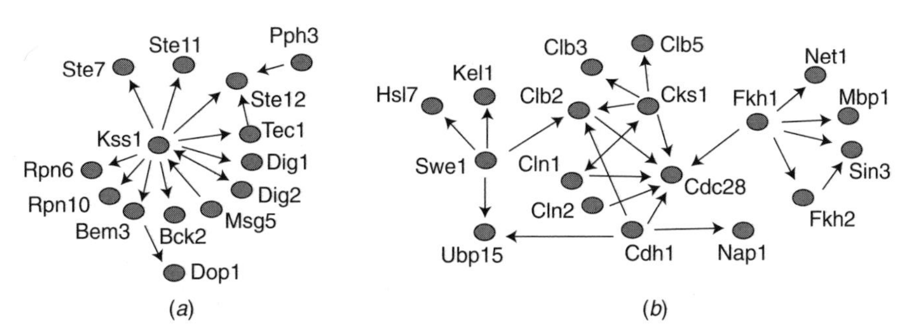

Figure 7.4. Two kinase-based signaling networks. (*a*) Interaction diagram for Kss1 complexes. (*b*) Interaction diagram for Cdc28 and Fkh1/2 complexes. Arrows point from the bait protein to the interaction partner. Blue arrows indicate known interactions; red arrows indicate new interactions. Not all detected interactions with all components are shown, nor are connections necessarily direct. [Reproduced from Ho et al. (2002) with permission from Nature Publishing Group.]

help attenuate the upstream Cdc42 Rho-type GTPase needed for mating (Fig. 7.4). In addition, several known components of the mating/filamentous growth pathway together with Ste11 and Ste7 and transcriptional regulators Ste12, Tec1, Dig1/Rst1, and Dig2/Rst2 (Fig. 7.4) were also identified. A network of interactions centered on the primary cyclin-dependent kinase (CDK) for cell division control were determined including the association between Cdc28 and Fkh1, one of the transcription factors that drives expression of mitotic cyclins CLB1/2 and other G2/M-regulated genes. A chief advantage of the HMS-PCI approach is its capacity to identify approximately threefold more literature-derived interactions per bait than each large-scale HTP-Y2H data set and about twofold more interactions when compared with the combination of both comprehensive HTP-Y2H data sets, which were used for comparison purposes (Ito et al., 2001; Uetz et al., 2000). Given that responses to biological signals are modulated and expressed through integrated and complex networks of interactions, and that the nature of the responses relies on the networks themselves, their state in various cell types and physiological conditions, MS-based methods such as HMS-PCI will be required to perform comprehensive investigations aimed at decoding the effects of disease, therapeutic intervention, aging, and so forth on global cellular responses. Conversely, such studies will benefit the fields of biological and medical sciences by enabling the discovery and validation of biologically relevant biomarkers and fostering their application in drug discovery, drug development, and diagnostics.

AFFINITY-BASED MASS SPECTROMETRY METHODS

In most of the above MS proteomics approaches the aim was primarily to detect and identify individual proteins or protein complexes in biological samples in a high-throughput fashion. These tasks are best achieved by submitting protein mixtures to proteolytic hydrolysis followed by 1-, 2- or 3D ESI LC/MS/MS or MALDI-TOF MS/MS analysis of the resulting peptides. Analysis of whole proteins is generally performed using MALDI, a technique in which the proteins are embedded in a solid medium by co-crystallization with a photoactive compound such as α-cyano-4-hydroxycinnamic acid (CHCA) or 3,5-dimethoxy-4-hydroxycinnamic acid (sinapinic acid). The sample is irradiated with laser light, which is absorbed by the matrix compound to desorb and ionize the embedded protein molecules. As the proteins do not fragment during desorption, MALDI is often referred to as being a "soft" ionization technique. Because MALDI is a pulsed technique, it is well suited for TOF instruments. Typically, MALDI is used on qTOF and on IT to take advantage of tandem MS capabilities, but quadrupole fields limit the mass range that one may need for particular proteomics studies.

A major limitation of MALDI-TOF is the presence of buffer components that inhibit direct analysis of the proteins within the matrix. Consequently, crude biological extracts must go through extensive separation and purification steps such as 2DE prior to analysis. Also, the limitations imposed by co-crystallization generally require the protein to be relatively pure prior to embedding in the matrix. Restrictions related to the conventional sample preparation approach to MALDI were in part removed using different strategies. For instance, Wei et al. (1999) introduced desorption–ionization on porous silica mass spectrometry (DIOS-MS), a matrix-free approach that employs porous silica to capture the molecules of interest. Because porous silica has the ability to absorb ultraviolet emission, it acts as an energy receptacle for the laser radiation. The structure and physicochemical properties of the porous silica surfaces are critical to optimal performance of DIOS-MS. These parameters are controlled by the selection of silicon and the electrochemical etching conditions. Hydrophobic porous surfaces were shown to generate better signals of analyte from aqueous-based media. Although the DIOS effective mass range is less than 18,000 Da, it is nevertheless amenable for the identification of proteins. Using either off-plate or on-plate tryptic digests Thomas et al. (2001) used DIOS-MS to identify bovine serum albumin (BSA) (Fig. 7.5*a*), an adenovirus penton protein (Fig. 7.5*b*), and β-lactoglobulin (Fig. 7.5*c*) based on their signature peptides. The low-mass signature fragment ions allow direct identification of the corresponding proteins using publicly available database searching algorithms. Furthermore, the mass information obtained was detailed enough to allow for the identification of the correct serotype (type II) of the adenovirus penton protein (Fig. 7.5*b* inset), which differs from a related serotype by only a few residues. Because DIOS is a matrix-free desorption technique, Thomas et al. (2001) also demonstrated that it has

Figure 7.5. DIOS mass spectra and database identification of (*a*) an 18-h solution tryptic digest of BSA (*b*), an on-plate tryptic digest of adenovirus penton protein at 37°C for 75 min, and (*c*) the 18-h solution tryptic digest of β-lactoglobulin. In (*c*), the intact protein (18 kDa) also was observed. Identification was from the PROFOUND protein search program (*http://129.85.19.192/prowl-cgi/ProFound.exe*) linked to the Swiss-Prot database, and search parameters included a 0.5-Da tolerance. [Reproduced from Thomas et al. (2001) with permission.]

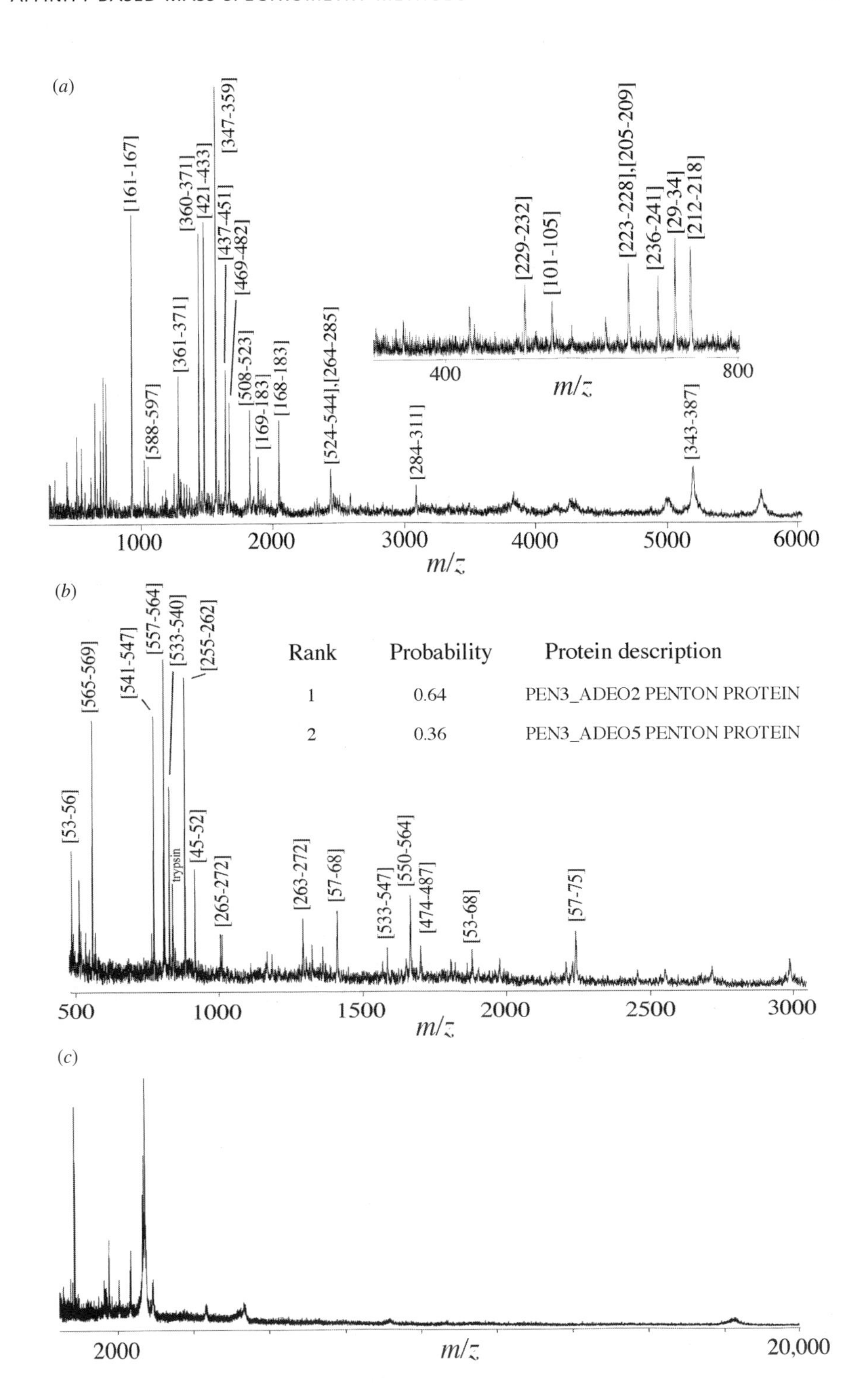
(a)
[161-167]
[588-597]
[361-371]
[360-371]
[421-433]
[347-359]
[437-451]
[469-482]
[508-523]
[169-183]
[168-183]
[524-544],[264-285]
[284-311]
[343-387]
[229-232]
[101-105]
[223-228],[205-209]
[236-241]
[29-34]
[212-218]
400
800
m/z
1000
2000
3000
4000
5000
6000
m/z
(b)
[53-56]
[565-569]
[541-547]
[557-564]
[533-540]
[255-262]
trypsin
[45-52]
[265-272]
[263-272]
[57-68]
[533-547]
[550-564]
[474-487]
[53-68]
[57-75]
Rank
Probability
Protein description
1
0.64
PEN3_ADEO2 PENTON PROTEIN
2
0.36
PEN3_ADEO5 PENTON PROTEIN
500
1000
1500
2000
2500
3000
m/z
(c)
2000
m/z
20,000

the potential for measuring low-molecular-weight synthetic compounds that could not be performed with MALDI due to matrix interference.

Several other alternatives have been developed to circumvent the problem of matrix-generated noise in the low-mass range, including that proposed by Lin and Chen (2002) where analytes are desorbed and ionized from a polymeric sol-gel-derived film in which 2,5-dihydroxybenzoic acid, the energy-capturing compound, is covalently bound. Taking advantage of the fact that the method is matrix interference free, small molecules and peptides such as bradykinin, insulin, and cytochrome C were measured at femtomole sensitivity, with a detection limit of 0.8 fmol in the case of cytochrome C.

Rapid mapping of protein extracts from a variety of biological sources has become feasible with the introduction of protein chips. A detailed review describing the theory and application of protein microchips integrating immunoassay principle and microarray technologies and capable of performing multiplexed and quantitative protein expression profiling is presented in Chapter 11. Proteinchips array technology based on surface-enhanced laser desorption/ionization mass spectrometry (SELDI/MS) is a series of analytical tools encompassing retentate chromatography, on-chip protein characterization, and multivariate analysis that allows for the generation and examination of patterns of protein expression and modification (Weinberger et al., 2000; Issaq et al., 2002, 2003; Fung et al., 2001). SELDI proteinchips are available in several defined chemistries including normal-phase silica, hydrophobic reverse phase, strong anion exchange, weak cation exchange, and immobilized metal affinity capture. Conversely, chips with biochemically modified surfaces incorporating covalently bound antibodies, enzymes, aptamers, phages, and so forth have been designed to promote specific biomolecular affinity interactions that capture individual or sets of proteins and peptides from complex biological extracts (Fig. 7.6). Typically, the proteinchip is incubated with a crude or purified preparation that contains the analyte molecules under investigation. Sample on-chip fractionation is achieved by retentate chromatography, and target analytes are bound by the affinity ligands on the chip. Residual material, including detergents, salts, and lipids that might interfere with conventional MALDI is removed through the application of a series of washes with appropriate solvents or buffers. Finally, a molecule that absorbs laser energy, the SELDI equivalent of the MALDI matrix, is added to the entire chip that is subsequently irradiated with a laser to measure the molecular weights of the bound proteins or peptides. Although the process is not well understood and may involve co-crystallization at some scale, the fact that the analyte is bound to the chip indicates that it does not have to be in solution simultaneously with the energy-absorbing molecule to be desorbed and ionized from the proteinchip surface.

This approach is particularly suitable to perform protein differential display studies (Fung et al., 2001) and was successfully applied in clinical and toxicological studies to detect and characterize biomarkers associated with a variety of carcinomas (Adam et al., 2001, 2002; Bischoff et al., 2002; Cazares et al., 2002; Petricoin et al., 2002; Srinivas et al., 2002; Wellmann et al., 2002), Alzheimer's disease (Davies et al., 1999; Xiang et al., 2002), as well as in drug-induced toxicity (Dare et al., 2002; He et al., 2003), and probiotic studies (Howard et al., 2000). Because of its physico-chemical

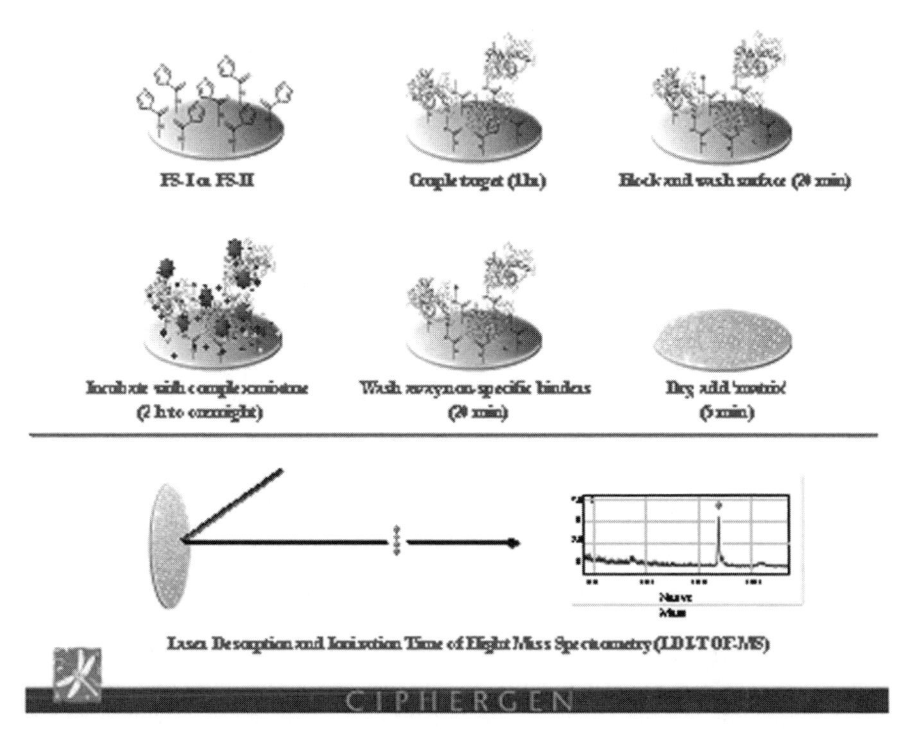

Figure 7.6. Principle of proteinchip SELDI TOF/MS. Chromatographic surfaces (e.g., reverse phase, ion exchange, immobilized metal affinity capture, etc.) or preactivated surfaces capable of forming covalent linkage are used to capture specific bait molecules such as antibodies, receptors, or oligonucleotides. Complex biological mixtures are deposited upon appropriate chromatographic arrays, each with different surface-interactive potentials. A gradient of wash conditions specific for each array surface is then applied to remove proteins of lower-binding potential. Matrix solution is ultimately applied and the arrays are read by SELDI-TOF/MS. [Reproduced with permission from Weinberger et al. (2000) with permission from Elsevier.]

features SELDI proteinchip can be used as a noninvasive method to perform protein expression profiling and clustering to ultimately establish statistically significant correlations and differences between protein patterns measured in disease and normal conditions. The main disadvantages of direct analysis of biological samples by SELDI are the preferential detection of low-molecular-weight proteins and the difficulty in measuring PTMs. This problem is partly circumvented by performing proteolytic digestion of the proteins combined with enzymatic dephosphorylation or any other relevant cleavage reaction prior to MALDI analysis. Using the SELDI approach, Vlahou et al. (2001) developed a noninvasive method for the diagnosis of transitional cell carcinoma (TCC) of the bladder. Analysis of urine specimens collected from patients with TCC, other urogenital diseases, and healthy donors using a strong anion exchange chip revealed several protein changes in the TCC group, including five prominent protein peaks and

TABLE 7.2. Sensitivity and Specificity of Multiple Biomarkers Panels[a]

Marker (kDa)	Sensitivity (%)	Specificity (N%)	Specificity (O%)	Specificity (All%)	PPV (%)	NPV (%)
3.3/9.5/100	83	71	63	67	54	90
3.3/44/85–92	83	71	63	67	54	90
3.3/9.5/85–92	87	71	60	66	54	91

[a] N, normal; O, other urogenital diseases; PPV, positive predictive value; NPV, negative predictive value.

seven protein clusters over the 3.3- to 133-kDa mass range, which were preferentially expressed in TCC patients. Representative mass spectra and gel views of these proteins are shown in Figure 7.7. When used individually, the five biomarkers showed TCC sensitivities and specificities ranging from 43 to 70 percent and from 70 to 86 percent, respectively. Combining these biomarkers and clusters increased the sensitivity for detecting TCC to 87 percent with a specificity of 66 percent (Table 7.2). This represents a significant improvement over the standard voided urine or bladder-washing cytology methods, which provided a sensitivity of only 33 percent. Frequency of most markers was observed to increase with progression from low-grade (I to II) to high-grade (III) and low-stage (Ta) to higher stage (T1–3) carcinomas. Although these results support the potential of the SELDI TOF/MS approach for the development of a highly sensitive urinary TCC diagnostic test, further large-scale clinical investigations will be required to validate these initial observations and identify the primary biomarkers to establish and confirm their biological and clinical relevance to TCC.

The traditional approach to the diagnosis of renal disease has been based upon the determination of blood urea nitrogen and serum creatinine levels, measurements of creatinine clearance, 24-h urine protein excretion, urinary sodium and chloride, and the like. In many cases, these methods are not sensitive enough to detect early renal disease, nor are they specific enough to provide information about the activity, etiology, or potential reversibility of the disease. Hampel et al. (2001) used SELDI proteinchip arrays to assess its applicability for protein profiling of urine and to demonstrate its use as a diagnostic tool to detect subtle failure in the kidney for a group of patients

Figure 7.7. Detection of five TCC-associated protein peaks in urine. Mass spectra (*top*) and respective gel views (*bottom*) of urine samples from four different TCC patients (C1–C4), two normals (N1–N2), and two patients with other urogenital diseases (B1 and B2). The average molecular mass of the five proteins identified to be unique or overexpressed in the TCC specimens are: UBC1, 3.352/3.432 kDa [(*a*), arrow] and occasionally 3.47 kDa [(*a*), arrowhead]; UBC2, 9.495 kDa [(*b*), arrow]; UBC3, 44.647 kDa [(*c*), arrow]; UBC4, 100.120 kDa; and UBC5, 133.19 kDa [(*d*), arrow]. Numbers in the mass spectra represent the observed mass of the marker in that particular sample. [Reproduced from Vlahou et al. (2001) with permission from the American Society for Investigative Pathology.]

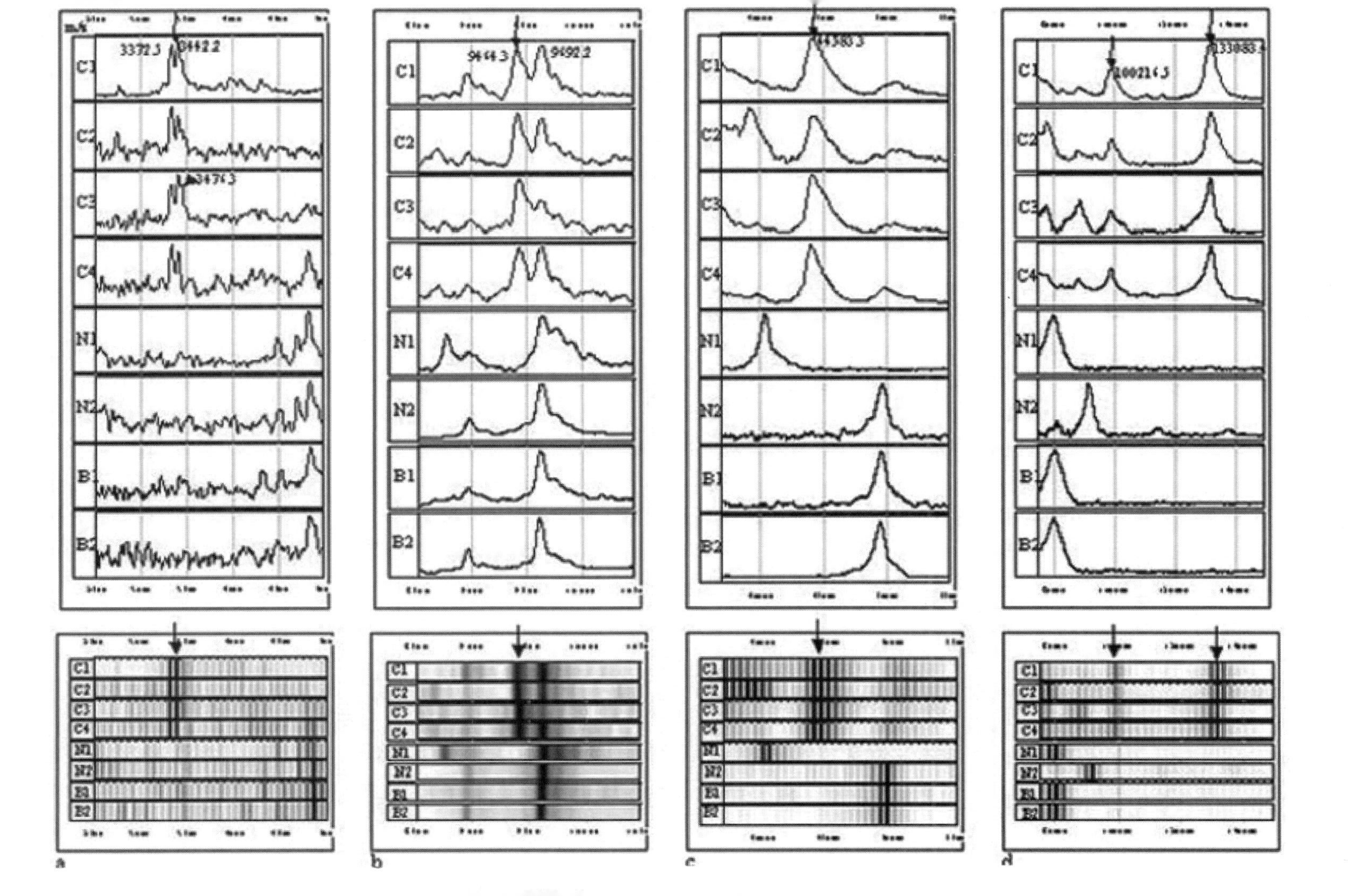
C1
C2
C3
C4
N1
N2
B1
B2
a
b
c
d

receiving Ioxilan, a radio-contrast medium. They first tested their approach in male Sprague-Dawley rats. Significant changes were observed in the abundance of urinary proteins of 9.9, 18.7, 21.0, and 66.3 kDa before and after intravenous administration of either Ioxilan or hypertonic saline solution as a control. Even in uncomplicated cases of radio-contrast medium infusion during cardiac catheterization, perturbations in the protein composition occurred but returned to baseline values after 6 to 12 h. Proteins with molecular masses of 9.75, 11.75, 23.5, and 66.4 kDa changed in abundance. For patients with impaired renal function, these changes were not reversible within 6 to 12 h. The protein with a molecular mass of 11.75 kDa, was identified as β-(2)-microglobulin, a known biomarker of renal dysfunction. Hence, SELDI is a promising and sensitive proteomic tool for the detection and characterization of trace amounts of proteins in urine. This approach can be used not only in pathologies such as arteriosclerosis, diabetes, and hypertension where disease progression and therapeutic intervention may have deleterious effects on the renal function but also in patients without renal complications where subtle drug-induced changes in urinary proteomics profiles might represent markers of impending nephropathy. The SELDI proteinchip technique has proven its efficacy to enhance the sensitivity and specificity of diagnostic tests and help in the discovery of individual or sets of proteomics biomarkers, particularly when used in combination with qTOF mass spectrometers that allow simultaneous biomarker identification, mapping, or clustering using differential protein expression data. A caveat of this technique is its relative inability to perform broad and quantitative screening of complex proteomes on a single proteinchip that can be performed with 2- or 3D LC/MS-based methods.

PROTEIN QUANTITATION

An important aspect of biomarker proteomics is to correlate the expression of proteins as a function of the cell or tissue state, or health versus disease states. A corollary objective of proteomics is to develop sensitive methods capable of performing the quantitative determination of protein fluxes and effluxes through the numerous biological networks involving considerable numbers of participant proteins and PTM analogs. Because cellular responses often entail amplification in signaling response or translation, quantitative methodologies effective in measuring proteins over large dynamic concentration ranges will be required. The quantitative dimension in MS-based proteomics is provided by the established technique of stable-isotope dilution. This method is based on the fact that (a) the mass spectrometer can differentiate pairs of chemically identical compounds of different stable-isotope composition (e.g., ^{1}H vs. ^{2}H, ^{12}C vs. ^{13}C, etc.) owing to their mass difference and (b) the ratio of intensities of the analyte pairs accurately correspond to the relative abundance of the two analytes in the sample. Yao et al. (2001) developed a modular shotgun proteolytic ^{18}O-labeling method whereby proteolysis is performed on two proteome mixtures. Two $H_2{}^{18}O$ atoms are incorporated into the carboxyl termini of all tryptic peptides during proteolytic digest of the first mixture, whereas proteins from the second mixture are cleaved similarly in the presence of $H_2{}^{16}O$. The sample are pooled and the ratios of peptide pairs differing by 4 Da are analyzed by high-resolution MS. Relative signal intensities of the paired peptides quantify the relative amounts of their precursor proteins. In other approaches, stable-isotope tags are introduced into proteins through metabolic labeling with heavy metals or amino acids (Conrads et al., 2002) or by attaching isotope-coded affinity tags through a variety of chemistries specific to the amino group (Munchbach et al., 2000) and serine hydrolases by reaction with a biotinylated fluorophosphonate probe (Liu et al., 1999). Greenbaum et al. (2000) engineered chemical probes that were used to broadly track activity of cysteine proteases throughout a defined model system for cancer progression. When used in combination with libraries of affinity probes, this approach provides a rapid means for obtaining detailed functional information without the need for prior purification and identification of targets. Methods with different enrichment procedures were developed for the analysis of phosphorylated proteins/peptides using chemical reactions (Oda et al., 2001; Zhou et al., 2001) or esterification combined with IMAC affinity capture (Ficaro et al., 2002). A well-established quantitative proteomic method is the isotope-coded affinity tag (ICAT) method developed by Gygi and Aebersold (Gygi et al., 1999a; Zhou et al., 2002). As depicted in Figure 7.8, the functional components of an ICAT reagent include a 2-iodoacetamido group with specific reactivity toward sulfhydryl residues, a light or heavily deuterated linker, and a biotin affinity tag. In a typical experiment where the aim is to compare the differential proteome expression in two different cell states, the following steps are performed: (a) free cysteinyl residues from a reduced protein sample representing one cell state are derivatized with the isotopically light form of the ICAT reagent and the same reaction is performed on the sample representing the second cell state using the heavily labeled reagent; (b) the two samples are combined and enzymatically digested to generate peptide fragments, where those bearing a sulfhydryl moiety are tagged; (c) the tagged peptides are isolated by avidin affinity chromatography; and (d) the isolated peptides are separated according to various schemes prior to analysis by capillary LC or LC CID (Griffin and Aebersold, 2001). Combination of the data obtained by MS and CID analyses of ICAT-labeled peptides allows for both the determination of the sequence identities of the constituent proteins in the sample as well as their relative amounts. Although only cysteine-containing proteins are measured by the ICAT method, the combination with 2D or 3D-LC separation and advanced MS sequencing/scanning strategies significantly increases the number of uniquely identified proteins (Gygi et al., 2002). In a study whose aim was to analyze changes in the proteome of *S. cerevisiae* induced by a shift from glucose to galactose in the carbon source, Smolka et al. (2002) demonstrated that postisolation isotopic protein labeling using the ICAT reagents is also compatible with 2DE and accurate protein MS quantitation.

These reports indicate that the ICAT approach can be advantageously used for biomarker discovery and quantitation. This was elegantly demonstrated by Griffin et al. (2003) who used abundance-ratio-dependent measurement for the quantitative profiling of androgen-regulated prostate cancer proteins in a human prostate cell line. MS data revealed a total of 306 ICAT-labeled peptides, of which 26 percent showed abundance differences considered to be significant, while 36 unique proteins were identified by CID analysis and sequence database searching. Also, Griffin et al. (2003) showed that this approach is notably more efficient than the data-dependent sequenc-

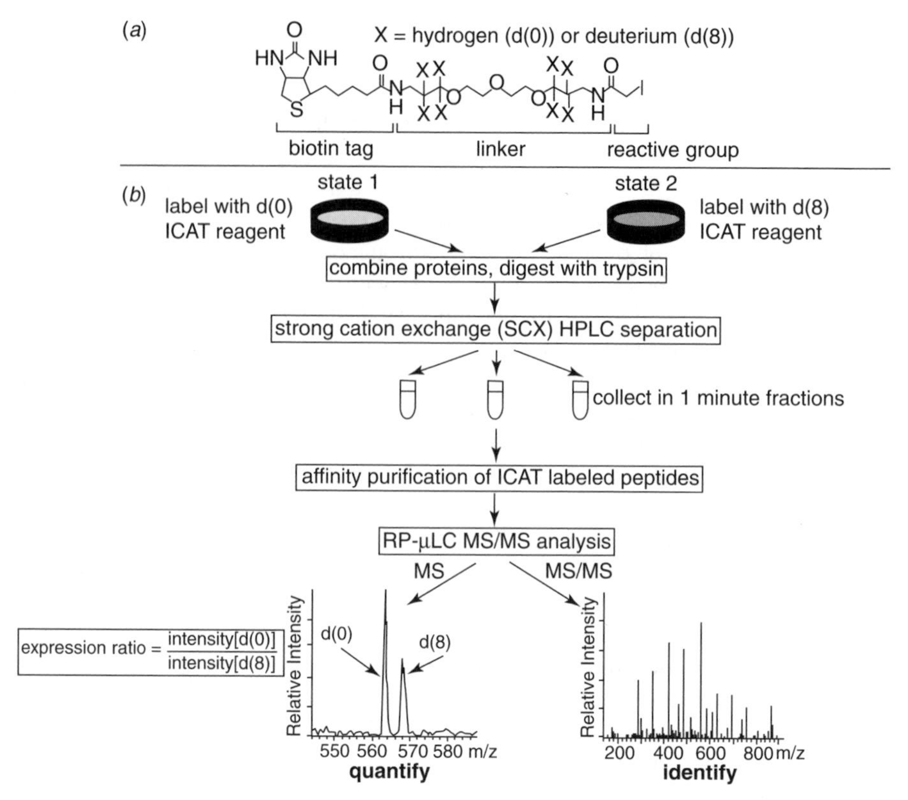

Figure 7.8. Global quantitative mass spectrometric analysis of protein expression. (*a*) Structure of the ICAT reagent. (*b*) Mass spectrometric analysis using selective protein labeling with the ICAT reagent and multidimensional chromatography. Equal amounts of total protein are isolated from cells existing in two different biological states and labeled with the d_0 or d_8 versions of the ICAT reagent. The proteins are mixed, enzymatically digested, separated by multidimensional chromatography, and analyzed by MS. Relative quantification of protein expression between the two states is accomplished by comparison of peak intensities of the isotopically different peptides, and identification is accomplished by selecting these peptides for MS/MS and subsequent sequence database searching with the generated CID spectra. [Reproduced from Griffin and Aebersold (2001) with permission from authors.]

ing method in detecting and analyzing differentially expressed peptides. This was substantiated by the difference in the percentage ratio of the number of identified peptides to the number of acquired CID spectra (peptide/CID spectra), which was 50 percent for the abundance-ratio-dependent method compared to 16 percent for the data-dependent method. Despite the large number of CID spectra recorded in the later method, only 2.6 percent resulted in the identification of a differentially expressed peptide compared to 44 percent for the abundance-ratio-dependent method. One of the main disadvantages of the ICAT method is that proteins without cysteine residues (~10 percent) cannot be recognized.

Cagney and Emili (2002) described an integrated strategy similar to ICAT called mass-coded abundance tagging (MCAT). In this procedure one of the samples is treated with *o*-methylurea at high pH to convert lysine residues to homoarginines. This is a selective reaction and essentially goes to completion. The amino terminus of the peptide is unaffected as well as the other side groups. The reaction causes an increase of 42 amu in the affected peptides. MCAT was shown to facilitate de novo peptide sequencing and relative quantitation of proteins in a single analysis. Product scanning reveals complete b and y ions at 42, 21, and 14 amu apart for singly, doubly, and triply, charged ions, respectively. Figure 7.9 shows a typical MCAT determination of the relative abundance of a 16-residue peptide in a complex biological sample. A full scan is performed in the LC-ESI/MS mode to monitor the relative intensity of related peptides, which are subsequently fragmented. Protein quantitation is subsequently performed by measuring the relative signal intensities of peptides with identical sequences but differing by the calculated lysine-specific mass tag. The method has been tested in a cell lysate (complex protein mixture) and is quite economical and simple.

Qiu et al. (2002) reported another alternative approach to ICAT termed acid-labile isotope-coded extractants (ALICE), a new class of chemically modified resins comprised of three elements: an acid-labile group covalently attached to the polymeric resin, a labeled linker containing 10 deuterium atoms, and a maleimido reactive group used as a bait to capture cysteine-containing peptides. Compared to ICAT, one apparent limitation of the ALICE method is that the cysteine-containing peptides are labeled after rather than before proteolytic digestion since complete substitution of all cysteine residues on proteins using ALICE is probably difficult to achieve due to the steric hindrance created by its bulky resin backbone. Conversely, the stability of the acid-labile bond allows extensive washing of the captured cysteine-containing peptides to remove contaminants.

Two alternatives to ICAT for quantitative proteomics were recently described. Tao et al. (2003) have devised a novel solid-phase isotope tagging (SPIT) technique for quantitative proteomics. Based on standard solid-phase peptide synthesis procedures, isotope tags and functional groups are bound to beads through which peptides are captured. The captured peptides are subsequently released by acid or photo cleavage with a stable isotope tag (heavy or light) incorporated at the N-terminus. The labeled peptides are then subjected to quantitative μLC-MS and CID analysis. This approach has been applied to several model protein mixtures as well as to peptides derived from immunopurified STE-12 protein complexes isolated from yeast in different states for the determination of relative abundance. The inherent redundant tagging of multiple peptides from one protein allows for better accuracy in quantification. The method may also be adapted for automated high-throughput experiments, as well as for quantitative analysis of PTMs.

Locke et al. (2003) described a method for isotopically differentiated derivatization of peptide amines with plurality of labels. The method was developed for the chemical derivatization of amines (lysines) by reaction with isotopically labeled aldehydes. The approach was tested with protein standards and then applied to model biological systems such as the bacterium *Aeromonas salmonicida* under different growth conditions and the fungi *Candida albicans* in either the yeast or hyphal form. Samples were

(*a*)

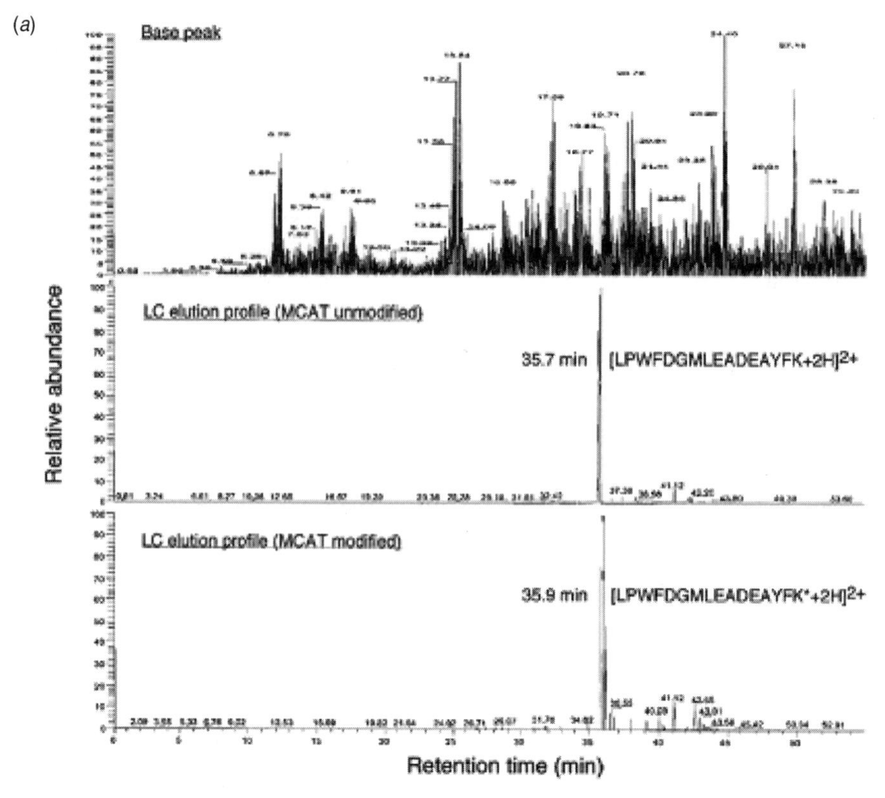
Base peak
LC elution profile (MCAT unmodified)
35.7 min
[LPWFDGMLEADEAYFK+2H]2+
LC elution profile (MCAT modified)
35.9 min
[LPWFDGMLEADEAYFK*+2H]2+
Relative abundance
Retention time (min)

(*b*)

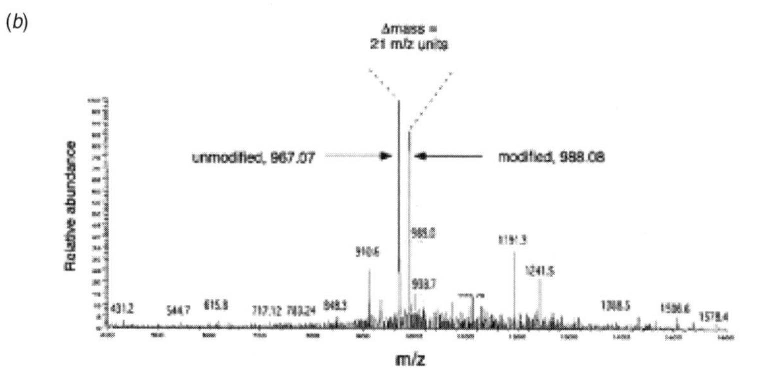
Δmass =
21 m/z units
unmodified, 967.07
modified, 988.08
Relative abundance
m/z
431.2
544.7
615.8
717.12
783.24
848.1
910.6
988.0
998.7
1191.3
1241.5
1388.5
1506.6
1578.4

(*c*)

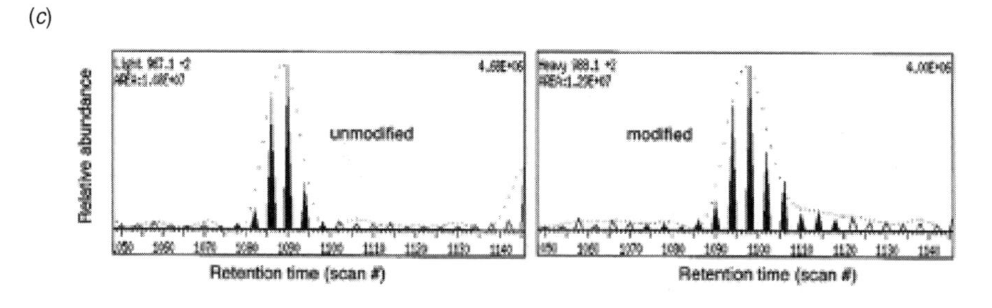
unmodified
modified
Relative abundance
Retention time (scan #)
Retention time (scan #)

treated with isotopic formaldehyde and isotopic cyanoborohydride in either H_2O or D_2O. The samples were then combined, digested with trypsin, and finally analyzed by LC-MS. The method involves labeling of lysine residues, whose relative abundance is higher than cysteine (ICAT reagents) resulting in increased sensitivity. Initial experiments indicate that labeled peptides are chromatographically indistinguishable under routine HPLC conditions, enhancing accurate quantitation. Labeled peptides are easily analyzed by CID with no interference in performing sequencing experiments.

Recently, absolute quantification (AQUA) was suggested as an ideal solution to define the state of all proteins within a given proteome (Regnier et al., 2001). This may involve the synthesis of over a million isotopically labeled peptides as internal standards to cover the global proteome (one or two for each protein). Once a linear relation can be derived between the signature peptide and the protein, this information can be used to determine protein concentration even in PTM species present within any organelle of a given cell. This technique was applied to PTM complex protein mixtures (Gerber et al., 2002).

DISEASE MECHANISMS AND DIAGNOSTICS

Historically, anatomical, histological, or molecular biomarkers have been associated with the presence and severity of specific diseases or physiological states. Advances in diagnosis and patient treatment based on molecular knowledge of disease are positively impacting on both therapy and prognosis. Understanding the dynamic nature of proteomes in cells or tissues represents an immense challenge emphasizing the fact that novel technologies will be required to bridge expression and functional proteomics, particularly as they correlate to disease. In that context the introduction of MS-based technologies has driven significant advances in disease proteomics research over the

Figure 7.9. MCAT determination of the relative protein abundance in complex biological mixture. (*a*) Ion chromatograms recorded for the base peak (*top*) and unmodified doubly charged peptide ion $[LPWFDGMLEADEAYFK+2H]^{+2}$ (*middle*), and its corresponding *O*-methylisourea (MCAT)-modified form (*bottom*). When mixtures of untreated and MCAT-treated protein digests are resolved by reverse-phase LC, MCAT-modified peptides elute with a modest delay relative to their respective unmodified forms (35.9 vs. 35.7 min, respectively, in this example). (*b*) Depending on charge, the *m/z* peaks observed for pairs of unmodified or modified peptides during MS become offset by 42, 21, or 14 *m/z* units (for singly, doubly, or triply charged ions, respectively). In this example, the peaks recorded for the unmodified (967.07 *m/z*) and modified (988.08 *m/z*) sister peptides are offset by 21 *m/z* units, indicating a +2 charge. The peptides were then independently selected and analyzed by MS/MS for protein identification. (*c*) The relative abundance of individual proteins is determined by comparing the ion chromatograms of each sister peptide and calculating the ratio of signal intensities by integrating the area under the curve. [Reproduced from Cagney and Emili (2002) with permission from *Nature Biotechnology*.]

past few years. Indeed, considerable efforts have been devoted to the discovery and validation of clinically relevant biomarkers using MS-based proteomics approaches, particularly in the field of cancer (Alalya et al., 2000, Simpson and Dorow, 2001) and cardiovascular disease (Van Eyk, 2001).

In exploring the immune response induced in human cancer, a creative approach for the identification of cancer biomarkers termed SEREX, based on the determination of autoantibodies against tumor proteins, has been introduced by Pfreundschuh and coworkers (Sahin et al., 1995; Türeci et al., 1997). This approach allows for the identification of specific cancer antigens that have relevance to the etiology and mechanistic aspects of cancer as well as to diagnosis and therapy (Hanash, 2003). Representative examples of this approach are the discovery of antiannexins I and II autoantibodies in lung cancer patients using 2DE and MS (Brichory et al., 2001) and the proteomic study by Prasannan et al. (2000) demonstrating that patients with neuroblastoma frequently develop an antigenic response against β-tubulin isoforms. The occurrence of these neuroblastoma-specific antibodies cannot only serve as a diagnostic biomarker, but also in the design of immunotherapies to overcome problems resulting from variable expression of a particular antigen resulting from tumor heterogeneity. Over the past 2 to 3 years, SELDI-TOF-MS was introduced as a unique MS-based proteomic tool capable of broadening the scope and breadth of molecular diagnostics in the various fields of clinical medicine. Science and technology have had an impact on applications in biomarker discovery as illustrated by developments in prostate cancer diagnosis. Decades ago, prostate volume was the sole clinical biomarker of both benign prostatic hypertrophy (BPH) and prostate cancer. Detection of prostate cancer was improved by the determination of prostate-specific antigen (PSA) level in serum (Barry, 2001). However, there is still uncertainty about its specificity in differentiating prostate cancer from BPH. Using a novel affinity capture protein biochip combined with SELDI TOF/MS, Xiao et al. (2001) demonstrated that serum prostate-specific membrane antigen (PSMA) may be a more effective biomarker than PSA for differentiating benign from malignant prostate disease. Furthermore, a study by Dhanasekaran et al. (2001), who identified prostate-cancer-related genes, demonstrated that the correlation of genomic and proteomics data could accelerate the discovery and validation of clinically relevant prostate cancer biomarkers. Such integrated investigations will also provide new insights in the biochemistry and pathophysiology of prostate cancer, thus enabling the discovery of new drug targets and the development of specific and highly efficacious therapies.

Proteomics research is also investigating protein complexes in an effort to decipher the mechanistic aspects of their disruption in different biological processes and disease states. Although the application of protein–protein interactions and protein complexes to disease investigations is still limited and narrowly focused, this area of proteomics research is rapidly expanding. Vondriska and Ping (2002) reviewed recently the novel approaches that have been developed to study ischemic heart proteomics and highlighted how affinity methods, 2DE, and MS were combined to demonstrate that cardioprotection is induced by activation of myocardial protein kinase C (PKC), which is physically associated with at least 36 other proteins with a diversity of functions. This type of proteomics application provides a solid basis for leading the design and discovery of new cardiac drugs and therapeutic strategies. Likewise, an elegant study by Borodovsky et al. (2002) using a chemical proteomics approach to quantitatively determine enzyme activity across normal and diseased tissue further illustrates the potential of MS-based proteomics methods in disease-related functional proteomics. These authors designed and synthesized a series of novel deubiquitinating enzyme-specific probes allowing for structural identification by tandem MS, as well as direct demonstration of enzymatic activity for gene products whose functions were deduced from primary structure data. Borodovsky et al. (2002) reported 23 active deubiquitinating enzymes in EL4 cells, including tumor suppressor CYLD1. Such studies are helpful in correlating structural features, biological activities, and functions of the constituents of complex protein networks or complexes and assisting biologists and clinicians to decipher the molecular complexity between normal and disease states. Also, such investigations will create new opportunities in the field of drug discovery and development by providing highly specific insights about the potential sites of actions of new chemical entities in tissues, cells, or cellular organelles and their mechanism(s) of action and interaction with the proteome.

BIOMARKERS IN DRUG DEVELOPMENT

Deciphering and understanding the role that protein networks play in diseases will create enormous clinical opportunities as these pathways are the source of tomorrow's drug targets (Atkinson, 2001; Figeys, 2002; Lathia, 2002; Wagner, 2002). In that perspective, proteomics biomarkers can also be used as surrogate end points to predict clinical benefit in evaluating the effectiveness, safety, and toxicity profiles of new drug candidates. As we will describe below, the judicious use of molecular biomarkers can accelerate the drug development process and regulatory approval by generating, in both the preclinical and clinical phases, relevant information about the biological activity or toxicity associated with a lead compound or a drug candidate. This information can be utilized to make pertinent decisions about further development or early termination of a particular molecule. The expectation is that such molecular biomarkers would provide clinicians and researchers with highly specific and sensitive surrogates or clinical end points to guide crucial development decisions such as selection of patients, doses, and dose regimens as well as to demonstrate the efficacy of the therapeutic agent and investigate the molecular basis and severity of potential adverse events. Aside from those genetic differences predicting altered drug metabolism and the associated drug toxicity, there is no documented example where pharmacogenomics has been capable of identifying those patients likely to experience severe drug-induced adverse events. Given that genetics plays a major role in a patient's response to therapeutic agents, it is likely that patient's susceptibility to such adverse effects is reflected, most probably, in a subtle fashion through intricate proteomics expressions that are obscured by the expression of so-called established biomarkers. The need for comprehensive biomarker strategies to assist in the development of safe and efficacious therapeutic agents is demonstrated by the numerous drug withdrawals that have occurred over the past decade. The antidiabetic drug troglitazone (Rezulin) used for the management of type II diabetes mellitus and cerivastatin (Baycol) are typical examples. Troglitazone was a promising therapeutic agent because it was effective in many patients whose diabetes had not been controlled by other medication. Although it showed some degree of liver toxicity in approximately 2 percent of the 2500 patients tested during drug development trials (Gale, 2001), liver enzyme levels decreased to within the normal range following discontinuation of therapy, thus indicating that the observed liver toxicity was temporary and reversible. During the first 2 years after troglitazone's approval, about one million patients were treated with the drug, with 70 experiencing severe liver failure, including 60 deaths and 10 transplants, which led to troglitazone's recall. Likewise, the lipid-lowering drug cerivastatin was withdrawn from the market in 2001 due to numerous fatal rhabdomyolysis, a severe muscle adverse reaction that results in muscle cell breakdown. These episodes indicate that the use of the standard liver toxicity biomarkers alone are often insufficient to predict liver toxicity at the drug discovery or development stages in a large heterogeneous patient population. Liver toxicity has been defined as any increase by more than two- to threefold in serum glutamine pyruvic transaminase or conjugated bilirubin than the normal blood levels. Given the intrinsic complexity of the underlying biochemical events and the role environmental and genetic factors have on hepatic function, more comprehensive proteomics approaches will be required to foster significant advances about the etiology and underlying biochemical mechanisms of drug-induced liver toxicity and define how they can be leveraged in the development of safer drugs.

In the same perspective, nephrotoxicity has long been related to certain drug classes such as the aminoglycoside antibiotics gentamicin, puromycin, and amikacin and immunosuppressive agents including cyclosporin and tacrolimus. Similarly, clinicians do not know the extent of renal damage that may occur with commonly used drugs such as nonsteroidal antiinflammatory drugs (NSAIDs), cyclooxygenase-2 inhibitors (COX-2 inhibitors), acetaminophen, and common lipid-lowering drugs such as atorvastatin and simvastatin or combinations thereof. Proteomics provides a unique approach to examine the extent of drug-induced nephrotoxicity by characterizing and validating suites of biomarkers that could be used as a reference protein signature of kidney injury to classify responses of new compounds with similar pharmacological or toxicological phenotype in drug safety evaluation trials. A recent study by Charlwood et al. (2002) reported an elegant proteomic approach for the investigation of the effect of increasing doses of gentamicin on protein expression in the rat kidney cortex. The authors treated three experimental groups of rats with gentamicin at increasing doses of 10, 30, and 70 mg/kg/day by intraperitoneal injection on four consecutive days. After necropsy, kidney cortex samples were analyzed by two-dimensional polyacrylamide gel electrophoresis (2D-PAGE) and were clustered into three groups: the control (Fig.7.10*a*), a 30-mg dose (Fig. 7.10*b*), and a 70-mg dose (Fig 7.10*c*). To identify the proteins of interest, the corresponding spots were excised from the gel, digested with trypsin, and the resulting peptide fragments analyzed by MALDI TOF-MS or nanospray TOF-MS. The most striking change between samples is the presence of five isoforms of α-2-microglobulin, which were found to be up-regulated in the latter samples compared to controls. Interestingly, α-2-microglobulin isoforms were not all up-regulated in a linear fashion with respect to the dose of gentamicin. Some isoforms were strongly up-regulated in both 30- and 70-mg treated samples, whereas others were less up-

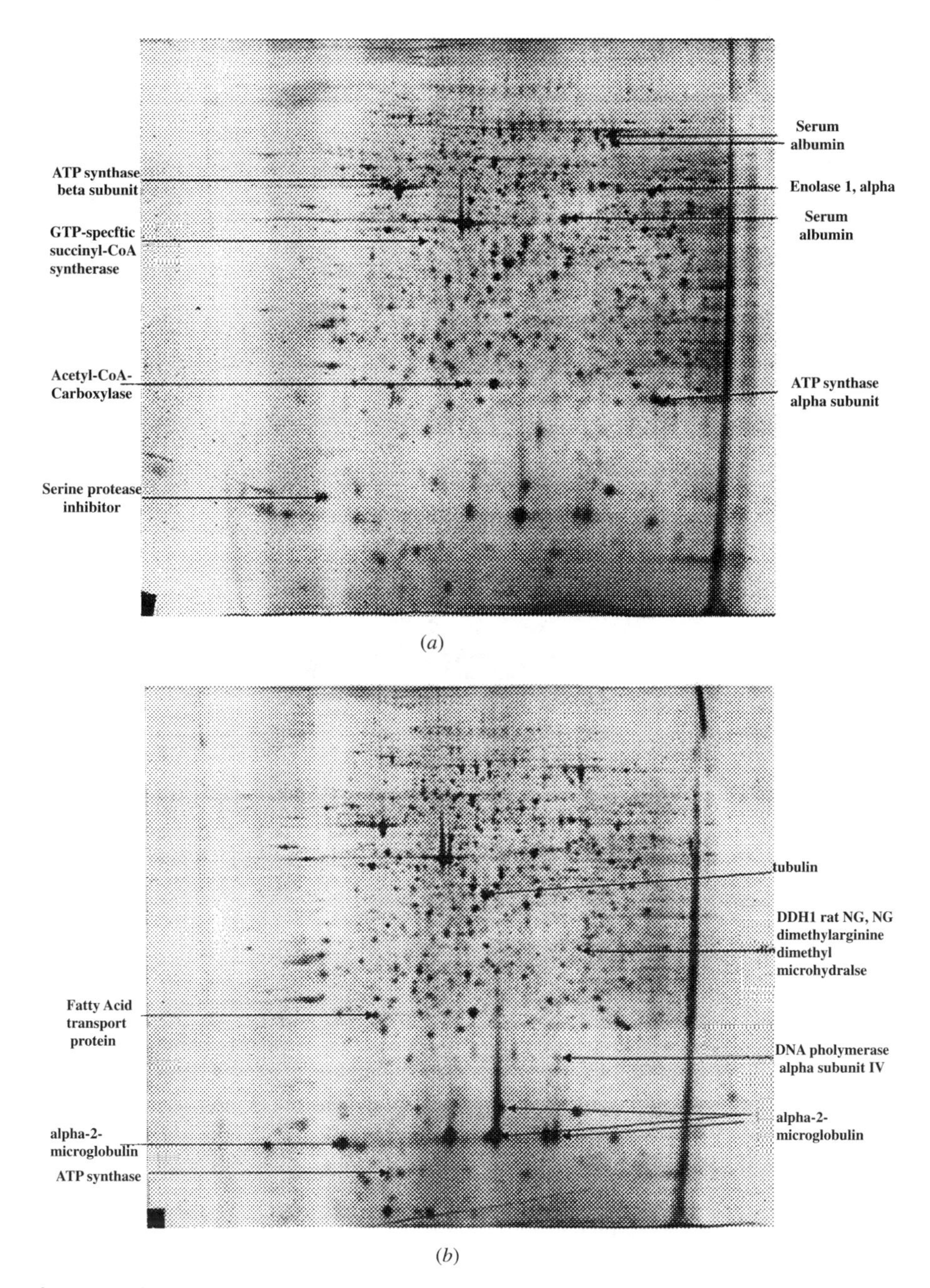

Figure 7.10. (*a*) Two-dimensional PAGE gel stained with Sypro Ruby of control rat kidney cortex. Proteins highlighted in pink are down-regulated in the middose samples and proteins highlighted in violet are down-regulated in the high-dose samples. (*b*) Two-dimensional PAGE gel stained with Sypro Ruby of kidney cortex from a rat treated with 30 mg/kg gentamicin (middose). Proteins highlighted in pink are up-regulated in the middose samples compared to the control.

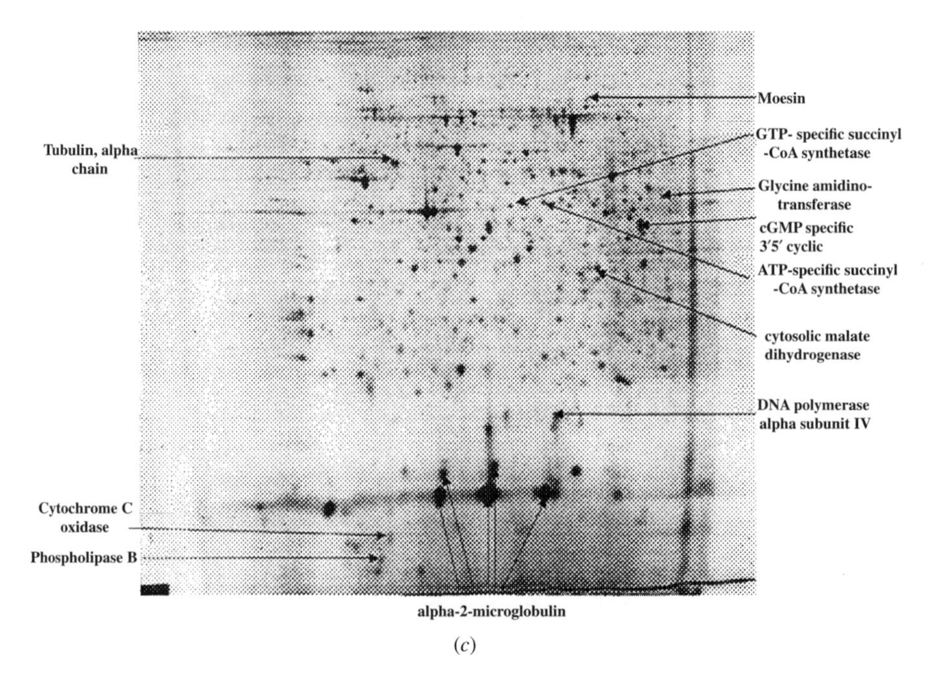

(*c*)

Figure 7.10. (*c*) Two-dimensional PAGE gel stained with Sypro Ruby of kidney cortex from a rat treated with 70 mg/kg of gentamicin (high dose), which showed nephrosis. Proteins highlighted in violet are up-regulated in the high-dose samples compared to control. (*a*)–(*c*) IEF range was pH 4–7 (left to right), SDS-PAGE was 12% acrylamide (top is high MW and bottom is low MW). Approximately 100 g of protein was loaded onto the gel. [Reproduced from Charlwood et al. (2002) with permission from the American Chemical Society.]

regulated in the 30-mg treated samples. More than 20 other proteins identified in this study are associated with four distinct biochemical pathways: gluconeogenesis and glycolysis, fatty acid transport and utilization, the citric acid cycle, and stress response. These results are in agreement with known mechanisms in gentamicin-induced nephrotoxicity, primarily through impairment of energy production and mitochondrial dysfunction. Hence, profiling of biomarker expression would be useful for selecting compounds for drug development and for investigating the mechanistic aspects of drug-induced nephrotoxicity.

The discovery and validation of biologically relevant biomarkers is a multifaceted endeavor that parallels the complexity of the drug development process itself. For example, the development of a cancer drug starts with a crucial in vivo efficacy study in which the human tumor cell line that showed the highest in vitro sensitivity to the candidate molecules is used as a xenograft in a subcutaneous implant in nude mouse models. The efficacy study is followed by an interspecies investigation of absorption, distribution, metabolism, and excretion (ADME) variation of the candidate drug's behavior in dogs, mice, and rats. Finally, a series of pharmacokinetics (PK) and safety studies are performed in at least two species to investigate qualitative and quantitative organ toxicities and their reversibilities. The latter investigations are conducted to establish the starting doses and escalating schedules for future human clinical trials. There have been limited applications of proteomic investigations in drug development. Thus far, functional proteomics studies have been used essentially for identification of regulated targets in specific pathways. In that perspective, the mitogen-activated protein kinase (MAPK) pathway represents an attractive target for therapeutic intervention. The development of pharmacological inhibitors of this pathway, particularly those exhibiting in vivo activity, remains an area of intense interest (Dent and Grant, 2001). In a well-designed proteomics study Lewis et al. (2000) identified 25 targets of the MAPK extracellular signal regulated kinase 5 (MAPK/ERK5) by combining functional proteomics with selective activation and inhibition of map kinase kinase (MKK1/2) with the aim of identifying cellular targets regulated by the MKK/ERK cascade. This study represents a classical application of functional proteomics concerning the identification of regulated targets of a distinct signal transduction pathway. Also, it demonstrates the utility of this discovery-based strategy in elucidating novel MAP kinase pathway effectors while providing a comprehensive approach not only for monitoring global molecular responses following activation of signal transduction pathways but also for reporting altered protein PTM and expression.

CONCLUSION

Because of its intrinsic technical flexibility, the variety of its ionization modes, the ability to perform rapid peptide sequencing, the inherent specificity of detection and ease of interfacing with a variety of front-end techniques such as affinity capture and LC systems, MS is a core component of proteomic biomarker discovery. As previously mentioned in this chapter, the development of new MS-based techniques has allowed for the detection, identification, and quantitation of several proteins in a variety of functional conditions. Conversely, this has accelerated the general understanding of complex biological processes, particularly the role of proteins in the etiology and progression of diseases. As pointed out by Aebersold and Mann (2003), this new paradigm has a number of important implications and poses so far inadequately addressed challenges for every aspect of experimental biology, namely experimental design and data analysis visualization and storage to name a few. It is recognized that further advances in technology will be required to develop proteomic biomarker discovery platforms allowing for the characterization of all biologically relevant biomarkers in a given biological sample. Such methods would provide a complete description of the biomarker proteins, their PTMs and degradation products, their interactions with other proteins, their cellular, subcellular, and tissue distribution, as well as their temporal quantitative variation in various biological, physiological, and therapeutic conditions.

Hence, bridging genomics and proteomics from a systems biology perspective will allow biologists to discovery and validate the optimum drug targets, select the most promising drug leads, and validate the best clinical or diagnostic biomarkers. This underscores the fact that proteomics biomarkers can be advantageously utilized to increase the efficiency and quality of preclinical and clinical development by providing assessment and screening tools allowing clinicians and researchers to make intelligent decisions at each of the critical stages of the drug development process. This approach can be used in the evaluation of the relative efficacy of a series of synthetic drug candidates, protein molecules, or peptidomimetic compounds that can block a receptor or inactivate a metabolic cascade in a known disease pathway. In such proof-of-concept studies, the measurement of a protein molecule (biomarker) or a series of proteins whose concentration(s) in whole-blood/plasma/serum change downstream from the blocked receptor/pathway may serve as indicator(s) in determining whether the tested compounds act through the hypothetical mechanism of action.

Biomarker discovery will ultimately be incorporated into a complete picture that begins with the discovery stage and translates into application. Biomarkers should provide earlier efficacy measurements, entailing a greater understanding of the mechanism of disease progression. For example, molecular changes observed in tumors may serve as signatures of malignancy in body fluids [blood, serum, plasma, urine, cerebral spinal fluid (CSF), tears, etc.] that may precede clinical cancer diagnosis. One can predict that MS will remain, in the foreseeable future, a core technology in biomarker discovery and characterization, and its capabilities enhanced by advances in microfluidic systems and nanodevices as they will enable increased selectivity and specificity of sample preparation processes.

REFERENCES

Adam, B.L., Qu, Y., Davis, J.W., Ward, M.D., Clements, M.A., Cazares, L.H., Semmes, O.J., Schellhammer, P.F., Yasui, Y., Ziding, F., and Wright, G.L. (2002). Serum protein fingerprinting coupled with a pattern-matching algorithm distinguishes prostate cancer from benign prostate hyperplasia and healthy men. *Cancer Res.* **62**:3609–3614.

Adam, B.L., Vlahou, A., Semmes, O.J., and Wright, G.L. (2001). Proteomics approaches to biomarker discovery in prostate and bladder cancer. *Proteomics* **10**:1264–1270.

Aebersold, R., and Mann, M. (2003). Mass spectrometry-based proteomics. *Nature* **422**:198–207.

Alalya, A.A., Franzen, B., Auer, G., and Linder, S. (2000). Cancer proteomics: From identification of novel markers to creation of artificial learning models for tumor classification. *Electrophoresis* **21**:1210–1217.

Atkinson, Jr., A.J. (2001). Physiological and laboratory markers of drug effect. In: Atkinson, Jr., A.J., Daniels, C.E., Dedrick, R.L., Grudzinskas, C., and Markey, S.P. (Eds.). *Principles of Clinical Pharmacology*, Academic, New York.

Barry, M.J. (2001). Prostate-specific-antigen testing for early diagnostic of prostate cancer. *N. Engl. J. Med.* **344**:1373–1377.

Baxter, R.C. (2001). Signaling pathways involved in antiproliferative effects of IGFBP-3. *Mol. Pathol.* **54**:145–148.

Biomarkers Definitions Working Group (2001). Biomarkers and surrogate endpoints, preferred definitions and conceptual framework. *Clin. Pharmacol. Therapeutics* **69**:89–95.

Bischoff, S.R., Kahn, M.B., Powell, M.D., and Kirlin, W.G. (2002). SELDI-TOF-MS analysis of transcriptional activation protein binding to response elements regulating carcinogenesis enzymes. *Intl. J. Mol. Sci.* **3**:1027–1038.

Blanchard, A.P., Kaiser, R.J., and Hood, L.E. (1996). High-density oligonucleotide arrays. *Biosens. Bioelectron.* **11**:687–690.

Borodovsky, A., Ovaa, H., Kolli, N., Gan-Erdene, T., Wilkinson, K.D., Ploegh, H.L., and Kessler, B.M. (2002). Chemistry-based functional proteomics reveals novel members of the deubiquitinating enzyme family. *Chem Biol* **9**:1149–1159.

Brichory, F.M., Misek, D.E., Yim, A.M., Krausse, M.C., Giordano, T.J., Beer, D.G., and Hanash, S.M. (2001). An immune response manifested by the common occurrence of annexins I and II autoantibodies and high circulating levels of IL-6 in lung cancer. *Proc. Natl. Acad. Sci.* **98**:9824–9829.

Cagney, G., and Emili, A. (2002). De novo peptide sequencing and quantitative profiling of complex protein mixture using mass-coded abundance tagging. *Nature Biotech* **20**:163–170.

Cazares, L.H., Adam, B.L., Ward, M.D., Nasim, S., Schellhammer, P.F., Semmes, O.J., and Wright, G.L. (2002). Normal, benign, preneoplastic, and malignant prostate cells have distinct protein expression profiles resolved by surface enhanced laser desorption/ionization mass spectrometry. *Clin. Cancer Res.* **8**:2451–2452.

Chapman, J.R. (2000). Mass spectrometry of proteins and peptides. In: Chapman, J.R. (Ed.). *Methods of Molecular Biology*™ *Series*, Walker JM (ed), Humana Press, Totowa, NJ, Vol. 146.

Charlwood, J., Skehel, J.M., King, N., Camilleri, P., Lord, P., Bugelski, P., and Atif, U. (2002). Proteomic analysis of rat kidney cortex following treatment with gentamicin. *J. Proteome Res.* **1**:73–82.

Collins, F.S., and McKusick, V.A. (2001). Implications of the human genome project for medical sciences. *JAMA* **285**:540–544.

Conrads, T.P., Issaq, H.J., and Veenstra, T.D. (2002). New tools for quantitative phosphoproteome analysis. *Biochem. Biophys. Res. Commun.* **290**:885–890.

Dare, T.O., Davies, H.A., Turton, J.A., Lomas, L., Williams, T.C., and York, M.J. (2002). Application of surface-enhanced laser desorption/ionization technology to the detection and identification of urinary parvalbumin-α: A biomarker of compound-induced skeletal muscle toxicity in the rat. *Electrophoresis* **23**:3241–3251.

Davies, H., Lomas, L., and Austen, B. (1999). Profiling of amyloid β peptide variants using SELDI proteinchips® arrays. *Biotechniques* **27**:1258–1261.

Dent. P., and Grant, S. (2001). Pharmacologic interruption of the mitogen-activated extracellular-regulated kinase/mitogen-activated protein kinase signal transduction pathway. *Clin. Cancer Res.* **7**:775–783.

Dhanasekaran, S.M., Barette, T.R., Ghosh, D., Shah, R., Varambally, S., Kurachi, K., Pienta, K.J., Rubin, M.A., and Chinnaiyan, A.M. (2001). Delineation of prognostic biomarkers in prostate cancer. *Nature* **412**:822–826.

Eng, J., McCormack, A.L., and Yates, J.R. (1994). An approach to correlate tandem mass spectral data of peptides with amino acid sequences in a protein database. *J. Am. Soc. Mass Spectrom.* **5**:976–989.

Fenn, J.B., Mann, M., Meng, C.K., Wong, S.F., and Whitehouse, C.M. (1989). Electrospray ionization for mass spectrometry of large biomolecules. *Science* **246**:64–71.

Ficarro, S.B., McCleland, M.L., Stukenberg, P.T., Burke, D.J., Ross, M.M., Shabanowitz, J., Hunt, D.F., and White, F.M. (2002). Phosphoproteome analysis by mass spectrometry and its application to *Saccharomyces cerevisiae*. *Nature Biotech.* **20**:301–305.

Figeys, D. (2002). Proteomics approaches in drug discovery. *Anal. Chem.* **74**:413a–419a.

Figeys, D. (2001). Two-dimensional gel electrophoresis and mass spectrometry for proteomics studies: State-of-the-art. In: Rehm, H.J., and Reed, G. (Eds.) in cooperation with Puhler, A., and Stadler, P. *Biotechnology, Vol. 5b: Genomics and Bioinformatics*, 2nd ed. Wiley, New York, pp. 243–268.

Fung, E.T., Thulasiraman, V., Weinberger, S.R., and Dalmasso, E.A. (2001). Protein biochips for differential profiling. *Curr. Opin. Biotechnol.* **12**:65–69.

Gale, E. (2001). Lessons from the glitazones: A story of drug development. *Lancet* **357**:1870–1875.

Gerber, S.A., Rush, J., Stemmann, O., Steen, H., Kirschner, M.W., and Gygi, S.P. (2002). Absolute quantification of cell cycle regulatory proteins and phosphorylation states: The AQUA strategy for protein profiling. Proceedings from the 50th ASMS Conference, Orlando, Florida, 2–6 June 2002.

Görg, A. (2000). Advances in 2D gel techniques. In: Mann, M., and Blackstock, W. (Eds.). *Proteomics: A Trend Guide*. Elsevier, London, pp. 3–6.

Greenbaum, D., Medzihradszky, K.F., Burlingame, A., and Bogyo, M. (2000). Epoxide electrophiles as activity-dependent cysteine protease profiling and discovery tools. *Chem. Biol.* **7**:569–581.

Griffin, T.J., and Aebersold, R. (2001). Advances in proteome analysis by mass spectrometry. *J. Biol. Chem.* **276**:45497–45500.

Griffin, T.J., Lock, C.M., Li, X.J., Patel, A., Chervetsova, I., Lee, H., Wright, M.E., Ranish, J.A., Chen, S.S., and Aebersold, R. (2003). Abundance ratio-dependent proteomic analysis by mass spectrometry. *Anal. Chem.* **75**:867–874.

Gygi, S.P., Rist, B., Griffin, T.J., Eng, J., and Aebersold, R. (2002). Proteome analysis of low-abundance proteins using multidimensional chromatography and isotope-coded affinity tags. *J. Proteome Res.* **1**:47–54.

Gygi, S.P., Rist, B., Gerber, S.A., Turecek, F., Gelb, M.H., and Aebersold, R. (1999a). Quantitative analysis of protein mixtures using isotope coded affinity tags. *Nature Biotechnol.* **17**:994–999.

Gygi, S.P., Rochon, Y., Franza, B.R., and Aebersold, R. (1999b). Correlation between protein and mRNA abundance in yeast. *Mol. Cell. Biol.* **19**:1720–1730.

Hampel, D.J., Sansome, C., Sha, M., Brodsky, S., Lawson, W.E., and Goligorsky, M.S. (2001). Toward proteomics in Uroscopy: Urinary protein profiles after radiocontrast medium administration. *J. Am. Soc. Nephrol.* **12**:1026–1035.

Hanash, S. (2003). Harnessing immunity for cancer marker discovery. *Nature Biotechnol.* **21**:37–38.

He, Q.Y., Yip, T.T., Li, M., and Chiu, J.F. (2003). Proteomics analyses of arsenic-induced cell transformation with SELDI-TOF proteinchip® technology. *J. Cell. Biochem.* **88**:1–8.

Ho, Y., Gruhler, A., Heilbut, A., Bader, G.D., Moore, L., Adams, S.L., Millar, A., Taylor, P., Bennett, K., Boutilier, K., Yang, L., Wolting, C., Donaldson, I., Schandorff, S., Shewnarane, J., Vo, M., Taggart, J., Goudreault, M., Muskat, B., Alfarano, C., Dewar, D., Lin, Z., Michalickova, K., Willems, A.R., Sassi, H., Nielsen, P.A., Rasmussen, K.J., Andersen, J.R., Johansen, L.E., Hansen, L.H., Jespersen, H., Podtelejnikov, A., Nielsen, E., Crawford, J.,

Poulsen, V., Sørensen, B.D., Matthiesen, J., Hendrickson, R.C., Gleeson, F., Pawson, T., Moran, M.F., Durocher, D., Mann, M., Hogue C.W.C., Figeys, D., and Tyers, M. (2002). Systematic identification of protein complexes in *Saccharomyces cerevisiae* by mass spectrometry. *Nature* **415**:180–183.

Hoogland, C., Sanchez, J.C., Wather, D., Baujard, O., Tonella, L., Hochstrasser, D.F., and Appel, R.D. (1999). Two-dimensional electrophoresis resources available from ExPASy. *Electrophoresis* **20**:3568–3571.

Horvath, S.J., Firca, J.R., Hunkalipper, T., Hunkalipper, M.W., and Hood, L.E. (1987). An automated DNA synthesizer employing deoxy-nucleoside 3′ phosphoramidites. *Methods Enzymol.* **154**:314–326.

Howard, J.C., Heinemann, C., Thatcher, B.J., Martin, B., Gan, B.S., and Reid, G. (2000). Identification of collagen-binding proteins in *Lactobacillus* spp. with surface-enhanced laser desorption/ionization–time of flight proteinChip technology. *Appl. Environ. Microbiol.* **66**:4396–4400.

Issaq, H.J., Conrads, T.P., Prieto, D.A., Tirumalai, R., and Veenstra, T.D. (2003). SELDI-TOF MS for diagnostic proteomics. *Anal. Chem.* **75**:149a–155a.

Issaq, H.J., Veenstra, T.D., Conrads, T.P., and Felschow, D. (2002). The SELDI TOF MS approach to proteomics: Protein profiling and biomarker identification. *Biochem. Biophys. Res. Commun.* **292**:587–592.

Ito, T., Chiba, T., Ozawa, R., Yoshida, M., Hattori, M., and Sakaki, Y. (2001). A comprehensive two-hybrid analysis to explore the yeast protein interactome. *Proc. Natl. Acad. Sci.* **98**:4569–4574.

Jones, S., and Thornton, J.M. (1996). Principles of protein–protein interactions. *Proc. Natl. Acad. Sci. USA* **93**:13–20.

Karas. M., and Hillenkamp, F. (1988). Laser desorption ionization of proteins with molecular masses exceeding 10,000 daltons. *Anal. Chem.* **60**:2299–2301.

King, R.A., Rotter, J.I., and Motulsky, A.G. (2002). *The Genetic Basis of Common Diseases*, 2nd ed. Oxford Monograph on Medical Genetics No 44, Oxford University Press, New York.

Kwok, P.Y. (2001). Methods for genotyping single nucleotide polymorphisms. *Ann. Rev. Genomics Hum. Genet.* **2**:235–258.

Lathia, C.D. (2002). Biomarkers and surrogate endpoints: How and when might they impact drug development? *Disease Markers* **18**:83–90.

Lewis, T.S., Hunt, J.B., Aveline, L.D., Jonscher, K.R., Louie, D.F., Yeh, J.M., Nahreini, T.S., Resing, K.A., and Ahn, N.G. (2000). Identification of novel MAP kinase pathway signaling targets by functional proteomics and mass spectrometry. *Mol. Cell* **6**:1343–1354.

Lin, Y.S., and Chen, Y.C. (2002). Laser desorption/ionization time-of-flight mass spectrometry on sol-gel derived 2,5-dihydroxybenzoic acid film. *Anal. Chem.* **74**:5793–5798.

Link, A.J., Eng, J., Schieltz, D.M., Carmack, E., Mize, G.J., Morris, D.R., Garvik, B.M., and Yates, J.R. III (1999). Direct analysis of protein complexes using mass spectrometry. *Nature Biotechnol.* **17**:676–682.

Lipton, M.S., Pasa-Tolic, L., Anderson, G.A., Anderson, D.J., Auberry, D.L., Battista, J.R., Daly, M.J., Fredrickson, J., Hixson, K.K., Kostandarithes, H., Masselon, C., Markville, L.M., Moore, R.J., Romine, M.F., Shen, Y., Stritmatter, E., Tolic, N., Udseth, H.R., Venkateswaran, A., Wong, K.K., Zhao, R., and Smith, R.D. (2002). Global analysis of the *Deinococcus radiodurans* proteome by using accurate mass tags. *Proc. Natl. Acad. Sci. USA* **99**:11049–11054.

Liu, Y., Patricelli, M.P., and Cravatt, B.F. (1999). Activity-based protein profiling: The serine hydrolases. *Proc. Natl. Acad. Sci, USA* **96**:14694–14699.

Locke, S.J., Pinto, D.M., and Rowland, E. (2003). Quantitative proteomics via isotopically differentiated derivatization. Proceedings from the 51st ASMS Conference, Montreal, Canada, 8–12 June 2003.

Mann, M.N., Hendrickson, R.C., and Pandey, A. (2001). Analysis of proteins and proteomes by mass spectrometry. *Annu. Rev. Biochem.* **70**:437–473.

McDonald, W.H., and Yates, J.R. III (2002). Shotgun proteomics and biomarker discovery. *Disease Markers* **18**:99–105.

McDonald, W.H., Ohi, R., Miyamoto, D.T., Mitchison, T.J., and Yates, J.R. III (2002). Comparison of three directly coupled HPLC MS/MS strategies for identification of proteins from complex mixtures: Single-dimension LC-MS/MS, 2-phase MudPIT, and 3-phase MudPIT. *Intl. J. Mass. Spectrom.* **219**:245–251.

Mullis, K., Faloona, F., Scharf, S., Saiki, R., Horn, G., and Erlich, H. (1986). Specific enzymatic amplification of DNA in vitro: The polymerase chain reaction. *Cold Spring Harb. Symp. Quant. Biol.* **51**:Pt1:263–273.

Munchbach, M., Quadroni, M., Miotto, G., and James, P. (2000). Quantitation and facilitated de novo sequencing of proteins by isotopic N-terminal labeling of peptides with a fragmentation-directing moiety. *Anal. Chem.* **72**:4047–4057.

Oda, Y., Nagasu, T., and Chait, B.T. (2001). Enrichment analysis of phosphorylated proteins as a tool for probing the phosphoproteome. *Nature Biotechnol.* **19**:379–882.

O'Farrell, P.H. (1975). High resolution two-dimensional electrophoresis of proteins. *J. Biol. Chem.* **250**:4007–4021.

Pandey, A., and Mann, M. (2000). Proteomics to study genes and genomes. *Nature* **405**:837–846.

Perrot, M., Sagliocco, F., Mini, T., Monrobot, C., Scheider, U., Shevchenko, A., Mann, M., Jeno, P., and Boucherie, H. (1999). Two-dimensional gel protein database of *Saccharomyces cerevisiae* (update 1999). *Electrophoresis* **20**:2280–2298.

Petricoin, E.F., Ardekani, A., Hitt, B.A., Levine, P.J., Fusaro, V.A., Steinberg, S.M., Mills, G.B., Simone, C., Fishman, D.A., Kohn, E.C., and Liotta, L.A. (2002). Use of proteomics pattern in serum to identify ovarian cancer. *Lancet* **359**:572–577.

Prasannan, L., Misek, D.E., Hindewrer, R., Michon, J., Geiger, J.D., and Hanash, S.M. (2000). Identification of β-tubulin isoforms as tumor antigens in neuroblastoma. *Clin. Cancer Res.* **6**:3949–3956.

Qiu, Y., Sousa, E.A., Hewick, R.M., and Wang, J.H. (2002). Acid-labile isotope-coded extractants: A class of reagents for quantitative mass spectrometric analysis of complex protein mixtures. *Anal. Chem.* **74**:4969–4979.

Regnier, F., Amini, A., Chakraborty, A., Geng, M., Ji, J., Riggs, L., Sioma, C., Wang, S., and Zhang, X. (2001). Multidimensional chromatography and the signature peptide approach to proteomics. *LCGC* **19**:200–213.

Sahin, U.O., Türeci, O., Schmitt, H., Cochlovius, B., Johannes T. Schmits, R., Stenner, F., Luo, G., Schobert, I., and Pfreundschuh, M. (1995). Human neoplasms elicit multiple specific immune responses in the autologous host. *Proc. Natl. Acad. Sci. USA* **98**:11810–11813.

Sali, A., Glaeser, R., Earnest, T., and Baumeister, W. (2003). From words to literature in structural proteomics. *Nature* **422**:216–225.

Scheele, G.J. (1975). Two-dimensional analysis of soluble proteins. *Biochemistry* **250**:5375–5385.

Simpson, R.J., and Dorow, D.S. (2001). Cancer proteomics: From signaling networks to tumor markers. *Trends Biotech.* **19**:S40–48.

Smith, L.M., Sanders, J.Z., Kaiser, R.J., Hughes, P., Dodd, C., Connell, C.R., Heiner, C., Kent, S.B.H., and Hood, L.E. (1986). Fluorescence detection in automated DNA sequence analysis. *Nature* **321**:674–679.

Smolka, M., Zhou, H., and Aebersold, R. (2002). Quantitative protein profiling using two-dimensional gel electrophoresis, isotope-coded affinity tag labeling, and mass spectrometry. *Mol. Cell Proteomics* **1**:19–29.

Srinivas, P.R., Verma, M., Zhao, Y., and Srivastana, S. (2002). Proteomics for cancer biomarker discovery. *Clin. Chem.* **48**:1160–1169.

Taniguchi, T. (1965). A new device for determining urinary metanephrine and normetanephrine and its clinical applications. *Med. J. Osaka Univ.* **15**:365–387.

Tao, A.W., Ranish, J., and Aebersold, R. (2003). Quantitative analysis of protein complexes using solid-phase isotope tags (SPIT). Proceedings from the 51st ASMS Conference, Montreal, Canada, 8–12 June 2003.

Thomas, J.J., Shen, Z., Crowell, J.E., Finn, M.G., and Siuzdak, G. (2001). Desorption/ionization on silicon (DIOS): A diverse mass spectrometric platform for protein characterization. *Proc. Natl. Acad. Sci.* **98**:4932–4937.

Türeci, O., Sahin, U., and Pfreundschuh, M. (1997). Serological analysis of human tumor antigens: Molecular definition and implications. *Mol. Med. Today* **3**:342–349.

Uetz, P., Giot, L., Cagney, G., Mansfield, T.A., Judson, R.S., Knight, J.R., Lockshon, D., Narayan, V., Srinivasan, M., Pochart, P., Qureshi-Emili, A., Li, Y., Godwin, B., Conover, D., Kalbfleisch, T., Vijayadamodar, G., Yang, M., Johnston, M., Fields, S., and Rothberg, J.M. (2000). A comprehensive analysis of protein–protein interactions in *Saccharomyces cerevisiae*. *Nature* **403**:623–627.

Valentinis, B., and Baserga, R. (2001). IGF-1 receptor signalling in transformation and differentiation. *Mol. Pathol.* **54**:133–137.

Van Eyk, J.E. (2001). Proteomics: Unravelling the complexity of heart disease and striving to change cardiology. *Curr. Opin. Mol. Therapeut.* **3**:546–553.

Vlahou, A., Schellhammer, P.F., Mendrinos, S., Patel, K., Kondylis, F.I., Gong, L., Nasim, S., and Wright, Jr., G.L. (2001). Development of a novel proteomic approach for the detection of transitional cell carcinoma of the bladder in urine. *Am. J. Pathol.* **158**:1491–1502.

Vondriska, T.M., and Ping, P. (2002). Functional proteomics to study the protection of the ischaemic myocardium. *Expert. Opin. Therapeut. Targets* **6**:563–570.

Wagner, J.A. (2002). Overview of biomarkers and surrogate endpoints in drug development. *Disease Markers* **18**:41–46.

Washburn, M.P., Wothers, D., and Yates, J.R. III (2001). Large-scale analysis of yeast proteome by multidimensional protein identification technology. *Nature Biotechnol.* **19**:242–247.

Washburn, M.P., and Yates, J.R. III (2000). New methods of proteome analysis: Multidimensional chromatography and mass spectrometry. In: Mann, M., and Blackstock, W. (Eds.). *Proteomics: A Trend Guide*. Elsevier, London, pp. 3–6.

Wei, J., Burlak, J.M., and Siuzdak, G. (1999). Desorption-ionization mass spectrometry on porous silica. *Nature* **399**:243–246.

Weinberger, S.R., Dalmasso, E.A., and Fung, E.T. (2000). Current achievements using proteinchip® array technology. *Curr. Opin. Chem. Biol.* **6**:86–91.

Wellmann, A., Wollscheid, V., Lu, H., Ma, Z.L., Albers, P., Schutze, K., Rohde, V., Behrens, P., Dreschers, S., Ko, Y., and Wernert, N. (2002). Analysis of microdissected prostate tissue with proteinchip® arrays—a way to new insights into carcinogenesis and to diagnostics tools. *Intl. J. Mol. Med.* **9**:341–347.

Wilkins, M.R., Pasquali, C., Applel, R.D., Ou, K., Golaz, O., Sanchez, J.C., Yan, J.X., Gooley, A.A., Hughes, G., Humphery-Smith, I., Williams, K.L., and Hochstrasser, D.F. (1996). From proteins to proteomes: Large scale protein identification by two-dimensional electrophoresis and amino acid analysis. *Biotechnology* **14**:61–65.

Wolters, D., Washburn, M.P., Yates, J.R. III (2001). An automated multidimensional protein identification technology for shotgun proteomics. *Anal. Chem.* **73**:5683–5690.

Xiang, Z., Ho, L., Shrishailam, Y., Zhao, Z., Pompl, P., Kelley, K., Dang, A., Qing, W., Telpow, D., and Pasinetti, G.M. (2002). Cyclooxygenase-2 promotes amyloid plaque deposition in a mouse model of Alzheimer's disease neuropathology. *Gene Expr.* **10**:271–278.

Xiao, Z., Adam, B.L., Cazares, L.H., Clements, M.A., Davis, J.W., Schellhammer, P.F., Dalmasso, E.A., and Wright, G. (2001). Quantitation of serum prostate-specific membrane antigen by a novel protein biochip immunoassay discriminates benign from malignant prostate disease. *Cancer Res.* **61**:6029–6033.

Yao, X., Freas, A., Ramirez, J., Demirev, P.A., and Fenseleau, C. (2001). Proteolytic ^{18}O labeling for comparative proteomics: Model studies with two serotypes of adenovirus. *Anal. Chem.* **73**:2836–2842.

Zhou, H., Watts, J.D., and Aebersold, R. (2001). A systematic approach to the analysis of protein phosphorylation. *Nature Biotechnol.* **19**:375–378.

Zhou, H., Ranish, J.A., Watts, J.D., and Aebersold, R. (2002). Quantitative proteome analysis by solid-phase isotope tagging and mass spectrometry. *Nature Biotechnol* **20**:512–515.

8

INDUSTRIAL-SCALE PROTEOMICS ANALYSIS OF HUMAN PLASMA

Keith Rose

GeneProt Inc., Meyrin/Geneva, Switzerland

INTRODUCTION

The scope of this chapter is the industrial-scale approach to the proteomics analysis of plasma. To begin with, we will briefly discuss the terms *proteomics*, *plasma*, and *industrial scale*. In the context of this chapter, the word *proteomics* refers to a broad protein discovery process—determining which polypeptides and proteins are present in

Industrial Proteomics: Applications for Biotechnology and Pharmaceuticals, edited by Daniel Figeys
ISBN 0-471-45714-0

a sample and in which relative amounts. This aspect of the proteomics endeavor has been referred to as discovery proteomics, expression proteomics, or analytical proteomics. Other aspects such as functional proteomics are not addressed in this chapter.

Plasma is one of the most important biofluids for proteomics analysis. As the term implies, a biofluid is a fluid associated with life. For humans, the term *biofluid* covers the common fluids such as blood (and sweat and tears), plasma, serum, urine, saliva, milk, cerebrospinal fluid (CSF), seminal plasma, and less common ones such as synovial fluid, blister exudate, and nipple aspirate. The term is less likely to be employed to cover certain fluids produced by medical intervention where an exogenous fluid is introduced such as bronchial lavage, hemodialysate, and so forth. In all there are about 40 fluids that may be obtained from the human body; they all contain at least some protein; and some are very rich indeed in protein (e.g., plasma and serum). Biofluids of all kinds are useful sources of proteins associated with health and disease (Kennedy, 2001). Of these fluids, plasma and serum are undoubtedly the ones most generally examined, and clinical chemists around the world perform millions of analyses each day to quantitate proteinaceous disease markers (antibodies, metabolic enzymes, troponins, C-reactive protein, D-dimer, prostate-specific antigen, carcinoembryonic antigen, etc.) in addition to electrolytes and the usual small metabolites.

Plasma is obtained by centrifugation of blood to remove the red cells, and filters are used to remove the white cells; fibrinogen is present in plasma. Serum is obtained through the natural process of coagulation of blood during which the red cells become trapped in an insoluble mass of fibrin, and fibrinogen releases quantities of soluble fibrinopeptides (Anderson and Anderson, 2002). The coagulation process, which is a major proteolytic event, is difficult to control. Also, the clot may release proteins into the serum. So while serum is a more stable fluid than plasma, and serum banks are widely available at many medical facilities, serum may be considered as the debris left after the bomb (proteolysis-driven coagulation) has gone off. Plasma, on the other hand, while less stable (it is ready to "detonate"), is closer to blood from which the red and white cells have been removed. It is thus not surprising that the Human Proteome Organization *www.hupo.org* has as a major objective the proteomics of plasma rather than serum.

Plasma presents a formidable challenge to proteomics analysis due to the enormous range of protein concentrations (Fig. 8.1). This range begins at almost millimolar levels for albumin and reaches down to low femtomolar levels for proteins such as TNF (tumor necrosis factor), and lower still for "leakage" proteins—ones present due to dying cells that release their contents into the circulation, and so are potentially useful as markers of disease. As indicated in Figure 8.1, the more abundant proteins (labeled common) are generally less interesting (except when differentially expressed in disease and control) than those present at lower concentrations (nanomolar and below, region of interest in Fig. 8.1). Of course, proteins present at subpicomolar levels are also very interesting but are difficult to identify in a discovery experiment at the subfemtomole per milliliter level, which represents current state of the art (Shen et al., 2004). As we shall see, by using larger volumes industrial-scale proteomics addresses this issue of low concentration proteins.

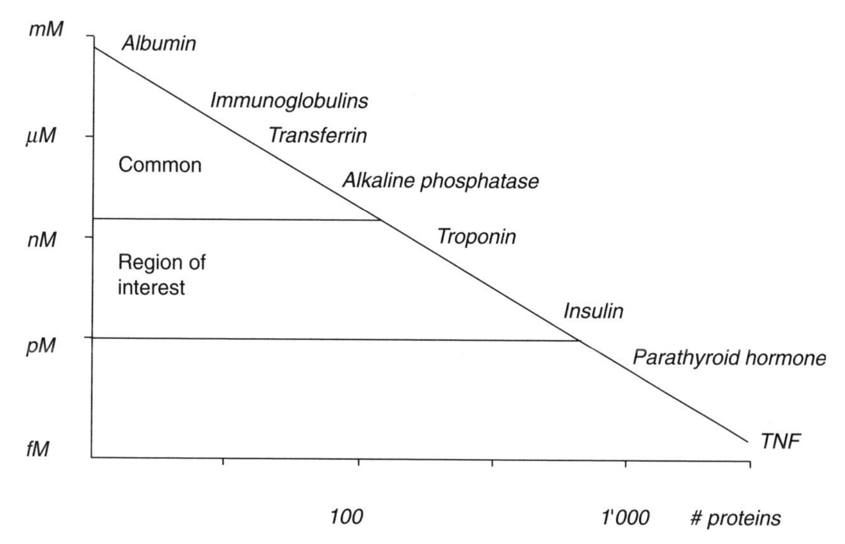

Figure 8.1. Abundance and dynamic range of plasma proteins (taken from Rose et al., 2004). Plasma concentrations of known proteins are shown in a log-log relationship with the number of protein species expected at the various concentration ranges. The dynamic range of known protein species spans at least 11 to 12 orders of magnitude.

Analysis is complicated by the presence of potential proteolytic activity, the very wide range of physicochemical properties of proteins (size, charge, hydrophobicity), the fact that some proteins self-associate or associate with other proteins or with lipids, and the sheer number of protein species present (Anderson and Anderson, 2002; Adkins et al., 2002; Anderson et al., 2004). Due to the very low concentration of some proteins, it is necessary to use large volumes of plasma in order to have sufficient quantities of such proteins to identify them in an exploratory proteomics experiment. Hence the need for an industrial-scale approach at the initial discovery stage of a proteomics investigation (Rose, 2003; Rose et al., 2004). Of course, once new and interesting proteins (or groups of proteins) have been discovered, it is possible and indeed preferable to develop rapid, low-cost and highly sensitive analytical assays for these analytes, and array techniques are being developed for this purpose.

DEFINITION OF INDUSTRIAL SCALE

An industrial-scale approach (Rose, 2003) has the goal of as comprehensive an analysis as possible. In order to identify proteins of lower abundance, relatively large sample volumes are required (Fig. 8.1), and in view of the effort (time and resources) required to exhaustively analyze large volumes, only a small number of samples can be analyzed in such depth. When analyzing liter volumes of plasma, the number of samples may be only one or two—for example, disease and control, treated and nontreated, young and old. This raises the question of sample pooling, discussed below, and of reproducibility.

In contrast to industrial-scale as just defined, a high-throughput approach has the goal to rapidly analyze a large number of individual samples for relatively few features of interest, although protein array techniques are increasing the number of proteins that may be analyzed at a time. This high-throughput approach, which normally employs much automation, has sometimes also been referred to as "large scale" or "industrial scale," but for the purposes of this chapter, the term *industrial scale* refers to the larger volume, proteome-wide approach. At GeneProt, for example, we study samples of <1, 10, 500, and even 2500 mL. While there is no difficulty to obtain 10 mL or even 100 mL of plasma from a single adult individual, the larger volumes are generated through a pooling step (see below).

SAMPLE SELECTION, COLLECTION, AND OPTIONAL POOLING

While it is valuable, industrial-scale proteomics analysis is time-consuming and expensive, so sample selection is crucial in order to extract maximum value from the study. Normally, two states are chosen for comparison: healthy versus diseased, treated versus nontreated, young versus old, and so forth. Through the pooling of carefully selected samples, spurious differences due to variation among individuals may be diluted and averaged out while preserving the more consistent and therefore relevant differences. Pooling also has the advantage of providing a larger sample for industrial-scale analysis, especially important when infant subjects are to be studied, or when studying diseases where very little blood can be drawn without compromising the patient. It is equally important to obtain control samples matched for age, sex, ethnic origin, and other parameters (smoking, alcohol consumption, medication including automedication, and many others). After selection based on these criteria, extensive clinical chemistry and protein electrophoresis should be performed on all individual samples, from the disease set and the control set, to exclude co-morbidity (people suffering from other diseases besides the one that is the subject of the study). Particularly in the case of elderly subjects, as many as 30 percent of people presenting as healthy controls may in fact be suffering from an unsuspected disorder. Quite often, the control subjects are not true matched healthy controls recruited from the general population; they tend to be people who consulted for an unrelated disorder. Careful clinical screening prior to pooling is thus essential and also offers an opportunity to exclude samples from seropositive individuals [hepatitis B and C, human immunodeficiency virus (HIV)], if it is thought this could affect results or for safety reasons. It should not be forgotten that large volumes are processed and that pathogens may become concentrated in certain fractions. In the case of plasma, as opposed to serum, which may be available from a serum bank, it is normally necessary to arrange for specific collection in the context of a prospective study. Such a study requires careful elaboration of a protocol, including inclusion and exclusion criteria, for submission to the Institutional Review Board (Ethics Committee) in the case of human subjects. The protocol should include very detailed specifications on the collection devices and handling procedures to be employed, including sample transport and shipping. The information is stored in a Laboratory Information Management System (LIMS), and all samples are bar-coded (Allet et al., 2004). Procedures must be in place to preserve anonymity of sample donors. The better controlled the collection process the higher the quality of the results of the study. In order to dilute out the small individual differences that still occur even when the above precautions are taken, it is recommended that plasma samples from at least 50 subjects be pooled. In our work at GeneProt we have on occasions pooled samples from more than 2000 individuals, but 50 is more usual. In a time-dependent study, where samples are taken from the same individuals prior to and after an event (e.g., treatment), there is less difficulty from individual variation of course. It is not generally practical to locate a sufficient number of consenting pairs of identical twins of which only one member has the disease, but where possible this would be an ideal situation.

DEPLETION OF ABUNDANT PROTEINS

Albumin is by far the most abundant protein in plasma, being responsible for more than 50 percent of the protein mass in normal individuals. Together with immunoglobulins, transferrin, fibrinogen, complement components, apolipoproteins, and a few other proteins, the top 20 or so proteins are responsible for about 99 percent of the protein mass in plasma (Anderson and Anderson, 2002; Tirumalai et al., 2003). The question thus arises, should plasma be depleted in abundant proteins in order to facilitate the analysis of the rarer ones, and if so, how? The answer has usually been affirmative, but there are recent signs of controversy (Petricoin and Liotta, 2004; Tirumalai et al., 2003). Tirumalai et al. (2003) took pains to try to avoid loss of small proteins that may have been bound to larger ones by having a denaturing agent present during ultrafiltration to recover the small proteins. For depletion, adsorption to dye columns (Ahmed et al., 2003), antibodies (Wang et al., 2003), and immobilized peptide ligands (see Shaw and Riederer, 2003) have been used, and these techniques are scalable. There have been attempts to remove the abundant proteins (which are generally >30 kDa) by ultrafiltration, also a scalable technique, successfully (Tirumalai et al., 2003) or otherwise (Georgiou et al., 2001). A device known as the Gradiflow has also been described for this purpose (Rothemund et al., 2003). Commercial kits are available for the depletion through adsorption of several proteins, such as the kits proposed by Agilent, Applied BioSystems, Bio-Rad, Sigma-Aldrich, and others. Some of the kits are very expensive, and, while very useful for small sample volumes, they are not practical for industrial-scale purposes. To remove albumin, we successfully use a laboratory prototype gel based on a peptidic compound linked to an agarose matrix from Amersham Biosciences. This high-capacity medium is very appropriate for industrial-scale use, and it may be recycled hundreds of times with complete retention of properties; besides albumin and certain fragments of albumin, some traces of hydrophobic proteins (apolipoproteins) were retained (Rose et al., 2004).

SMALLER PROTEINS VERSUS LARGER ONES

Once the question of removal or not of abundant proteins has been settled, the question arises whether to study the smaller proteins or the larger ones, or both. Ultrafiltration has been used for size separation of proteins of plasma (see above), and this technique is used on an industrial scale for other applications (food processing, bioprocess engineering). Other techniques include gel filtration (Rose et al., 2004) and chromatography on restricted access media, although these latter are mostly used for extracting small-molecule drugs rather than small proteins from plasma. Gel filtration can be performed on a very large scale, and inclusion of urea or acetonitrile in the elution buffer tends to avoid aggregation and does not compromise an ion exchange step that may follow. Nonetheless, given the small ratios of loading volume to bed volume necessary to preserve resolution, gel filtration is time and resource consuming for industrial-scale proteomics of plasma, and there is a risk (at least theoretical) of losses by adsorption to media. Rose et al. (2004) describe an industrial-scale process for the analysis of smaller proteins of liter volumes of plasma comprising depletion, gel filtration, ion exchange chromatography, and two steps of reverse-phase chromatography.

GEL-BASED PROTEIN SEPARATION

In the case of larger proteins, a decision must be taken concerning the separation approach: two dimensional (2D) or not 2D, that is the question. Gel electrophoresis has a loading limit of normally less than about 1 mg per gel if resolution is to be preserved. Large-scale operations with plasma can easily provide 20 g of protein per liter of plasma even after depletion of albumin and IgG. Only 500 mg of these 20 g is of low molecular weight, so with 19.5 g of higher molecular weight protein, direct gel separations are not practical for proteome-wide industrial-scale proteomics at this scale! While even 10 mL plasma is far in excess of what can be applied to a gel, after depletion and fractionation it becomes possible to run a number of gels and exploit the unique separating power of the 2D technique. When enough sample is available, parallel separations based on different enrichment strategies such as group-selective extractions (glycoproteome, antibody-based phosphoproteome, metal chelate chromatography, etc.), or free-flow electrophoresis or other separation techniques can be performed, and the resulting fractions applied to gels. No results of very large scale sample processing for gel analysis have been published yet, but see Pieper et al., 2003. While 2D gels offer a very useful image of the proteins present, even before identification, there are several difficulties associated with 2D gel-based separations besides the technical difficulty of obtaining reproducible results: proteins with extreme isoelectric point, those that are very hydrophobic, and those of very low molecular weight. The reproducibility issue has been addressed through the use of the difference gel electrophoresis technique (Van den Bergh and Arcens, 2004), and the throughput issue is being addressed by approaches such as the molecular scanner (Binz et al., 2004).

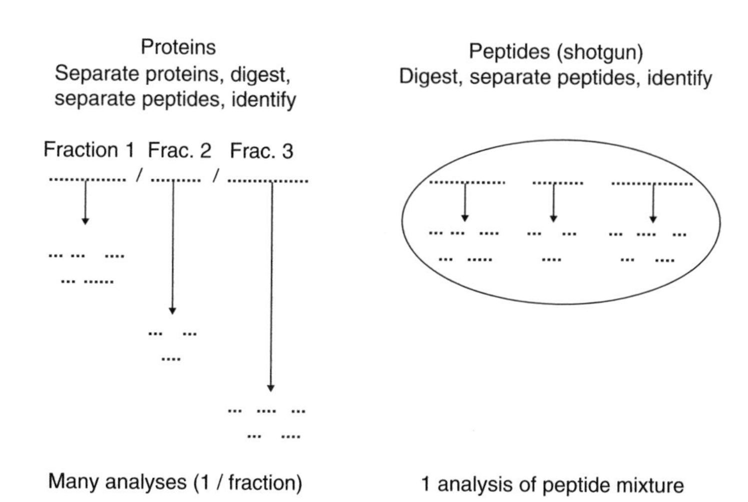

Figure 8.2. Scheme depicting the protein separation approach (separation of intact proteins prior to digestion, *left*) and the peptide separation approach (digestion of the protein mixture and separation only at the peptide level, *right*).

SEPARATION OF INTACT PROTEINS VERSUS SEPARATION OF PEPTIDES

Equally important as the decision on whether to use 2D gels for the larger proteins is the general question of whether to separate the proteins as intact species and then follow by digestion to facilitate identification or to perform digestion up-front and follow by separation of the peptides in order to identify the proteins present.

Figure 8.2 depicts these two approaches. On the left the preseparation of proteins is depicted. The rows of dots separated by slashes represent separated protein fractions (of course, in reality there are many more than 3 fractions, and there are generally many proteins in each fraction not just the one shown here). Each fraction is then digested (downward-pointing arrows) to peptides, and each digested fraction is analyzed individually (100 fractions necessitate 100 analyses, 1000 fractions necessitate 1000 analyses).

Depicted on the right of Figure 8.2 is the so-called shotgun approach, also associated with the term MudPIT (multidimensional protein identification technology, Wolters et al., 2001). In this approach, the proteins in the plasma sample are digested without prior fractionation of the proteins (except sometimes for depletion or isolation of a low-molecular-weight fraction; see Tirumalai et al., 2003). The resulting very complex mixture of peptides is then fractionated, usually by combinations of ion exchange chromatography and reversed-phase high-performance liquid chromatography (HPLC). Both of these approaches shown in Figure 8.2 may in principle be applied at an industrial scale. As currently practised, the MudPIT approach is not applied at an industrial scale to plasma or serum analysis, although multidimensional chromatography of peptides (those present naturally in a fluid) has been applied to very large volumes of hemodialysate (see, e.g., Schulz-Knappe et al., 1997).

The advantages of the shotgun analysis approach are speed (as there are fewer steps), lower sample consumption (as losses are lower), relatively easy separations (separating peptides rather than proteins), and more suitable for higher throughput. It should be noted that some proteins with extreme properties (e.g., extremely hydrophobic or highly charged) might be lost during the protein separation approach, and loss of the protein entails a missed identification. However, it is possible to imagine that if the alternative (digestion and peptide separation) approach is adopted, there might exist some peptides with more normal properties that would not be lost during separation and so would lead to a positive identification of the parent protein. The main disadvantage of the shotgun approach is the loss of the correlation between a protein fraction and an identification.

This is important since a given peptide may (and normally does) come from several protein species present in the sample. These protein species may be separated, and often are in the case of 2D gel approaches, but are not separated in the shotgun approach. This state of affairs is represented in Figure 8.3, which depicts three forms of a given protein: a long form, a short form, and a modified form represented by the small circle. A given peptide present in all three forms of the protein is represented by a row of x's. On the left panel of Figure 8.3, which applies to a 2D gel approach or to a chromatographic separation approach, the three forms of the protein are shown as appearing in separate fractions (this will not always be the case and will depend on the degree of fractionation at the protein level that is employed). Let us suppose that the given peptide is present in all three forms of the protein (represented by the row of x's) and is identified in each of the three fractions. It is most important to realize that one (at least) of the three forms of the protein represented may be present in different relative amounts in the case of disease and control samples, and this differential expression may be detected through the protein separation approach shown on the left panel of Figure 8.3, even if globally (when all three forms are considered) there may be little or no difference in expression level. If the differential expression can be associated to the truncated form of a posttranslationally modified form, this can be very helpful for the biological interpretation of the results.

In contrast, the right panel of Figure 8.3 depicts the shotgun approach, where the three forms of the protein are not separated and are digested (along with the other proteins present of course, not shown for clarity). The peptide represented by the row of x's is thus just identified, but its presence cannot be associated with the three forms of the protein represented. In the case of differential expression between disease and control samples of a minor form of the protein, this would not be easy to detect.

Figure 8.4 shows the separation schemes we have used at GeneProt. Where possible, corresponding steps are aligned horizontally in the figure. The 1-mL process involves separation of peptides, whereas the larger, industrial-scale processes involve separation of proteins prior to digestion. The numbers of final fractions are indicated, and the number of small proteins found. It is important to note that the 2500-mL study was performed with Esquire 3000 ion traps, whereas the smaller studies were performed using the upgraded 3000+ machines (hence the + on the top line of Figure 8.4).

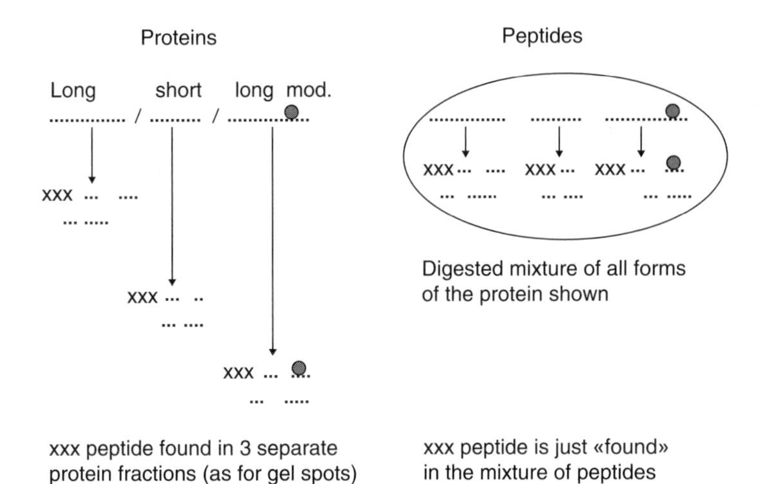

Figure 8.3. Differential expression of a particular form of a protein (the protein is identified through the peptide xxx) is in principle detectable in the protein separation approach (*left*) but only with difficulty in the peptide separation approach (*right*).

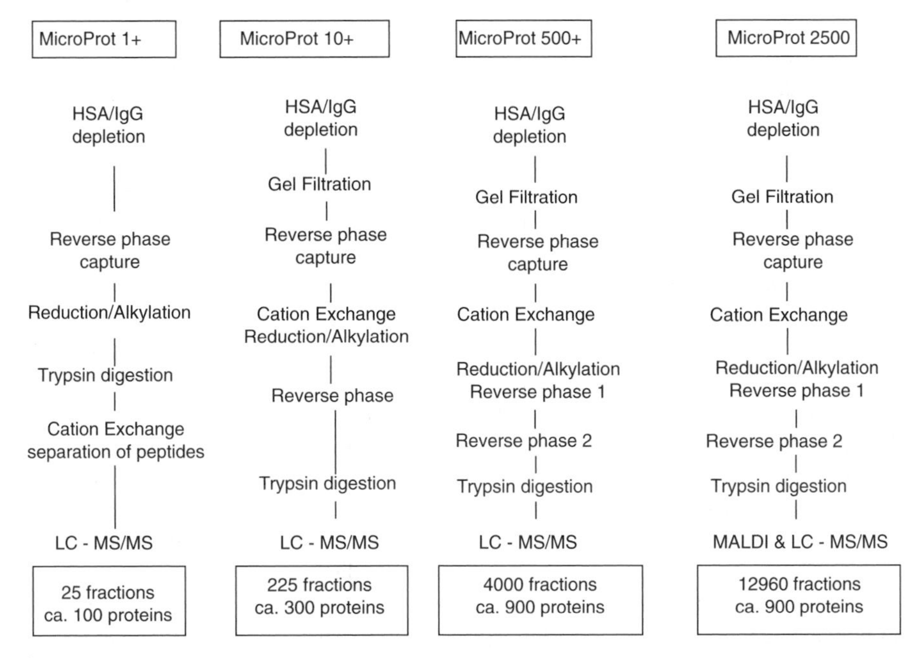

Figure 8.4. Separation schemes for large and small volumes of plasma or serum. Where possible, corresponding steps are aligned horizontally. The 1-mL process involves separation of peptides, whereas the larger scale processes involve prior separation of proteins.

MASS SPECTROMETRY

There are two main ionization methods used in proteomics (see Chapter 1): matrix-assisted laser desorption ionization (MALDI) mass spectrometry and electrospray ionization (ESI) mass spectrometry. Both lend themselves well to industrial-scale proteomics. Besides ionization mode, choice of mass analyzer is critical. The most widely used ones for proteomics analysis are: ion trap, time-of-flight (TOF), and various hybrid analzers such as the QTOF and the tandem TOF-TOF analyzers. Some groups (financially well-endowed) work with Fourier transform ion cyclotron resonance machines (FT-ICR; Page et al., 2004). The high-resolution, mass precision, sensitivity, and dynamic range of FT-ICR instruments is unsurpassed, and they may be used for the so-called top-down approach, where intact proteins are fragmented within the mass spectrometer. This enables a great deal of information to be obtained from a small sample volume (Page et al., 2004). So far, FT-ICR machines have not been used for industrial-scale work. Both major ionization methods (MALDI and ESI) have been interfaced to the analyzers mentioned above. This is not the place to discuss the relative merits of particular combinations. Nevertheless, it should be noted that all are relatively costly, ranging (depending on configuration, software, interfaces such as micro-HPLC) from a few hundred thousand dollars to a million or so. This difference becomes very significant when multiple instruments are required for industrial-scale or high-throughput use. While ESI was used mainly to obtain tandem mass spectrometry data (MS/MS) for sequence-specific fragmentation and MALDI used mainly for intact mass measurements of peptides, the newer MALDI TOF-TOF instruments (available from Applied Biosystems and Bruker Daltonics) also offer sequence-specific fragmentation. This feature is especially useful for the molecular scanner (Binz et al., 2004).

For the industrial-scale analysis of intact smaller proteins, mass spectrometry is generally performed online by LC-ESI-MS and offline by MALDI-TOF-MS or MALDI-TOF-TOF-MS. Liquid chromatography (LC) MS/MS may also be performed offline using devices such as the Nanomate from Advion. It is not easy to extract mass information on intact components from fractions containing many protein components, particularly online during a chromatographic separation process. There is a narrow time window (elution of the chromatographic peak), and the multicharging inherent to ESI complicates the signal obtained. MALDI analysis is not affected by these latter two difficulties but is affected by suppression phenomena and dynamic range. While all these difficulties may be addressed by techniques such as peak parking, nanospray, reversed-phase preconcentration, and the like, these are not always applicable to 24/7 operation simultaneously on multiple machines in a robust and reproducible manner. A main advantage of offline analysis, and which is practical for industrial-scale work, is that portions of sample may be retained for follow-up studies or repeat analysis.

After digestion, fragments are analyzed by LC-ESI-MS/MS and MALDI-TOF-MS and may also be analyzed by MALDI-TOF-TOF mass spectrometry. Since the species being analyzed are peptides rather than proteins, separation and analysis is easier, but there are of course many more peptides than proteins. In the case of LC-ESI-MS/MS or LC-MALDI-TOF (or TOF-TOF) analysis, the additional LC step helps to compensate for the increase in number of species for analysis. During the industrial-scale analysis of liters of plasma (Rose et al., 2004), more information was obtained from LC-ESI-MS/MS than from MALDI, but there were nevertheless peptides that were identified by MALDI that were not identified by ESI-MS/MS. These tended to be the earlier-eluting (hydrophilic) peptides that were obscured by the front during reversed-phase chromatography.

Quantitation (reviewed by Sechi and Oda, 2003) and posttranslational modifications (reviewed by Mann and Jensen, 2003) are very important but have not yet been addressed for human plasma on an industrial scale in a publication.

BIOINFORMATICS

As important as the mass spectrometric hardware is the software. This is particularly true in the case of industrial-scale operation. When there is an experienced operator behind each of the one or two machines in a smaller laboratory, manual intervention at the acquisition stage and the data processing stage can compensate for imperfections with the software. For industrial-scale operation, this is not efficient. An operator should check, calibrate, and launch a set of machines that then run for the rest of the day and all through the night, with samples being automatically injected from a cooled tray.

Data acquisition software is supplied by the manufacturer, but this can be improved (e.g., peak detection algorithms) to extract more data and more reliable data from the signals being generated. Once the raw data have been acquired, postprocessing software is applied, and here again it can be improved (e.g., charge assignments). The identification software sold with commercial instruments is usually based on the MASCOT identification engine (Matrix Sciences) or on SEQUEST (Thermo Finnigan), but several others are available. This identification software is used to compare the MS/MS spectra acquired with those expected from computer analysis of databases containing either protein sequence data (SwissProt), EST sequence data, or Genomic sequence data (Human Genome Project data or Celera data). At GeneProt, after manually checking well over 300,000 MS/MS spectra we developed a new algorithm, OLAV (Colinge et al., 2003) and applied it to a set of 1.6 million such spectra generated during industrial-scale analysis of human plasma (Rose et al., 2004). Postprocessing followed to enhance the quality of the scoring of each identification and an integration step then checked the consistency of the global identification across the different databases used. Annotation (automatic and manual) was performed on the validated identifications to emphasize important features and further characterize the observed proteins or fragments of proteins. The hardware required to operate in industrial-scale mode (processing and online storage) is quite impressive since terabytes of data are produced during analysis.

CONCLUDING REMARKS

Industrial-scale proteomics involves large sample volumes and so requires extensive separation prior to final analysis by mass spectrometry. As mass spectrometers and associated equipment and software become faster and more sensitive, it will become possible to obtain more and more information from a given sample volume. Our own work with the small proteins of plasma and serum has shown that similar results may be obtained with the Esquire 3000+ ion trap (Bruker Daltonics) from 500 mL as we obtained with the (less sensitive) Esquire 3000 machine from 2500 mL. The new model HCT is more sensitive still, and similar improvements are being made continuously also by the other manufacturers. Operation at an industrial scale will always offer more information, but at a cost in sample volume (important issue for, e.g., tears), time, and expense. The working scale is chosen to balance these factors.

Acknowledgments

Colleagues at GeneProt, Geneva, Professor J.-D. Tissot, CHUV, Lausanne, and Dr. Martine Michel of the Blood Transfusion Service of Geneva Cantonal Hospital, and all plasma donors.

REFERENCES

Adkins, J.N., Varnum, S.M., Auberry, K.J., Moore, R.J., Angell, N.H., Smith, R.D., Springer, D.L., and Pounds, J.G. (2002). Towards a human blood serum proteome: Analysis by multidimensional separation coupled with mass spectrometry. *Mol. Cell. Proteomics* **1**:947–955.

Ahmed, N., Barker, G., Oliva, K., Garfin, D., Talmadge, K., Georgiou, H., Quinn, M., and Rice, G. (2003). An approach to remove albumin for the proteomic analysis of low abunbance biomarkers in human serum. *Proteomics* **3**:1980–1987.

Allet, N., Barillat, N., Baussant, T., Boiteau, C., Botti, P., Bougueleret, L., Budin, N., Canet, C., Carraud, S., Chiappe, D., Christmann, N., Colinge, J., Cusin, I., Dafflon, N., Depresle, B., Fasso, I., Frauchiger, P., Gaertner, H., Gleizes, A., Gonzalez-Couto, E., Jeandenans, C., Karmime, A., Kowall, T., Lagache, S., Mahé, E., Masselot, A., Mattou, H., Moniatte, M., Niknejad, A., Paolini, M., Perret, F., Pinaud, N., Ranno, F., Raimondi, S., Reffas, S., Regamey, P.-O., Rey, P.-A., Rodriguez-Tomé, P., Rose, K., Rosselat, G., Saudrais, C., Schmidt, C., Villain, M., and Zwahlen, C. (2004). *In vitro* and *in silico* processes to identify differentially expressed proteins. *Proteomics* **4**(8):2333–2351.

Anderson, N.L., and Anderson, N.G. (2002). The human plasma proteome: History, character, and diagnostic prospects. *Mol. Cell. Proteomics* **1**:845–867.

Anderson, N.L., Polanski, M., Pieper, R., Gatlin, T., Tirumalai, R.S., Conrads, T.P., Veenstra, T.D., Adkins, J.N., Pounds, J.G., Fagan, R., and Lobley, A. (2004). The human plasma proteome: A non-redundant list developed by a combination of four separate sources. *Mol. Cell. Proteomics* **3**(4):311–326.

Binz, P.-A., Müller, M., Hoogland, C., Zimmermann, C., Pasquarello, C., Corthals, G., Sanchez, J.-C., Hochstrasser. D.F., and Appel, R. (2004). The molecular scanner: Concept and developments. *Curr. Opin. Biotechnol.* **15**(1):17–23.

Colinge, J., Masselot, A., Giron, M., Dessigny, T., and Magnin, J. (2003). OLAV: Towards high-throughput tandem mass spectrometry data identification. *Proteomics* **3**:1454–1463.

Georgiou, H.M., Rice, G.E., and Baker, M.S. (2001). Proteomic analysis of human plasma: Failure of centrifugal ultrafiltration to remove albumin and other high molecular weight proteins. *Proteomics* **1**:1503–1506.

Kennedy, S. (2001). Proteomic profiling from human samples: The body fluid alternative. *Toxicol. Lett.* **120**:379–384.

Mann, M., and Jensen, O.N. (2003). Proteomic analysis of post-translational modifications. *Nature Biotechnol.* **21**:255–261.

Page, J., Masselon, C.D., and Smith, R.D. (2004). FTICR mass spectrometry for qualitative and quantitative bioanalyses. *Curr. Opin. Biotechnol.* **15**(1):3–11.

Petricoin, E.F., and Liotta, L.A. (2004). SELDI-TOF based serum proteome pattern diagnostics for early detection of cancer. *Curr. Opin. Biotechnol.* **15**(1):24–30.

Pieper, R., Gatlin, C.L., Makusky, A.J., Russo, P.S., Schatz, C.R., Miller, S.S., Su, Q., McGrath, A.M., Estock, M.A., Parmar, P.P., Zhao, M., Huang, S.-T., Zhao, J., Wang, F., Esquer-Blasco, R., Anderson, N.L., Taylor, J., and Steiner, S. (2003). The human serum proteome: Display of nearly 3700 chromatographically separated protein spots on two-dimensional electrophoresis gels and identification of 325 distinct proteins. *Proteomics* **3**:1345–1364.

Rose, K. (2003). Industrialization of proteomics: Scaling up of proteomics processes. In: Cooper, D.N. (Ed.). *Encyclopedia of the Human Genome*. Macmillan, London, Vol. 3, pp. 435–439.

Rose, K., Bougueleret, L., Baussant, T., Böhm, G., Botti, P., Colinge, J., Cusin, I., Gaertner, H., Gleizes, A., Heller, M., Jimenez, S., Johnson, A., Kussmann, M., Menin, L., Menzel, C., Ranno, F., Rodriguez-Tomé, P., Rogers, J., Saudrais, C., Villain, M., Wetmore, D., Bairoch, A., and Hochstrasser, D. (2004). Industrial-scale proteomics: From litres of plasma to chemically synthesized proteins. *Proteomics* **4**(7):2125–2150.

Rothemund, D.L., Locke, V.L., Liew, A., Thomas, T.M., Wasinger, V., and Rylatt, D.B. (2003). Depletion of the highly abundant protein albumin from human plasma using the Gradiflow. *Proteomics* **3**:279–287.

Schulz-Knappe, P., Schrader, M., Ständker, L., Richter, R., Hess, R., Jürgens, M., Forssmann, W.-G. (1997). Peptide bank generated by large-scale preparation of circulating human peptides. *J Chromatogr. A* **86**:213–217.

Sechi, S., and Oda, Y. (2003). Quantitative proteomics using mass spectromstry. *Curr. Opin. Chemical Biol.* **7**:70–77.

Shaw, M.M., and Riederer, B.M. (2003). Sample preparation for two-dimensional gel electrophoresis. *Proteomics* **3**:1408–1417.

Shen, Y., Jacobs, J.M., Camp, D.G., Fang, R., Moore, R.J., and Smith, R.D. (2004). Ultra-high-efficiency strong cation exchange LC/RPLC/MS/MS for high dynamic range characterization of the human plasma proteome. *Anal. Chem.* **76**(4):1134–1144.

Tirumalai, R.S., Chan, K.C., DaRue, A.P., Issaq, H.J., Conrads, T.P., and Veenstra, T.D. (2003). Characterization of the low molecular weight human serum proteome. *Mol. Cell. Proteomics* **2**:1096–1103.

Van den Bergh, G., and Arcens, L. (2004). Fluorescent two-dimensional difference gel electrophoresis unveils the potential of gel-based proteomics *Curr. Opin. Biotechnol.* **15**(1):38–43.

Wang, Y.Y., Cheng, C.P., and Chan, D.W. (2003). A simple affinity spin tube filter method for removing high-abundant common proteins or enriching low-abundant biomarkers for serum proteomic analysis. *Proteomics* **3**:243–248.

Wolters, D.A., Washburn, M.P., and Yates, J.R. (2001). An automated multidimensional protein identification technology for shotgun proteomics. *Anal. Chem.* **73**:5683–5690.

9

CHEMICAL GENOMICS: TARGETS ON DISPLAY

Steve Doberstein

Five Prime Therapeutics, Inc., S. San Francisco, California

Philip W. Hammond and René S. Hubert

Xencor, Monrovia, California

INTRODUCTION: CHEMICAL GENETICS AND CHEMICAL GENOMICS

For most of human history, medically important compounds have been identified primarily by direct observation of the effects that complex mixtures of natural products have on human disease. Both Eastern and Western medical traditions have relied on nature to supply the lion's share of clinically active compounds. The isolation of those compounds from their natural sources became the task of the pharmaceutical chemistry industry in the nineteenth and early twentieth centuries, followed by the development of commercially viable synthetic routes for their in vitro production. This

Industrial Proteomics: Applications for Biotechnology and Pharmaceuticals, edited by Daniel Figeys
ISBN 0-471-45714-0

tradition is exemplified by the development of aspirin as a human therapeutic. Aspirin has its roots in the use of birch bark as an analgesic, from which salicylic acid was first purified in 1828. This was followed by the development of a synthetic route to acetylsalicylic acid in 1869. The molecular targets of aspirin (cyclooxygenases) were not discovered until the 1970s, more than a century later. The determination of this class of proteins as the primary targets allowed the development of a second generation of more specific cyclooxygenases-2 (COX-2) inhibitors that bypass some of the side effects of aspirin.

Many more drugs were discovered and developed based on their phenotypic effects in simple animal and cell-based models of disease as well as during clinical evaluation in humans. Nearly all antibiotics were discovered based on their ability to elicit a simple phenotype, death, when applied to bacterial cultures. The thiazolidinediones were discovered in screens for compounds that would lower blood glucose in rabbits. Most neuroactive compounds have been identified in animal-based screens due to the lack of sufficient mechanistic understanding of important diseases such as schizophrenia and depression. More recently, sildenafil citrate (Viagra) was developed as a treatment for erectile dysfunction after researchers noted its physiological effects on male patients during its testing as a therapy for cardiovascular disease.

However, since the discovery of the statins in the early 1970s, much pharmaceutical research and development has centered on the screening of diverse libraries of small molecules against specific protein targets selected on the basis of a therapeutic hypothesis tying the protein to the disease state. Although target-based drug discovery has proven versatile and broadly applicable, shortcomings in our understanding of human molecular physiology are demonstrated by the relatively high rate of projects that fail in the clinic due to inappropriate target selection (Drews, 1999). The process of target validation, in which the biology of the proposed therapeutic target is matched to the pathophysiology of the disease state, has lagged behind the tools of high-throughput screening and medicinal chemistry. This challenge is brought to the forefront by the complete sequencing of the human genome. While all possible protein drug targets can now be enumerated, resulting in a "target glut," clearly designating a small set of specific proteins as valid targets in disease pathogenesis remains a major challenge.

Difficulties encountered in target validation have led to the development of a screening process that combines features from both phenotype-based and target-based paradigms. By applying the tools of modern cell and molecular biology, high-throughput cell-based screening, and combinatorial chemistry, the best attributes of both phenotype- and target-based approaches can be applied to drug discovery. Chemical genetics, in which very large-scale phenotype-based screening of compound libraries is applied to sophisticated cell-based models of disease, has emerged as a powerful new approach to drug discovery. The use of phenotype-based screens raises confidence that the target being modulated is therapeutically relevant, and is "druggable" (amenable to inhibition or activation by a small molecule) in the context of the living cell. In addition, high-throughput screening of large compound libraries results in the early identification of relatively potent compounds, consequently requiring less refinement in the optimization phase.

However, the full integration of these techniques into drug discovery requires that the protein target of the active compound be identified. Knowledge of the protein target allows the full armamentarium of medicinal chemistry to be applied to lead optimization, resulting in development of the safest, most potent, specific, and bioavailable drug possible. The process of determining the protein target has variously been called *chemical genomics* and *chemogenomics* with several different and overlapping meanings. For the purposes of this review, we use the following definition. Given an observable phenotype of a cell or organism elicited by exposure to a chemical compound, chemical genomics is the process of assigning that phenotype to the inhibition or activation of a specific protein.

The full potential of high-throughput screening of small molecules for pharmaceutical effect will not be realized until the supporting field of chemical genomics matures. Substantial progress has been made toward achieving this goal with the development of several new technologies. In this chapter, we briefly review various methods in chemical genomics with a focus on protein display technologies, a powerful set of tools in which the functionality of proteins is coupled with the ease of manipulation and sensitive detection of deoxyribonucleic acid (DNA) replicons.

CHEMICAL GENOMICS TECHNIQUES

General Introduction

Many methods that have been developed to identify the molecular targets of small molecules have yielded positive results, yet none appears to be totally generalizable. By and large, the current methods of chemical genomic target identification fall into two categories: (a) in vivo and cell-based methods driven by changes in phenotype and (b) those driven by the tools of biochemistry and molecular biology based on detecting binding of the compound directly to its protein target. Although this chapter is focused on the latter, we will begin with a brief overview of the former.

Phenotype-Based Chemical Genomics Methods

In vivo model organism genetics has been proposed as a generalizable and rapid tool for ligand/target identification (Matthews and Kopczynski, 2001; Nislow and Giaever, 2003; Margolis and Duyk, 1998). *Drosophila*, *C. elegans*, and *Saccharomyces* have all been used successfully to identify the mechanism of action of small-molecule ligands. Typically, the researcher identifies the effects generated by administration of a compound to normal, wild-type animals, then performs genetic screens for mutations that cause resistance or hypersensitivity to that effect. Among the genes thus identified should be the target of the compound, as well as components of any signaling pathway involved. While this approach has in the past been time-consuming and labor-intensive, the advent of large-scale mutant collections and genome-wide messenger ribonucleic acid (mRNA) knockdown methods (short interfering (si)RNA, antisense, and ribozyme libraries; Miyagishi et al., 2003) has reduced the time required for cloning of the mutant (target) locus.

The model organism approach is of particular utility in cases where the model organism is closely related phylogenetically to the organism of commercial use. For example, yeasts are very well suited for antifungal target identification while *Drosophila* has shown utility in identifying the mechanism of action of insecticides.

Perhaps the greatest drawback of genome-wide model organism approaches for human therapeutics is the difficulty of clearly translating sequence homologs between organisms. Two of the most chemically tractable target classes are the nuclear hormone receptors and the seven-transmembrane G protein carpled receptors (GPCR) families; the compositions of many chemical compound libraries are designed to affect these targets. Therefore, one might expect a large number of those targets to emerge from the model organism approach. However, unambiguously identifying the ortholog of the model organism target within the human genome is nearly impossible in many cases. This issue is minimized in pathways that are very well conserved. For example, homologs in the kinase cascade that controls the eukaryotic cell cycle are relatively straightforward to translate between organisms.

The advent of large-scale mRNA knockdown methods in mammalian cell culture should be helpful in overcoming model genetics shortcomings. These methods reduce the amount of protein made from a specific mRNA in response to an administered oligonucleotide. This allows cells to be grown in the presence of normal protein expression and avoids the pitfalls of toxic mutations or knockouts in a model organism. Another advantage is that the cells used for observation of the desired phenotype can be the same ones used to validate the target. In such experiments the elimination of the target mRNA should either mirror the effect of a small-molecule inhibitor or abrogate the effect of a small-molecule potentiator.

Biochemical and Display Methods

Introduction and Background. Underlying many chemical genomics technologies is the hypothesis that proteins that interact with the active compound can generally be identified using affinity-based binding and capture techniques. Traditionally, in vitro biochemical methods have been applied to the identification of interacting proteins from a cell or tissue lysate. Specific techniques used successfully in the past include chemical cross-linking of a modified compound to interacting proteins, co-purification of targets with radiolabeled ligands, and identification of interacting proteins by affinity chromatography.

These techniques, while useful in some cases, are difficult to generalize to many interactions. First, the interaction needs to be one of high affinity (and low off-rate) so that the target protein can be efficiently purified. Second, sufficient quantities of high-purity material must be obtained in order to identify the protein using peptide sequencing or mass spectrometry. These particular shortcomings may be magnified in a chemical genetics context, as the compounds identified in a primary chemical genetic screen may be of relatively low potency and affinity compared to compounds that have been optimized, either by medicinal chemistry or natural selection.

An alternative approach to target identification involves the use of hybridization arrays to measure the relative levels of different mRNAs in the transcriptome (the full complement of genes being expressed at a given time). Within this complex set of information is likely to be an mRNA encoding the target(s) of the compound of interest. The challenge is identifying which mRNA encodes the target among the many transcripts available in a given cell at a particular time. After exposure to the compound of interest, the relative amounts of many different mRNAs are assayed simultaneously using a hybridization array. A basic assumption is made that exposure will alter the amount of the target protein mRNA as a means of feedback compensation, or that the pathway involved can be deduced from changes in the pattern of mRNA expression. mRNAs whose levels change either up or down are identified as candidates for further investigation. This assumption has not yet been rigorously proven, and it is widely observed that the expression of many nontarget genes changes in response to compound administration. Another limitation is that only targets represented on the chip can be identified. Although this approach has attained widespread acceptance, its generality remains to be determined.

The display technologies provide a functional bridge between the two extremes of mRNA and protein characterization. They all have in common the formation of a durable and selectable linkage between phenotype (protein) and genotype (mRNA or DNA) (Fig. 9.1). Methods include biological linkage wherein a self-replicating unit such as a phage both displays the protein and packages the DNA, chemical linkage in which a covalent bond is formed between an mRNA and its encoded protein (mRNA display), and mechanical linkage as in microbead display.

Figure 9.1. Description of phenotype–genotype linkages used in various display systems. A fusion protein is shown bound to an immobilized drug with the tag (light gray ribbon) symbolically linked to the nucleic acid encoding it.

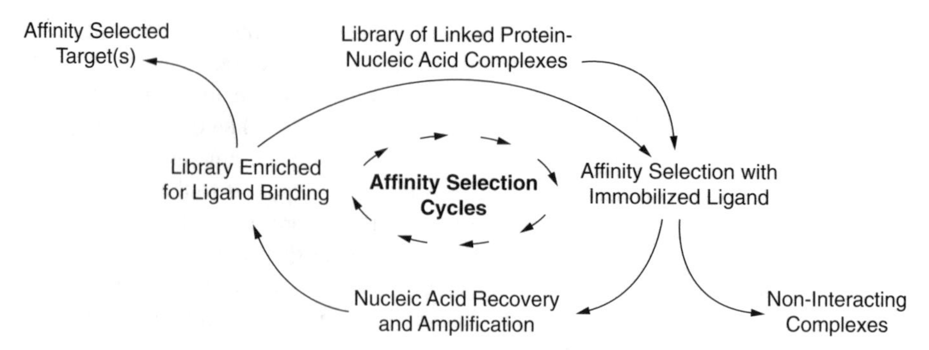

Figure 9.2. Main affinity selection steps in most display technologies.

The display of a single protein is in and of itself primarily a clever trick of molecular engineering. The major utility arises from the ability to produce complex libraries of displayed proteins, each linked to its corresponding nucleic acid sequence. From such a library, a subset with a specific attribute, such as binding to a small molecule, can be isolated, amplified, and identified. This is generally accomplished through several rounds of selection (Fig. 9.2). Briefly, a ligand is immobilized to a conveniently manipulated support such as a bead or a microtiter well. The display library is then incubated with the ligand to allow binding. After subsequent washing to remove unbound library members, the retained members are recovered by elution and replicated to produce an enriched library. The method of replication depends on the type of display. Commonly, a phage or plasmid is amplified by infection or transformation, while in the case of mRNA and ribosome display polymerase chain reaction (PCR) is used. After several rounds of selection, sufficient enrichment is obtained to allow individual binding members of the library to be identified by cloning and sequencing. The number of rounds of selection required will vary depending on the complexity of the initial library and the rate at which enrichment occurs. In the examples below, three to six rounds of selection were required for identification of specific binding proteins.

Display methods were initially developed using libraries of stochastically generated peptides and complex libraries of scaffolded binding proteins such as single-chain antibodies. These libraries have proven very useful in the generation of both research reagents and antibody-based pharmaceuticals. In addition, the application of these libraries to the identification of the target protein of taxol was demonstrated (Rodi et al., 1999). In this case the target protein Bcl-2 was inferred from homology to the sequence of a selected peptide, rather than by a positive identification as is obtained from a complementary DNA (cDNA) library.

The underlying methods for affinity selection have proven extensible to chemical genomics. For this application the normal cellular complement of proteins (rather than random peptides) are displayed. Specifically, methods have been developed for the robust display of cDNA. Although for many applications full-length cDNA is desirable, for display methods this is not necessarily the case. Because most display methods rely on a fusion with another protein through some constant region sequences, the presence of either the native start or stop codon (depending on the display method) can be problematic. Therefore, in general, random priming is the preferred method with the result being a library of protein fragments of varying sizes and reading frames.

A random-primed library will necessarily contain a number of members too small to produce intact folded proteins or independently folding protein domains. Typical small domains such as SH3 and SH2 are 60 to 120 amino acids in length. Size-based purification, such as by gel extraction, may be required to bias this library toward larger proteins, and correspondingly away from short peptides. Therefore, the utility of simple, full-length libraries should not be underestimated.

During the construction of libraries, a conserved region is usually added to allow for subsequent amplification and cloning of the library en masse. One group (Hammond et al., 2001) has described the incorporation of "tags" into the 5′ untranslated region (UTR) of libraries during cDNA synthesis. The resulting libraries can then be mixed and treated as a single population for purposes of selection. Subsequent sequencing of an individual clone will reveal the actual tissue or cell source of the initial template.

General Considerations in Choosing a Display Method. Once a display library has been constructed, it is ready for use in a selection designed to isolate the target(s) of the compound of interest. When choosing a display method, a number of variables should be taken into consideration (Table 9.1).

A high-valency interaction allows for avidity to supplement simple affinity during the selection process. In a case where the compound/protein interaction is presumed to be high affinity, a low-valency method would be preferred. However, if low-affinity interactions are being investigated, or an exhaustive list of interacting proteins is desired, avidity may produce an interaction that is stable to the manipulations of the selection methodology. mRNA display is inherently a monovalent method, whereas phage display can be either high (>100) or low (<1) valency, depending on the phage system selected. In plasmid display there is the potential for exact control of valency by engineering the desired number of binding sites into the display vector.

Another important consideration is library size, with the premise that bigger is better (Gold, 2001). Phage display libraries are effectively limited by the efficiency of bacterial transformation to <109 distinct members. This limitation is overcome by the methods of ribosome display and mRNA display. These methods scale relatively well and therefore allow for the formation of libraries of >1013 members. Larger libraries theoretically allow for the display of all mRNAs present in a single cell (~400,000) as well as all possible N- and C-terminal truncations. The ability to examine truncated versions of proteins is especially useful for those proteins whose mRNAs encode membrane anchors or secretion signals that might interfere with the function of a mature soluble protein. In addition, protein architecture is often modular. Larger proteins may consist of several small, independently folding, domains. Therefore, the display of protein fragments allows the identification of the specific protein domain involved in the interaction with ligand.

Although in vitro methods such as mRNA and ribosome display allow for large libraries, they have a downside in that libraries are amplified by PCR (or reverse transcription (RT)-PCR), which may introduce bias depending on fragment size, GC

TABLE 9.1. Summary of Most Prevalent Chemical Genomics Display Technologies

Display Technology	Library Size	Valency	Linkage Type	Linkage to Protein	Host (P)rokaryote (E)ukaryote	Posttranslational Modifications	Development Stage
Phage	$<10^9$	1 to >100	Viral capsid protein	N-ter	P	No	HTP complex systems
mRNA	$>10^{13}$	1	Covalent	C-ter	E in vitro	No	Complex model systems
Ribosome	$>10^{11}$	1	Noncovalent	C-ter	P/E in vitro	No	Complex model systems
Plasmid	$<10^{11}$	1 to ~10	Noncovalent or covalent	N or C-ter	P/E	Yes	Simple model systems
Bacterial/yeast	10^9–10^8	>100	Cell	N or C-ter	P/E	Yes	Simple model systems
Microbead	10^7	>100	Bead	N or C-ter	P/E in vitro	No	Simple model systems

content, and other physical attributes of the PCR template. Methods such as plasmid and phage display depend on bacterial growth for amplification and so might not be expected to have this issue. However, subtle growth advantages for plasmids with small or no insert have been observed, and these aberrant plasmids can overrun the system in a few cycles of selection (Hubert, unpublished results).

Many proteins require posttranslational modification for their function, and the display technologies have mixed abilities to accommodate requirements for such modifications (Table 9.1). As many as one-third of all human proteins are phosphorylated, many of them at multiple sites. Additional modifications include acetylation, sulfation, myristylation, and glycosylation.

Translation in bacteria would not be expected to generate posttranslational modifications. In one phage display example discussed below (Jin et al., 2002), a non-phosphorylated protein was bound by the ligand used in affinity selection while the phosphorylated version was not. However, should the reverse be true and a binding interaction require phosphorylation, it would have been missed in the selection. The methods for which translation occurs inside a eukaryotic or preferably mammalian cell would be expected to have the highest likelihood for correct modification. Both mRNA display and ribosome display can be performed using a mammalian cell lysate, and therefore some modification might be expected. Indeed, activating phosporylation of a protein kinase was observed in mRNA display (Hammond et al., unpublished results).

One demonstrated solution was the posttranslational modification of an entire library (Cujec et al., 2002). For their study using mRNA display, the library was phosphorylated en masse after its preparation using the v-Abl kinase. Only phosphorylated substrates of the v-Abl kinase were subsequently isolated by affinity selection. One could similarly imagine the incubation of a library with a more complex cocktail of enzymes or even a cell lysate to allow various posttranslational modifications to occur.

Compound Immobilization. In the specific case of chemical genomics, the ligand used for affinity selection is the lead compound identified in phenotypic screening. An important point to note is that the actual compound tested in an activity assay is generally not suitable for this purpose. Some chemical derivatization may be required to provide a means of immobilization. Often, the introduction of a reactive chemical moiety such as an amine, thiol, or carboxylic acid is sufficient for subsequent derivatization. A suitably derivatized compound may then be conveniently immobilized through either a direct chemical coupling or via an affinity tag such as biotin.

This derivatization may necessitate that some information on the structure–activity relationship (SAR) be generated. Such information may be provided by a comparison of related members of a synthetic compound library or by systematic modification of a natural product. However, compounds identified from libraries from natural products may require difficult and/or complex chemistries for derivatization and immobilization. In some cases, the ligand is simple enough in molecular structure that derivatization at multiple convenient sites and pooling for affinity selection is sufficient. Chemical compound libraries generated by directed combinatorial synthesis allow for the introduction of a uniform immobilization handle (Blackwell et al., 2001; Clemons et al., 2001). In fact, much chemical synthesis is performed from starting materials immobilized on beads. This may provide a natural handle for subsequent immobilization of the compound.

Types of Display

PHAGE DISPLAY. Phage display, invented more than 10 years ago, is the most thoroughly developed of the display methods (reviewed in Rhyner et al., 2002). Although a number of significant advancements have been made, the underlying scheme has remained the same. A DNA library of peptides or proteins is cloned into the phage as a fusion with one of the coat proteins. The choice of coat protein dictates the number of copies of each library member that will be displayed on an individual phage particle, ranging from <1 to >100. For the filamentous phages, the coat proteins gIIIp and gVIIp are used by cloning a library in-frame between the signal sequence and the coat protein, thereby generating an N-terminal fusion. However, for cDNA libraries that contain members that begin and end in different reading frames, due to random priming, this presents a problem. Although the library protein may start in-frame, a shift prior to translation of the coat protein leads to a nonviable phage. The result is very low display efficiency. This problem can be circumvented through the use of a C-terminal fusion to gVIp. Alternatively, a leucine zipper method was developed (Crameri and Suter, 1993). In this case, the phage coat protein is fused to one-half of the zipper while the library is independently translated as a C-terminal fusion with the other half. The protein fusion to be displayed then forms a complex with the phage particle after its assembly. In order to stabilize the zipper, a disulfide pair can be engineered between the two halves.

Another limitation of filamentous phage display is the requirement that the fusion protein be secreted through the periplasm of the bacteria. Many proteins are poorly secreted and therefore inhibit the assembly of viable phage. To circumvent this problem, a phage display system was developed using the lytic phage T7 (Rosenberg et al., 1996). In this case the phage is completely assembled in the cytoplasm, eliminating the need for secretion. This is the system that has proven most useful for chemical genomics.

Example 1: FK506. T7 phage display was applied in affinity selection using the immune suppressant drug FK506 as the ligand (Sche et al., 1999). This drug is known to bind to a number of proteins containing a so-called FK-binding domain, including the well-characterized FKBP12 (Van Duyne et al., 1993). The starting library was prepared from human brain cDNA by random priming during first-strand synthesis. Second-strand synthesis was performed such that mRNA 5′ UTRs were included in the library clones and translated as part of the final product. The resulting libraries should contain multiple copies of an individual protein that are heterogeneous at the C-terminus, but relatively homogeneous at the N-terminus, N-terminally truncated fragments being the result of incomplete cDNA synthesis. The size of the library was 3.3×10^6 individual phage plaques. After 6 rounds of selection, the FKBP12 gene product was identified in 5 out of 16 clones by PCR analysis. These clones were all identical and contained the complete coding sequence of FKBP12. Fortunately, the absence of any stop codons in the 5′ UTR allowed the entire gene sequence to be correctly translated.

Example 2: Doxorubicin. In a second example, the small molecule drug doxorubicin was used as the ligand (Jin et al., 2002). In contrast to the FK506 selection reported above, the protein target of doxorubicin was not known prior to the selection. Four rounds of affinity selection were performed using biotinylated ligand and a cDNA library containing 10^9 members derived from human liver. After selection, 20 phages were characterized by sequencing. Two out of the 20 contained an open reading frame (ORF) of 216 amino acids. The remainder encoded ORFs of <20 amino acids and were therefore deemed background.

The selected ORF was identified as a C-terminal fragment of the protein hNopp140. Further characterization demonstrated a K_d of 4.5 μM that was specific for a nonphosphorylated protein complex. Verification that hNopp140 is indeed the biologically relevant target of doxorubicin will require additional validation as discussed below. This example provides a clear case wherein posttranslational modification or lack thereof is a crucial factor in binding. In this instance, the nonphosphorylated protein was bound to the ligand used in selection. However, should a binding interaction require phosphorylation, it would likely be missed with this phage display method.

MESSENGER RNA DISPLAY. Messenger RNA display produces the most compact display complex of the various methods (reviewed in Takahashi et al., 2003). To produce the display complex, a DNA library is prepared with homogeneous 5′ and 3′ termini incorporated during cDNA synthesis. The 5′ region encodes a promoter for RNA transcription as well as a UTR and Kozak sequence for translation initiation. The 3′ region need only be devoid of a stop codon. mRNA transcribed from such a library is then derivatized at the 3′ end to incorporate the translational chain-terminating antibiotic puromycin as part of a DNA linker. Both enzymatic and photochemical methods of adding the puromycin have been described. When these derivatized mRNAs are translated using an in vitro translation system, typically rabbit reticulocyte lysate, the absence of a stop codon causes the translation machinery to pause when it reaches the end of the message. This paused complex has a sufficiently long lifetime to allow the puromycin moiety to enter the ribosome and covalently couple to the C-terminus of the nascent polypeptide chain. After subsequent purification of the complex, a cDNA copy of the mRNA is generated as a means of stabilizing the RNA and preventing the selection of a structured RNA aptamer. The resultant complex thereby formed contains a protein fragment stably linked to the mRNA that encoded it.

The libraries used for mRNA display were prepared differently from those reported for plasmid display. For mRNA display, cDNA synthesis was randomly primed for both first- and second-strand synthesis. The resulting libraries consequently have a higher complexity corresponding to the variability at the N-terminus as well as at the C-terminus. However, the relatively large library size achievable with mRNA display makes this additional diversity manageable.

Example 3. The selection of proteins that recognize FK506 as a ligand using mRNA display has been reported (McPherson et al., 2002). This conveniently allows for a direct comparison to the results obtained by phage display (see above). The mRNA display selection used a starting library pool of ~1.8×10^{11} members assembled from individual tissue-specific libraries prepared from human liver, kidney, and bone marrow. After 3 rounds of selection, 23 out of 24 sequences cloned from the dominant PCR product were FKBP12. Among these sequences were several that contained slightly varied 5′ and 3′ sequences indicating that they arose from independent priming events during construction of the library. The higher number of unique clones observed using mRNA display may reflect the relatively larger size of the starting library. A subsequent selection using only the FK506 mRNA for preparation of a single-gene library was used to more accurately map the minimal active domain. These studies corroborated the results of the full library selection in that any significant truncation at either N- or C-terminus produced an inactive protein.

RIBOSOME DISPLAY. Ribosome display (reviewed in Schaffitzel et al., 1999) has been successful in protein–protein interaction studies and may have utility for chemical genomics. However, it lacks published successes in the identification of targets for small-molecule ligands from complex cDNA libraries. This technology makes the phenotype–genotype linkage through stable complexes of mRNA, ribosome, and nascent polypeptide. This ternary complex is formed by variously stalling translation through the use of chloramphenicol, chilling, and omitting the stop codon. The complex is further stabilized using high concentrations of magnesium to ensure its survival during ligand binding and washing steps. Correspondingly, the complex can be destabilized during elution using magnesium chelation by ethylenediaminetetraacetic acid (EDTA). Most published applications of ribosome display describe the generation of high-affinity single-chain antibody variable regions. However, a chemical genomics application has been demonstrated (Takahashi et al., 2002) wherein active dihydrofolate reductase variants were identified using immobilized methotrexate. This example only showed enrichment from a two-component mixture, and full validation of ribosome display for chemical genomics awaits the isolation of protein targets from complex libraries.

PLASMID DISPLAY. Plasmid display is conceptually and mechanistically simple in design. Essentially, the library is encoded on a plasmid as a fusion with a protein that binds to a DNA element on that same plasmid. Thus, the fusion protein is translated and bound to its cognate plasmid within the confines of a cell. Optimally, each cell will carry a single plasmid and express a single fusion protein that binds to that plasmid in the cytosol. This requires careful control of the multiplicity of infection to prevent scrambling. Plasmid display permits expression in native host cells, thereby allowing for correct posttranslational modification and possibly for protein complex formation. It also has the potential to work in many contexts including bacterial, mammalian, yeast, and insect cells.

Plasmid display linkage can take place through either noncovalent or covalent interactions as long as the interaction is stable to the conditions of selection. For example, the NF-kB p50 and lac repressor DNA binding proteins form noncovalent linkages, while covalent attachments are possible with protein fusions of the viral proteins Rep68 (Snyder et al., 1990) or P2A (Fitzgerald, 2000).

Pioneering plasmid display experiments used cDNA libraries of relatively short peptides fused to the lac repressor protein (Cull et al., 1992). This library was used to identify peptide sequences binding to a monoclonal antibody for dynorphin B. A significant improvement was made with the development of fusions to the NF-kB p50 protein, which binds to plasmid with higher affinity (Speight et al., 2001). In addition, the NF-kB-fusion protein library contained longer protein fragments. In proof-of-principle experiments selecting for amylose binding proteins from a 105-member library prepared in *Escherichia coli*, a malE fragment was enriched to near homogeneity in three rounds of selection. However, as exciting as this technology is, the scarcity of publications describing successful selections limits its acceptance in the chemical genomics community.

Cell Surface Display. Display on the surface of bacterial or yeast cells (Samuelson et al., 2002) holds promise for chemical genomics applications, but acceptance still awaits the publication of examples showing the identification of protein targets selected using a small-molecule ligand. These display systems are similar to phage display in that the phenotype–genotype relationship consists of an engineered cell surface protein that is linked to the microorganism carrying the gene for this displayed protein. However, unlike phage display, the cell surface display organism is self-replicating. Affinity selection procedures enrich for the bacteria or yeast displaying the drug-binding protein without the need to retransform or amplify the coding DNA. Another advantage is the ability to screen cells for ligand binding by fluorescence-activated cell sorting (FACS; Francisco et al., 1993; Olsen et al., 2000), thereby circumventing the typical binding, washing, and elution steps required in affinity selection protocols. Using this technique, relatively small differences in affinity between displayed protein fragments are detectable based on fluorescent signal intensity. Such small differences are much more difficult to distinguish using selection methods.

Potential disadvantages of bacterial and yeast display systems include steric hindrance of drug binding caused by the complex environment of the cell surface and transfection-limited library sizes. To date, these display systems have been successful in applications primarily focused on protein–protein interactions such as epitope mapping (Lu et al., 1995), antibody affinity maturation (Francisco et al., 1993), peptide–major histocompatability complex (MHC) engineering (Brophy et al., 2003), and T cell receptor engineering (Kieke et al., 1999). Conceptually, bacterial and yeast display systems should be amenable to chemical genomics applications, and such demonstrations are eagerly awaited.

MICROBEAD DISPLAY. Microbead display, also known as in vitro compartmentalization, is an in vitro analog of cell surface display (IVC; reviewed in Griffiths and Tawfik, 2000). The phenotype–genotype connection is made through a microbead complex that is created within a water-in-oil emulsion. The compartmentalization is designed such that each compartment contains only one microbead. In vitro transcription-translation of the cDNA tethered to a single microbead within the compartment produces an epitope-tagged protein that binds to that same bead through an antitag antibody. This technique has evolved rapidly from proof-of-concept studies in model systems to the incorporation of selection methods using FACS (Sepp et al., 2002) and applications in enzyme selection (Griffiths and Tawfik, 2003). However, published chemical genomics applications are lacking.

Detection of Intracellular Drug–Protein Interactions. The display methods described above are all basically extracellular display. Potentially powerful alternatives can be described as intracellular display. Within the cellular environment, proteins are posttranslationally modified, compartmentalized, and accessible to binding partners and cofactors so that intracellular display ligands are exposed to physiologically relevant forms of target proteins.

One such methodology is the yeast three-hybrid system (Licitra and Liu, 1996; reviewed in Henthorn et al., 2002). It is a modification of the familiar two-hybrid system wherein a heterodimeric ligand that includes the compound of interest serves as a bridge between the protein fusions containing DNA-binding and transactivation domains. Among the limitations of the three-hybrid system are potential toxicity of the test compound to the yeast, the requirement for passage of the heterodimeric ligand complex across the cell membrane, and the relatively high affinity of protein for ligand that may be required for detection (nanomolar or subnanomolar K_d). All three of these limitations are circumvented by the use of an in vitro system such as those described above. In addition, the three-hybrid system is limited in the number of cells that can be transformed and screened. This is typically in the range of 106, several orders of magnitude lower than those attainable in vitro.

The primary challenges faced in developing a more robust system include: (1) introducing the cDNA library into the relevant cell type, (2) ensuring that drug enters the cell, (3) detecting the cells within the library in which the protein–drug interaction is occurring, and (4) recovering the cDNA sequence for the protein. Many recent technological advances may help address these issues. For example, mammalian cell transfection reagents have evolved to be highly efficient and suitable for many cell types. The ideal transfection protocol should introduce one plasmid per cell in order to reduce the scrambling of phenotype–genotype complexes.

Although early methods used cell survival as a readout for positive interactions, alternative systems are also being developed and have been applied to the detection of protein–small molecule interactions. Rapamycin has been used in proof-of-principle experiments to show that it can trigger protein splicing when coupled with the split intein system (Mootz and Muir, 2001). This system can logically be modified to splice together a selectable marker or a screening polypeptide that leads to FRET (fluorescence resonance energy transfer) or BRET (bioluminescence resonance energy transfer) as a result of drug binding. Another method is the protein–fragment complementation assay using dihydrofolate reductase (DHFR) that has been used to detect protein–protein interactions (reviewed in Michnick, 2001). This system may also be extensible to protein–ligand interactions.

DISCUSSION AND CONCLUSIONS

Current drug development efforts within the pharmaceutical industry are being refocused on cell-based, phenotype-driven screening. This process has been fueled by improvements in automated screening tools, high-content screening paradigms, and diversity-oriented parallel chemical synthesis. However, an unmet need remains for chemical genomics methods that will rapidly identify the relevant target of a bioactive compound.

The existing chemical genomics techniques of model organism genetics and cDNA display generate lists of potential targets that are either biologically or biochemically relevant, respectively. Neither approach will unambiguously reveal both aspects of a relevant target without further study. The true mechanistic target of a drug may be found at the intersection of these two lists. The key to reducing the chemical genomics bottleneck will be to efficiently use one class of techniques to rapidly reduce genomic-scale complexity (all possible targets) to a short list of potential targets, and then to rapidly analyze that list along the second axis. For example, while a display experiment may generate a handful of potential targets, subsequent gene-specific analysis, by a method such as siRNA, might be sufficient to identify the true target from that group.

As detailed above, significant progress has been made in the invention and early development of near-genome-wide chemical genomics display technologies. Early successes in proof-of-principle studies have generated great optimism for the future of this approach. Both phage display and mRNA display have yielded particularly promising results in the identification of target and potential target proteins from extremely complex cDNA libraries. Some classes of targets will prove particularly amenable to display such as soluble cytosolic and nuclear proteins and their subdomains.

However, several limitations remain for these approaches. Specifically, none of the current display technologies adequately address the issue of multispan transmembrane proteins, which represent a disproportionate number of therapeutically significant drug target classes. This is a class of targets for which an approach such as mRNA knockdown is more appropriate. In addition, many proteins require accessory factors for their activity. Although it is possible that a method such as plasmid display will allow for the assembly of functional multisubunit complexes, none has yet been demonstrated. On the whole, the widespread acceptance of chemical genomics awaits the publication of successful target identifications and the subsequent development of efficacious drugs using this approach.

The promise of chemical genomics is to enable the expedited development of safer and more effective drugs. Although it is clear that display methods are not a panacea for the chemical genomics challenge, the obvious potential of these approaches will necessitate their inclusion in any comprehensive drug discovery effort.

REFERENCES

Blackwell, H.E., Perez, L., Stavenger, R.A., Tallarico, J.A., Cope Eatough, E., Foley, M.A., and Schreiber, S.L. (2001). A one-bead, one-stock solution approach to chemical genetics: Part 1. *Chem. Biol.* **8**:1167–1182.

Brophy, S.E., Holler, P.D., and Kranz, D.M. (2003). A yeast display system for engineering functional peptide-MHC complexes. *J. Immunol. Methods* **272**:235–246.

Clemons, P.A., Koehler, A.N., Wagner, B.K., Sprigings, T.G., Spring, D.R., King, R.W., Schreiber, S.L., and Foley, M.A. (2001). A one-bead, one-stock solution approach to chemical genetics: Part 2. *Chem. Biol.* **8**:1183–1195.

Crameri, R., and Suter, M. (1993). Display of biologically active proteins on the surface of filamentous phages: A cDNA cloning system for selection of functional gene products linked to the genetic information responsible for their production. *Gene* **137**:69–75.

Cujec, T.P., Medeiros, P.F., Hammond, P., Rise, C., and Kreider, B.L. (2002). Selection of v-abl tyrosine kinase substrate sequences from randomized peptide and cellular proteomic libraries using mRNA display. *Chem. Biol.* **9**:253–264.

Cull, M.G., Miller, J.F., and Schatz, P.J. (1992). Screening for receptor ligands using large libraries of peptides linked to the C terminus of the lac repressor. *Proc. Natl. Acad. Sci.* **89**:1865–1869.

Drews, J. (1999). *In Quest of Tomorrow's Medicines*. Springer, New York.

Fitzgerald, K. (2000). In vitro display technologies—new tools for drug discovery. *DDT* **5**:253–258.

Francisco, J.A., Campbell, R., Iverson, B.L., and Georgiou, G. (1993). Production and fluorescence-activated cell sorting of *Escherichia coli* expressing a functional antibody fragment on the external surface. *Proc. Natl. Acad. Sci. USA* **90**:10444–10448.

Gold, L. (2001). mRNA display: Diversity matters during in vitro selection. *Proc. Natl. Acad. Sci. USA* **98**:4825–4826.

Griffiths, A.D., and Tawfik, D.S. (2003). Directed evolution of an extremely fast phosphotriesterase by in vitro compartmentalization. *EMBO J.* **22**:24–35.

Griffiths, A.D., and Tawfik, D.S. (2000). Man-made enzymes—from design to in vitro compartmentalisation. *Curr. Opin. Biotechnol.* **11**:338–353.

Hammond, P.W., Alpin, J., Rise, C.E., Wright, M., and Kreider, B.L. (2001). In vitro selection and characterization of Bcl-X(L)-binding proteins from a mix of tissue-specific mRNA display libraries. *J. Biol. Chem.* **276**:20898–20906.

Henthorn, D.C., Jaxa-Chamiec, A.A., and Meldrum, E.A. (2002). GAL4-based yeast three-hybrid system for the identification of small molecule-target protein interactions. *Biochem. Pharmaco.* **63**:1619–1628.

Jin, Y., Yu, J., and Yu, Y.G. (2002). Identification of hNopp140 as a binding partner for doxorubicin with a phage display cloning method. *Chem. Biol.* **9**:157–162.

Kieke, M.C., Shusta, E.V., Boder, E.T., Teyton, L., Wittrup, K.D., and Kranz, D.M. (1999). Selection of functional T cell receptor mutants from a yeast surface-display library. *Proc. Natl. Acad. Sci. USA* **96**:5651–5656.

Licitra, E.J., and Liu, J.O. (1996). A three-hybrid system for detecting small ligand-protein receptor interactions. *Proc. Natl. Acad. Sci. USA* **93**:12817–12821.

Lu, Z., Murray, K.S., Cleave, V.V., LaVallie, E.R., Ståhl, M.L., and McCoy, J.M. (1995). Expression of thioredoxin random peptide libraries on the *Escherichia coli* cell surface as functional fusions to flagellin: A system designed for exploring protein–protein interactions. *Bio/Technol.* **13**:366–372.

Margolis, J., and Duyk, G. (1998). The emerging role of the genomics revolution in agricultural biotechnology. *Nature Biotechnol.* **16**:311.

Matthews, D.J., and Kopczynski, J. (2001). Using model-system genetics for drug-based target discovery. *Drug Discov. Today* **6**:141–149.

McPherson, M., Yang, Y., Hammond, P.W., and Kreider, B.L. (2002). Drug receptor identification from multiple tissues using cellular-derived mRNA display libraries. *Chem. Biol.* **9**:691–698.

Michnick, S.W. (2001). Exploring protein interactions by interaction-induced folding of proteins from complementary peptide fragments. *Curr. Opin. Struct Biol.* **11**(4):472–477.

Miyagishi, M., Hayashi, M., and Taira, K. (2003). Comparison of the suppressive effects of antisense oligonucleotides and siRNAs directed against the same targets in mammalian cells. *Antisense Nucleic Acid. Drug Dev.* **13**:1–7.

Mootz, H.D., and Muir, T.W. (2002). Protein splicing triggered by a small molecule. *J. Am. Chem. Soc.* **124**(31):9044–9045.

Nislow, C., and Giaever, G. (2003). Chemogenomics: Tools for protein families and Chemical genomics: Chemical and biological integration. *Pharmacogenomics* **4**:15–18.

Olsen, M.J., Stephens, D., Griffiths, D., Daugherty, P., Georgiou, G., and Iverson, B.L. (2000). Function-based isolation of novel enzymes from a large library. *Nature Biotechnol.* **18**:1071–1074.

Rhyner, C., Kodzius, R., and Crameri, R. (2002). Direct selection of cDNAs from filamentous phage surface display libraries: potential and limitations. *Curr. Pharm. Biotechnol.* **3**:13–21.

Rodi, D.J., Janes, R.W., Sanganee, H.J., Holton, R.A., Wallace, B.A., and Makowski, L. (1999). Screening of a library of phage-displayed peptides identifies human bcl-2 as a taxol-binding protein. *J. Mol. Biol.* **285**:197–203.

Rosenberg, A., et al. (1996). T7Select Phage Display System: A powerful new protein display system based on bacteriophage T7. *inNovations* **6**:1–6.

Samuelson, P., Gunneriusson, E., Nygren, P.-A., and Ståhl, S. (2002). Display of proteins on bacteria. *J. Biotechnol.* **96**:129–154.

Schaffitzel, C., Hanes, J., Jermutus, L., and Pluckthun, A. (1999). Ribosome display: An in vitro method for selection and evolution of antibodies from libraries. *J. Immunol. Methods* **231**:119–135.

Sche, P.P., McKenzie, K.M., White, J.D., and Austin, D.J. (1999). Display cloning: Functional identification of natural product receptors using cDNA-phage display. *Chem. Biol.* **6**:707–716.

Sepp, A., Tawfik, D.S., and Griffiths, A.D. (2002). Microbead display by in vitro compartmentalisation: Selection for binding using flow cytometry. *FEBS Lett.* **532**:455–458.

Snyder, R.O., Im, D.-S., and Muzyczka, N. (1990). Evidence for covalent attachment of the adeno-associated virus (AAV) Rep Protein to the ends of the AAV. *J. Virology* **64**:6204–6213.

Speight, R.E., Hart, D.J., Sutherland, J.D., and Blackburn, J.M. (2001). A new plasmid display technology for the in vitro selection of functional phenotype-genotype linked proteins. *Chem. Biol.* **8**:951–965.

Takahashi, T.T., Austin, R.J., and Roberts, R.W. (2003). mRNA display: Ligand discovery, interaction analysis and beyond. *Trends Biochem. Sci.* **28**:159–165.

Takahashi, F., Ebihara, T., Mie, M., Yanagida, Y., Endo, Y., Kobatake, E., and Aizawa, M. (2002). Ribosome display for selection of active dihydrofolate reductase mutants using immobilized methotrexate on agarose beads. *FEBS Lett.* **514**:106–110.

Van Duyne, G.D., Standaert, R.F., Karplus, P.A., Schreiber, S.L., and Clardy, J. (1993). Atomic structures of the human immunophilin FKBP-12 complexes with FK506 and rapamycin. *J. Mol. Biol.* **229**:105–124.

10

BIOINFORMATICS FOR PROTEOMICS

Christian Ahrens, Hans Jespersen, and Soeren Schandorff

MDS Inc.—Denmark, Odense, Denmark

INTRODUCTION

Proteomics is the systematic study of the set of proteins encoded by the genome of an organism under investigation, also referred to as its proteome. In its broadest sense, the

Industrial Proteomics: Applications for Biotechnology and Pharmaceuticals, by Daniel Figeys
ISBN 0-471-45714-0

proteome covers all proteins and isoforms that can be generated by alternative splicing and/or posttranslational modifications in different tissues and at different times during development (Tyers and Mann, 2003).

A differentiating feature of proteomics technologies is their ability to study biological systems directly at the level of the molecules carrying out most cellular functions, the proteins. This provides the basis for important advantages over gene-centric technologies, including the ability to (i) determine the actual expression levels of proteins and their isoforms, which frequently do not correlate with those measured at the messenger ribonucleic acid (mRNA) level; (ii) identify posttranslational modifications, which often are important for protein function; (iii) study the interaction between proteins and thereby discover novel insights into pathways, networks, and functional complexes; and (iv) identify novel proteins missed by current gene prediction algorithms, as well as novel splice variants. In order to exploit these advantages, bioinformatics solutions have to provide the researcher with an integrated view of the experimental data in the context of a wealth of additional relevant information and at the level of detail outlined above.

Traditionally, reductionist biochemical analysis has involved the detailed study of individual proteins, which requires only a moderate level of bioinformatics support. Web sites that link to a wealth of sequence analysis tools for an in-depth analysis of the particular protein of interest are readily available (e.g., *www.expasy.org*). In contrast, large-scale parallel protein-based expression analyses aim to look at proteins on a global scale and in a high-throughput fashion. These studies generate massive amounts of data and therefore critically depend on advanced protein informatics and bioinformatics solutions for efficient data handling and control, as well as data analysis and interpretation. The term *bioinformatics* should be differentiated from *protein informatics*. Protein informatics broadly covers information-technology-based approaches and applications as they relate to the study and analysis of proteins. Bioinformatics is mainly concerned with the collection, organization, integration, and analysis of biological data and thus focuses on extracting knowledge from experimental data after it has been generated and integrated with additional types of information (gene expression data, interaction data, functional studies, annotation, etc.).

The industrial application of proteomics has lagged behind that of genomics for several reasons. First, proteins are inherently less amenable to large-scale studies than deoxyribonucleic acid (DNA) due to their widely varying biochemical characteristics, higher lability, and the lack of a technology for easy manipulation and amplification such as polymerase chain reaction (PCR). Second, both the large dynamic range of protein expression and the severalfold higher complexity of the proteome (see below) raise significant challenges for the experimental approach, the instrument sensitivity, and the bioinformatics setup. Finally, protein identification relies on matching mass spectra against protein databases, which are mainly derived from genomic information. The completion of the human genome sequence has thus been a prerequisite to fully enable human proteomics studies. Combined with significant improvements in instrument sensitivity and experimental fractionation technologies, these advances have enabled the widespread use of large-scale industrial proteomics analysis. The pharmaceutical and biotechnology industries expect proteomics to provide a significant number of higher quality and better validated drug targets than genomic technologies (Jain, 2003). The understanding of proteins in their respective functional context should enable the development of additional mechanism-of-action-based drugs and, by targeting alternative entry points in pathways, the development of drugs that exhibit fewer side effects.

The two main approaches currently employed for large-scale parallel expression proteomics analysis include gel-based technologies, which trace back to the advent of two-dimensional (2D) gel electrophoresis in the mid-1970s (O'Farrel, 1975) and gel-free analysis of complex protein mixtures, also referred to as shotgun proteomics (Link et al., 1999; Wolters et al., 2001). Shotgun proteomics has gained a lot of interest since it has enabled researchers to identify a significant number of proteins that are severely underrepresented in 2D gel-based proteomic analyses, for example, plasma membrane proteins. Notably, these include target classes of high relevance for the pharmaceutical and biotechnology industries such as transporters, ion channels, G-protein-coupled receptors (GPCRs), and other receptors. Protein microarray technology represents an additional approach for parallel protein-based analysis and is expected to play an important role in the near future (Kambhampati, 2003).

The bioinformatics requirements differ substantially for areas such as expression proteomics, structural proteomics, chemoproteomics, interaction proteomics, or analysis of protein microarrays and cannot be covered adequately in one chapter. We chose to focus on the protein informatics and bioinformatics tools required to enable large-scale shotgun proteomics, as well as general issues encountered in data integration and data analysis. While numerous commercial hardware and software solutions exist to support gel-based complex mixture analysis, there are currently no commercially available software tools that adequately address all the issues of shotgun proteomics approaches. Some of the major issues we discuss in this chapter include the need for (i) solutions to handle and track the massive amounts of data generated by semiautomated high-throughput setups, (ii) comprehensive databases to maximize the number of protein identifications, (iii) strategies to reduce the immense complexity of the proteome at the experimental level, as well as the data complexity at the protein informatics and bioinformatics level, and (iv) solutions to integrate other information resources with the experimental data enabling efficient prioritization of the results. Figure 10.1 illustrates this integration of experimental data and information from a large variety of external sources, which is a prerequisite for efficient data analysis and knowledge discovery.

RISE IN COMPLEXITY WHEN MOVING FROM GENOMICS TO PROTEOMICS DATA—THE CHALLENGE

As outlined before, the proteome covers all proteins and isoforms generated by alternative splicing and/or posttranslational modifications in different tissues and at different times during development. As such the proteome is much more complex than the genome, both with respect to the number of different functional players and their dynamic range of expression.

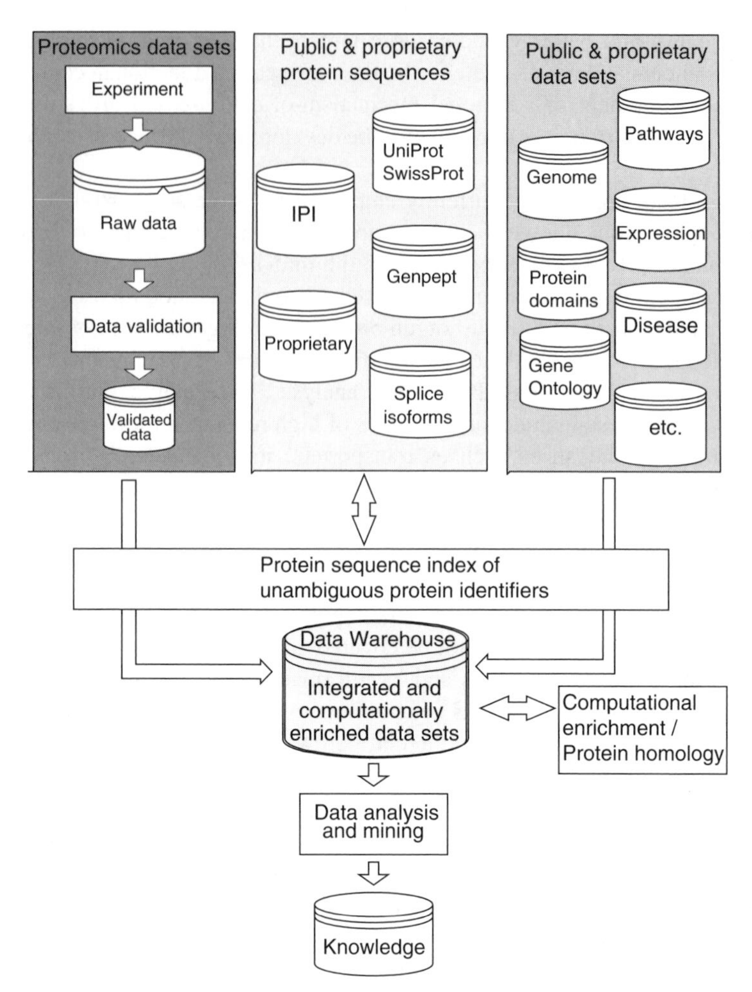

Figure 10.1. Schematic overview of a data warehouse integration strategy enabling efficient large-scale proteomic data analysis and knowledge discovery. Validated experimental data (*left side*) can be accurately integrated with a variety of preprocessed public and proprietary information (*right side*). This integration is based on the protein index (*middle*), an index of unambiguous protein identifiers. The data are computationally enriched using state-of-the-art protein analysis and prediction software, and the data becomes ideally suited for data analysis and mining of results from large scale proteomics experiments.

The human genome encodes roughly 35,000 to 40,000 genes, and it is estimated that between 40 to 50 percent of the genes have one to several splice isoforms (Mironov et al., 1999; Croft et al., 2000; Levanon and Sorek, 2003). When one considers the different posttranslational modifications on top of this, it is expected that the complexity of the proteome is at least an order of magnitude greater than that of the genome (Aebersold, 2003).

The dynamic range of protein expression varies from 6 orders of magnitude for normal cells to 10 orders of magnitude for the plasma proteome (Patterson, 2003). In plasma, serum albumin is the most abundant protein with 35×10^9 to 50×10^9 pg/mL, while interleukin-6 is only present with 0 to 5 pg/mL (Anderson and Anderson, 2002). This is significantly larger than the dynamic range of mRNA expression, which varies approximately between 1 copy/cell and 500,000 copies/cell. While the linear range of technologies that measure gene expression is close to the dynamic range of mRNA expression, current technologies for unbiased large-scale parallel protein analysis only have a linear range of 10^2 to 10^4. In stark contrast, technologies that measure the concentration of known analytes, such as immunoassays, can achieve a linear range of detection over 10 orders of magnitude. An increase of the linear range of protein expression by complex mixture analysis is intimately linked to efficient sample preparation and prefractionation steps that reduce sample complexity and/or remove abundant proteins, which prevent detection of lower abundance proteins (see below). The combination of current instrumentation with different prefractionation protocols (e.g., chromatography, immunoaffinity purification, isoelectric focusing), can achieve a linear range between 10^3 and 10^6 for large-scale parallel protein analysis (Anderson and Anderson, 2002).

Finally, the proteome is much more dynamic and constantly changing with respect to posttranslational modification state, changes in the interaction between proteins, their subcellular localizations, and others. One therefore looks at a proteomic snapshot of the functional state of the cell. The immense complexity of the proteome has important implications for the experimental approach as well as the bioinformatics requirements to represent this information.

COMPLEX MIXTURES AND MASS SPECTROMETRIC ANALYSIS—THE DATA

An overview of the steps involved in a generic setup for gel-free complex mixture analysis is shown in Figure 10.2 and includes contributions from biology/biochemistry, analytical, and protein informatics/bioinformatics functions, respectively. The biology/biochemistry function covers the steps prior to mass spectrometric analysis, which include biological sample handling (e.g., cell culture, metabolic labeling for quantitative protein expression analysis, tissue procurement from model organisms or clinical specimens), sample preparation (e.g., prefractionation, affinity purification, differential protein labeling of tissue samples by protein derivatization), and sample processing (digestion, separation). The analytical function performs the liquid chromatography-tandem mass spectrometry (LC-MS/MS) analysis and validates the data (supported by protein informatics). Validated data is stored in the experimental database and subsequently integrated with information from a variety of other sources for efficient data analysis and knowledge discovery. Additional information resources can include, for example, annotations, gene ontology classifications, protein domains, gene expression data, patent data, or data relevant to human disease (Fig. 10.1). The bioinformatics tools and solutions that aid in this knowledge discovery process are described in more detail below.

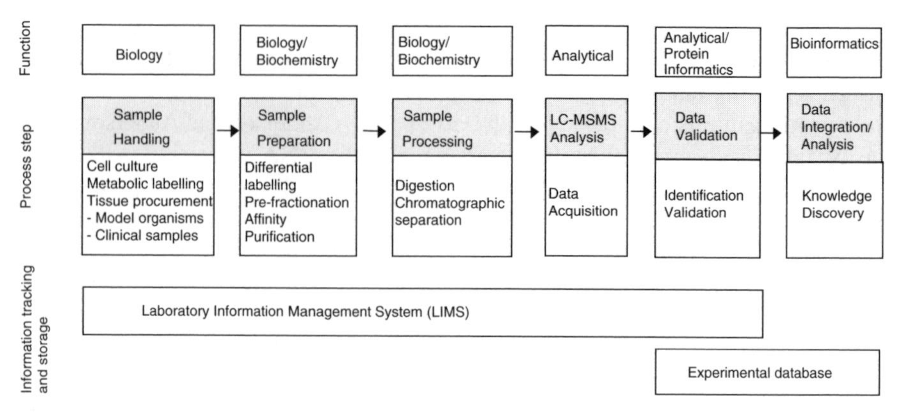

Figure 10.2. Overview of a generic gel-free complex mixture analysis process. Different functions within an organization (*upper panel*) carry out the various process steps (categorized and explained in some detail in the *middle panel*) sequentially from biological sample handling to bioinformatic data analysis. The contribution of the major information tracking and storage solutions is mapped onto this process (*lower panel*).

Peptides are the units of information in shotgun proteomics. The majority of gel-based versus gel-free complex mixture analysis approaches differ at the sample processing step, schematically illustrated in Figure 10.3. For gel-free complex mixture analysis, protein mixtures are mainly digested early during sample processing, before subsequent chromatographic fractionation steps are carried out (Fig. 10.3, lower panel). For complex soluble protein mixtures (e.g., serum or plasma proteins), the digestion step can also occur after chromatographic separation. In both cases, however, the experimentally identified peptides have lost the physical relation to the proteins they were originally derived from and represent independent units of information. This is an important difference from 2D gel-based parallel protein analysis where the proteolytic digestion step occurs after protein separation (Fig. 10.3, upper panel). Therefore, 2D gels have the advantage of a clearcut association between the identified digested peptides and the protein isoform excised from the respective gel spot and allow the identification of multiple modified isoforms of a given protein within the same gel.

Subsequently, proteins can either be identified by peptide mass fingerprinting (PMF), which produces a spectrum of intact peptides, or by tandem mass spectrometry, which involves fragmentation of individual peptides in order to determine their amino acid sequence. PMF is a robust and fast method, which is often used in combination with 2D gels on less complex samples. However, since the mass accuracy of the detection is inadequate to unambiguously identify a large number of proteins based on a series of peptide masses searched against a large database, it is not used for the analysis of complex samples. Protein identification for complex mixture analysis is therefore mainly based on tandem mass spectrometry (MS) analysis, which provides accurate structural information. The structural information is used to identify peptide hits, which may be associated with one or more proteins.

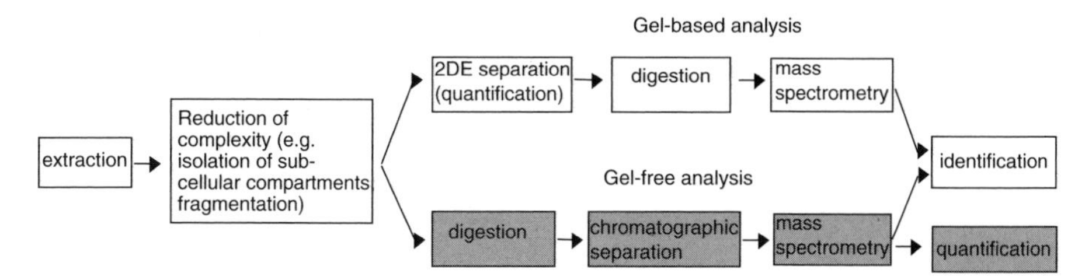

Figure 10.3. Overview over differences in the sample processing steps between gel-based and gel-free complex mixture analysis approaches. During gel-free complex mixture analysis (*lower panel*), the separation step occurs typically after protein digestion, while for gel-based analysis the digestion is carried out after the protein separation. Peptides derived from these approaches therefore have a different information content associated with them.

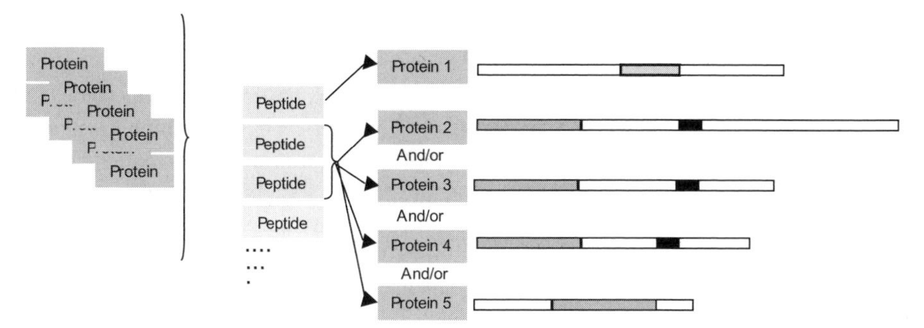

Figure 10.4. Schematic illustration of the consequences of the loss of the physical relation of peptides from proteins during gel-free complex mixture analysis. Sets of peptides are identified from proteolytic digestion of the proteins present in an experimental sample. The relation between peptide and proteins is lost. In the subsequent assignment of peptides to proteins (*right* in the figure) some peptides may unambiguously identify a specific protein sequence (see the peptide associated with protein 1), while other peptides may be associated with several proteins without discriminating, hence resulting in an ambiguous identification.

Since shotgun proteomics uses peptides as independent units of information (as outlined above), an important concept is illustrated in Figure 10.4. The list of identified proteins contains unambiguous identifications (protein 1, Fig. 10.4), as well as many cases where proteins or isoforms cannot be identified unambiguously based on the combination of the observed peptides, and thus a list of potential proteins is returned (proteins 2 to 5, Fig. 10.4). In cases where relevant proteins cannot be distinguished based on the peptide evidence, further validation of possible candidates has to be performed. This could involve the use of genomic technologies or, alternatively, additional database searches, for example, against predicted posttranslational modifications can be considered to gain additional peptide evidence. Any measure that improves the peptide coverage and thus increases the information content can help to distinguish between closely related proteins and to provide additional confidence in the protein identifications. The results of MS-based parallel protein analysis consist of long lists of protein identifications.

Since the proteome is continuously changing, an individual experiment only represents a snapshot in time. Ideally, these snapshots have to be compared over time or according to different experimental conditions and treatments to identify those proteins that change in activity, expression level, or modification. Quantitative differential protein analysis methods using differential metabolic labeling of cell lines have become widely used (Gygi et al., 1999; Ong et al., 2002). Their major benefit is that they reduce the vast number of protein identifications down to a manageable subset of differentially expressed proteins. The ability to quantitatively analyze proteins from tissues or clinical specimens using protein derivatization protocols (Olsen et al., 2004) has important implications. It will enable researchers to study disease processes on relevant clinical samples and promises to be a major driver for clinical proteomics studies.

Ideally, parallel protein expression analysis technologies aim to look at all proteins simultaneously to provide a systems biology view. Despite many technological advances and a current instrument sensitivity that enables protein detection down to the attomolar range (Martin et al., 2000), even studies on prokaryotes with a few thousand gene products have not yet accomplished this goal. Studies on human samples will only identify a fraction of the severalfold more complex human proteome. Adequate measures to prefractionate an experimental sample are critical to achieve a higher coverage of the human proteome (see below). Conceptually, protein microarrays (Kambhampati, 2003) could come closest to the goal of studying all proteins in parallel. However, key technology improvements and most of the reagents needed to study all proteins are yet to be generated.

There is a major difference between data generated by parallel protein-based versus parallel gene expression analysis. For the latter, the identities of the probes [mainly oligonucleotides or complementary DNA (cDNA) probes] that are deposited onto the gene array are known. Both qualitative information as to whether a specific gene is expressed or not as well as its relative expression level can be obtained by comparison to internal standards such as housekeeping genes. In contrast, protein identification from a complex mixture can only provide qualitative data as to whether a protein is present in the sample. The failure to identify a protein in a certain experiment is not absolute proof that it is not expressed; it could be expressed below the level of detection, concealed by abundantly expressed proteins, or it could even be expressed but lost in the prefractionation step.

AUTOMATION AND HIGH THROUGHPUT

A critical factor for the industrial success of genomics technologies such as DNA sequencing and gene expression analysis has been the ability to scale them to run in an automated, high-throughput fashion. In analogy, parallel protein-based expression proteomics approaches have to aim to run in a highly automated and robust fashion with minimal user interaction in order to provide value for the biotechnology and pharmaceutical industry. For 2D gel-based parallel protein analysis, integrated platforms are available that automate the steps from imaging analysis and robotic band/spot picking up to deposition of samples on targets for mass spectrometric analysis, thus achieving a high throughput (e.g., Ettan spot handling platform from Amersham Biosciences). The setup required for shotgun proteomics is simpler in the sense that it does not require the expensive imaging and robotics hardware and software. However, in lack of adequate commercial solutions, groups that rely on a high-throughput shotgun proteomics approach are forced to develop their own solutions.

The sample preparation and processing steps vary significantly depending on the respective experimental focus, and are less amenable to automation. Yet, commercially available kits could greatly improve the experimental reproducibility of, for example, the fractionation, especially for inexperienced users. The steps more amenable to automation include those that fall into the analytical, protein informatics, and bioinformatics categories, most importantly data handling, data validation, and data analysis. Performing shotgun proteomics in a high-throughput format requires efficient informatics solutions to handle and process the data and to store the large amounts of experimental data. Several commercial providers offer laboratory information management systems (LIMS) that track the data at various steps in the experimental process (e.g., LabVantage, Amersham Biosciences, Applied Biosystems, InnaPhase; Fig. 10.2). These commonly require customization according to the specific user needs, and several groups have developed solutions internally. Data validation is a critical step in the process and requires expert knowledge. It is therefore an important bottleneck in large-scale studies. Efforts to incorporate expert knowledge into a semiautomatic validation process promise to improve both the throughput and the data quality achieved by high-throughput analyses. Experimental data that passes user-defined quality control criteria can subsequently be stored in a database for experimental results. A link between the data storage and the LIMS system is important. Databases that store and accumulate the experimental information over time and enable comparison of different experiments are of high value.

Current semiautomated large-scale expression proteomics studies (both gel-free and gel-based approaches) create massive amounts of data. For shotgun proteomics, typically, a few thousand tandem mass spectra are acquired per fraction, and up to 50 percent of the tandem mass spectra have peptides assigned to them. These data have to be clustered over the different fractions of one experiment and over repetitions of the same experiment to collapse redundant protein identifications. Up to a few thousand proteins can be identified, which already poses the formidable challenge to prioritize among this long list of proteins and to identify the proteins most relevant to a particular biological question. It is of critical importance to integrate experimental data with the large body of public and proprietary biological knowledge and to filter out the relevant information enabling knowledge discovery (see below).

Today, the most significant bottleneck is no longer the generation of data but the lack of tools for efficient handling, analysis, prioritization, and interpretation of the data. This bottleneck will become even more pronounced with newer, more sensitive mass spectrometers, which produce severalfold more data. Bioinformatics can assist industrial proteomics by providing solutions that address several of the major bottlenecks. These include methods to (i) ensure robust protein identification and to provide an overview over potential alternative protein identifications, (ii) enable data quality control and provide a measure for the reproducibility over different experiments (statistics), (iii) provide tools to reduce data complexity, (iv) integrate the experimental data with publicly available bioinformatics and proprietary data in order to put it into a functional context and to enable various filtering capabilities, and (v) create tools that help to visualize and efficiently navigate the data.

ISSUES RELATED TO PROTEIN SEQUENCE DATABASES: COMPREHENSIVENESS VERSUS REDUNDANCY

A critical step in the protein identification process is the choice of protein sequence database. For the purpose of protein identification from shotgun proteomics, the database should be as comprehensive as possible to maximize the number of proteins that can be identified from various expression proteomics studies, while at the same time aiming to be nonredundant. Despite the availability of complete genome sequences, many small proteins and protein variants cannot be correctly predicted due to our incomplete understanding of gene structures and alternative splicing. Therefore, currently no database comes even close to representing the complete proteome. The comprehensiveness of the database has a big impact on the resolution at which cellular processes can be studied, and therefore the chances for discovery of a novel protein, splice variant, or even an allele with a single amino acid difference. Eliminating database redundancy greatly improves the analysis of the returned search results since the scientist is not drowning in long lists of hits to almost identical proteins, which is counterproductive to the goal of getting a fast overview of the identified proteins.

Historically, the preferred databases have been Swiss-Prot and national center for biotechnology information (NCBI)'s nrdb. Swiss-Prot has been the database of choice for researchers that prioritize to view identified proteins in the context of the detailed annotation and functional information that is stored in highly curated protein records. A drawback of Swiss-Prot is its lack of comprehensiveness, which can be overcome by adding the less curated but more inclusive TrEMBL database. Nrdb or taxonomic subsections thereof have been favored by researchers that prioritize a more comprehensive and nonredundant database over a database with extensive annotation. Nrdb includes a nonredundant set of protein sequences from the Swiss-Prot, PIR, PDB, and "GenPept" databases. Database redundancy can include protein sequences that are 100 percent identical over their entire length from the same organism (e.g., contributed by population studies) or from different species (homologs), as well as identity to partial protein sequences (see below). In nrdb, identical sequence entries are collapsed into one single protein sequence entry, whose header consists of concatenated descriptions of the collapsed entries. The first description in the header is the one that is displayed by most database search software using nrdb, including Mascot (Perkins et al., 1999). In December 2003, around 900 entries in nrdb were derived from identical sequences from multiple species, including one or more human proteins. In around 300 of these cases, the first description is not taken from a human entry. This can potentially mislead researchers to call an alternative human identifier of lower quality when analyzing a human sample and hence looking specifically for human proteins.

The redundancy issue is further complicated by the presence of many C-terminal, N-terminal, or internal fragments. Those fragments that arise from conceptual translation of partial gene sequences do not represent true biological processing events and introduce tryptic peptides in the database search routine that do not correspond to existing biological molecules. This can lead to incorrect identifications and impairs the database search statistics. On the other hand, fragments containing additional units of information due to, for example, allelic differences, may lead to complementary information. True redundancy should be avoided at the level of the database search, but this requires that specialized sequence databases be created, which is not yet common practice.

Since nrdb is not truly comprehensive, a number of alternative protein databases have emerged. The International Protein Index (IPI; *www.ebi.ac.uk/IPI*) is curated by the EMBL-European Bioinformatics Institute (EMBL-EBI) and provides indices for human, mouse, and rat proteomes. IPI is a top-level guide to Ensembl (Hubbard et al., 2002), SWISS-PROT/TrEMBL, and RefSeq (Pruitt and Maglott, 2001) with efficient cross referencing, minimal redundancy, and stable identifiers. Another recently introduced solution is the Universal Protein Knowledgebase (UniProt) (Apweiler et al., 2004), an initiative between the EBI, the Swiss Institute of Bioinformatics (SIB), and the Protein Information Resource (PIR). UniProt unifies information from Swiss-Prot, TrEMBL, and PIR-PSD. It furthermore offers a UniProt Archive (UniParc), which is currently the most comprehensive public protein sequence collection. On top of the main protein databases, it also includes sequences from fly (Flybase Consortium, 2003), worm (Harris et al., 2003), and from European, American, and Japanese patent records. Furthermore, databases that incorporate many additional sequences predicted from expressed sequence tag (ESTs) and high-throughput genome sequences are available, for example, the Alternative Splicing Database (Thanaraj et al., 2004). Of high potential value are commercially available gene collections from Incyte (LifeSeq) and Compugen (Genecarta). These curations of tens of thousands of transcripts derived from proprietary EST sequences that are not present in public databases offer the advantage to identify novel genes and splice isoforms not represented accurately by current gene prediction algorithms.

An important part of our data warehouse approach (Fig. 10.1) is a curated protein sequence index. This allows us to (i) control redundancy, (ii) include cross references to a variety of proteins sequence databases, (iii) provide stable identifiers for protein records, and (iv) track outdated and deleted sequences. The tracking of outdated database entries is very important to enable consistent analysis and reinvestigation of data. There is a huge turnover of identifiers mainly from predicted protein sequences. In NCBI's RefSeq database, for example, up to 50 percent of all predicted records change over the course of one year. Using our protein index, we have observed that more than 300,000 sequence entries have been replaced or deleted from NCBI Entrez during the last 2 years. However, close to 30 percent of these exact protein sequences are still present at Entrez and only referenced by another gene identifier (GI). This large flux illustrates the clear need to track these changes and control them at the database level rather than introducing this additional complexity to the scientists looking at long lists of protein identifications (see below). If one directly stores the identifiers (e.g., GIs) for identified proteins, some of these may later refer to a deleted record at the source database whereby important information may potentially be lost. Providing unique keys to sets of identical proteins is hence an important component of enabling control of redundancy and consistent analysis over time.

The most beneficial way of thinking about protein sequence databases for shotgun proteomics is to consider the units of information, the peptides. An optimal solution in this context would be to create preprocessed databases that contain only those proteins and isoforms representing existing biological molecules. Protein fragments that are based on translations of partial gene sequences but identical to full-length proteins should be removed. Conversely, isoforms or processed forms likely to exist but missing in the protein sequence database could be included. These comprise a large number of processed proteins, while the proteins present in the database are typically the nascent, full-length sequences. It was recently shown (Gevaert et al., 2003) that by using a theoretical preprocessing of the N-terminal sequences, 22 percent more matching N-terminal peptides could be found compared to just using nrdb, which lacks these processed proteins. Without the preprocessing, a high-quality experimental spectrum can thus not be matched to a biologically relevant peptide. Since very precise methods exist for prediction of, for example, signal peptides (see below), it is possible to identify full-length secreted proteins and to remove the signal peptide to include the processed isoform in the search database.

STRATEGIES TO REDUCE SAMPLE AND DATA COMPLEXITY

Due to the complexity of the proteome and the large dynamic range of protein abundance, a peptide mixture obtained from the digest of a total cell lysate is far too complex for an in-depth mass spectrometric analysis, and only the most abundant peptides can be identified. In order to increase the coverage of the proteome and to identify less abundant proteins, experimental approaches that reduce the actual sample complexity are of great importance. Since industrial proteomics applications create massive amounts of data, both protein informatics and bioinformatics strategies are important to reduce the data complexity. Protein informatics approaches can additionally contribute to the identification of less abundant proteins.

The complexity of the actual physical sample can only be reduced by experimental approaches (Fig. 10.5). Biochemical fractionation techniques can be used at the sample preparation stage to purify specific subcellular compartments, organelles, or even substructures within organelles such as the nucleolus (Fig. 10.2). Using fractionation techniques, we and others have achieved a larger coverage of organelle proteomes, for example, the mitochondrial (Mootha et al., 2003) and nucleolar proteome (Leung et al., 2003). In addition, the complexity of a protein or peptide mixture can be further reduced by chromatographic separation approaches, such as reverse-phase, ion exchange, size exclusion, and affinity chromatography. Examples include multidimensional protein identification technology, which combines ion exchange and reverse-

Category	Procedure	Benefit
Experimental		
Sample preparation & processing	Cell or tissue fractionation, subcellular fractionation	Limit analysis to specific compartments, Increase proteome coverage Reduce sample complexity Identify less abundant proteins
	Protein separation by biochemical methods	Focus analysis on proteins with specific properties
	Peptide separation by biochemical methods	Focus analysis on peptides with specific properties Introduce an additional parameter for peptide identification
Protein informatics		
Data acquisition	Removal of low quality spectra	Reduce search time and data volume
	Exclusion lists	Identify less abundant proteins Increase number of experimentally identified peptides Gain better coverage of proteins
Identification	Database search engines with improved peptide identification accuracy	Assign higher number of spectra Reduce number of false positives
	Tools for assessing database search results	Exclude low quality data
Data validation	Automated validation using expert knowledge	Automatically exclude low quality data
Bioinformatics		
Data integration	Standardized, pre-processed information in data warehouse	Achieve better data quality control, higher information content
	Tracking outdated protein identifiers	Enable consistent analysis over time
Data analysis	Clustering of proteins	Reduce complexity Exclude irrelevant protein groups
	Isoform distinction	Remove ambiguous protein identifications
	Filtering by annotation	Focus on relevant subset
	Applying statistics	Increase confidence of results

Figure 10.5. Overview of experimental, protein informatics and bioinformatics procedures and strategies that can be applied in order to reduce sample complexity, as well as data complexity and volume at various steps during a gel-free complex mixture analysis process.

phase chromatography (MudPIT; Washburn et al., 2001), and sucrose gradient centrifugation (Hanson et al., 2001). Alternatively, specific peptides can be isolated by lectin affinity chromatography (Bunkenborg et al., 2004), covalent chromatography (Wang and Regnier, 2001), combined fractional diagonal chromatography (COFRADIC; Gevaert et al., 2002), or affinity chromatography used for the isotope-coded affinity tag method (ICAT; Gygi et al., 1999). Phosphopeptides can be purified using antibodies specific for phosphorylated tyrosine (Pandey et al., 2000) or serine/threonine amino acid residues (Gronborg et al., 2002). The global enrichment for phosphorylated peptides using immobilized metal affinity chromatography (IMAC) columns has enabled the study of complex signaling pathways in yeast (Ficarro et al., 2002). Aliquots of individual fractions obtained from these separations are subjected to liquid chromatography tandem mass spectrometry (LC-MS/MS) analysis (Fig. 10.2).

Protein informatics strategies can contribute significantly to both a reduction of the data complexity and volume and to an increase in the coverage of the proteome. Importantly, they are also critical for data validation and data quality control. Approaches most relevant to industrial MS-based parallel protein analysis include identification and removal of bad spectra prior to database searching, selection of most intense precursor spectra for subsequent fragmentation in MS/MS mode, database search engines with a more robust protein identification capability, and tools that enable investigators to critically assess database search results. Typically, several thousand tandem mass spectra are generated for each fraction of a complex mixture analysis (the actual number depends on the instrument sensitivity). A significant portion of these spectra is of low quality and/or may not represent peptide fragmentations. These spectra should not be further analyzed. This consideration is especially relevant for industrial high-throughput approaches where a large number of mass spectrometers are run in parallel and massive amounts of data generated. In addition, when comprehensive protein databases or even the human genome sequence are used, the search time becomes a limiting factor, even for search engines that can run in parallel mode on a computer cluster (e.g., PepSea, Mascot). Efforts to automate the elimination of low-quality spectra thus are valuable to reduce search time and data volume, but dependent on the respective database search engine. This type of functionality is starting to be incorporated in mass spectrometry software packages (e.g., Agilent's Spectrum Mill MS Proteomics Workbench).

Peptide precursors that are derived from very abundant proteins will correspond to intense peaks in the survey spectrum and thus be preferentially selected for subsequent fragmentation in MS/MS mode over and over again, without increasing the number of different peptides and coverage of proteins. We therefore routinely apply another protein informatics approach to identify more and less abundant proteins: The same sample is analyzed three times and respective mass/charge ratios and retention times of peptides identified in the first run are cumulatively added onto an exclusion list. If peptides with the same mass/charge ratio and retention time are detected again, they will not be selected for an additional fragmentation. The same approach is taken in the second and third run, highly increasing the probability that less abundant peptide precursors are selected for fragmentation. Recent studies have reported increases in the number of protein identifications in an experimental sample by up to 50 percent, while simultaneously improving the peptide coverage by up to 30 percent (Kristensen et al., 2003). Importantly, the number of unambiguous protein identifications is increased because a larger number of independent units of peptide information are identified.

A very important and often overlooked area is that of validation of the database search results. Since this is a probabilistic approach, lists of possible peptides that match the experimental spectrum are returned. Different search engines use different matching algorithms and proprietary parameters and produce a varying proportion of peptide matches that are incorrectly assigned. This requires that validation tools are available that enable an expert to assess and validate the tandem mass spectra. It is apparent that such a manual validation is not feasible for industrial approaches. Therefore, an important step toward high-quality data acquisition in high-throughout mode is the automation of validation using expert knowledge.

The overall data set of assigned peptides has commonly changed after the validation step and the resulting list of protein identifications is different. These identifications have to be summarized over various fractions of one experiment (or repetitions of the same experiment) and can be compared to other experiments to provide a full overview. To support such tasks efficiently, a database solution for storage of the experimental results is the optimal solution (Kristensen et al., 2004). Due to the massive amounts of data that are being generated, subsequent data analysis and data mining greatly benefit from tools that can significantly reduce the data complexity. This can be achieved either by deploying statistical models for whether proteins are present in a sample (Nesvizhskii et al., 2003) with predictable sensitivity and false-positive identification error rates. Alternatively, we suggest implementation of expert rules into automated validation schemes with subsequent deployment of the right bioinformatic strategies (see Fig. 10.5) to bridge into the knowledge discovery process where the experimental data is investigated in the context of other publicly or proprietary available data (see below).

DATA WAREHOUSE INTEGRATION STRATEGIES TO FACILITATE LARGE-SCALE FUNCTIONAL PROTEOMICS ANALYSIS

We have chosen a data warehouse integration strategy to support the data analysis needs of our internal proteomics platform. In contrast to the federated approach, it requires a significant effort to generate software that download various data sets from different resources and further process them in an automated fashion. On the other hand, using a unified data model gives us complete control over the types of information we integrate. A data warehouse provides a platform for storing information collected from various sources in a customized format after processing and standardizing, while still keeping links back to the original information sources. By integrating these with the experimental data, a very efficient analysis of the huge data sets generated by large-scale shotgun proteomics approaches can be achieved (Fig. 10.1).

In the various databases of protein information, the same protein is often identified by different accession keys and often without comprehensive cross references to the accession keys of other databases. Even within the same database, the same protein can appear more than once with different accession keys and different annotation. To standardize information from various sources and incorporate it into the common framework of the data warehouse, we have introduced the concept of an unambiguous protein identifier. Any information that is incorporated into the data warehouse must refer, directly or indirectly, to such an identifier. The correlation between an external database key and the unambiguous protein identifier is obtained by sequence comparisons, disregarding any existing cross references provided by the external database. This makes it possible to obtain a comprehensive cross referencing while at the same time avoiding the propagation of cross-referencing errors.

A number of providers have realized the value of offering thorough and well-controlled annotations for the scientific community. The existing solutions range from sets of carefully annotated individual proteins to automated annotation projects of complete genomes, such as the Ensembl Genome Browser (*www.ensembl.org/*). Among the more comprehensive solutions in the category of carefully annotated protein databases are commercial databases, such as HumanPSD from Incyte, GeneCarta from Compugen, as well as UniProt/Swiss-Prot. These databases serve very broad scopes of scientific research and hence do not provide the format and the level of integration and standardization needed for large-scale proteomics research. Therefore, these solutions by themselves are not suitable without further processing. A data warehouse approach, for example, allows for the creation of the necessary consolidated data sets of annotation.

The type of annotation required to support efficient analysis of large-scale proteomics data is information that can be applied in an automated way to the data sets of thousands of protein identifications. The annotation should permit filtering, grouping, and sorting of complete data sets as well as the distinction of protein isoforms among all subsets of related proteins. Ideal for this purpose would be high-quality data in the form of compact/consolidated annotation information with little redundancy.

The reduction of the complexity of the overall data set can be achieved by grouping proteins based on common characteristics (e.g., proteins with transmembrane domains or proteins that contain a predicted signal peptide). Categories of annotation suitable for filtering include, for example, subcellular localization, conserved domains, selected categories of gene ontology, or pathway association. In contrast, the annotation required for a distinction of specific protein isoforms may include quite detailed information related to gene structures, including information on the affiliated genes, splice isoforms (where applicable), alleles, and whether the protein sequence represents a processed versus nascent protein or is a fragment. In certain cases it may moreover be relevant to have information related to RNA editing and sequencing errors/prediction errors and the evidence level for a protein sequence (e.g., predicted or cloned sequence) to assess whether a particular protein sequence is even likely to be real.

Even with an efficient data warehouse integration where annotation from public and proprietary data sets is consolidated, there still is a significant percentage of proteins that are poorly annotated. To deal with this we have introduced the strategy of computational enrichment, where state-of-the-art protein analysis and prediction software is run on every protein in the protein index to provide a comprehensive and consistent resource of complementary annotation. For example, information about specific types of subcellular localization, protein domains, and protein sequence similarity can be obtained from very accurate software packages. One such software tool is SignalP (Nielsen et al., 1997), which predicts signal peptides in secreted proteins. Its predictions have a very high specificity and sensitivity and can be used complementary to the annotation to divide huge data sets into proteins partly or entirely exposed on the exterior of the cell versus those that are not. Likewise, a number of transmembrane prediction methods have a high accuracy in correctly predicting the presence of one or several transmembrane segments (while being less specific in predicting the actual topology). Such tools, however, have to be chosen carefully since there are serious pitfalls associated with using tools that apply a large number of algorithms of varying accuracy to predict several subcellular localizations, for example, PSORT (Nakai and Horton, 1999). PSORT gives a good overview of the possible fate of a particular protein; but, since the sequence features associated with compartmentalization and transportation are still very poorly understood for many types of compartments, this kind of tool is not suitable for a large-scale prediction but rather for analysis of individual proteins by an expert user. Many prediction tools give a qualitative measure about the prediction accuracy (e.g., a probability). When predictions are to be incorporated in a data warehouse, it is often convenient to only store references to predictions above a certain threshold value. The actual prediction can then be rerun on demand if details are required. Annotation and prediction data can be standardized to fit a predefined set of codes, for example, GO terms, thereby allowing complex queries and enabling sorting and filtering. The data warehouse strategy facilitates an efficient bridging of experimental protein identifications to the subsequent bioinformatic analysis.

BRIDGING EXPERIMENTAL RESULTS TO BIOINFORMATIC ANALYSIS

The typical result of a single shotgun proteomics experiment is a long list with several hundreds of protein names and their identifiers. This list can number up to several thousand proteins when data from several experiments are combined. The ability to reduce this data complexity and to focus on a subset of proteins relevant to the respective biological question is a critical requirement for efficient data prioritization. To accomplish this, the mere list of protein names and identifiers has first to be turned into biologically meaningful objects. This can be achieved by matching detailed information from the data warehouse to the identified proteins and displaying it graphically. The bioinformatics system has to be able to provide the overview over the entire data set(s), while at the same time allowing to zoom in on any protein of interest and display additional information at an explicit level of detail. Many of the informatics solutions currently available for industrial proteomics research are not capable of adequately handling the complexity and redundancy issues associated with large-scale parallel protein-based analysis. We therefore describe solutions we have developed internally to solve these problems.

A first step in the reduction of data complexity involves the removal of redundant protein identifications, while recording their respective occurrences for statistics. Since the naming of proteins is very inconsistent with many aliases and synonyms in use, the unambiguous protein identifier (see below) is better suited for this purpose. However, this approach typically reduces the data set only slightly. As discussed above, nrdb is a commonly used database for the analysis of mass spectrometry data. It is nonredundant with respect to 100 percent identical sequences, yet contains many protein sequences from different species that differ by only one or a few amino acids. Experimentally identified peptides often cannot distinguish between these closely related proteins, nor between highly similar members of a protein family. Since the database searching is a probabilistic approach, independent database searches can return almost identical proteins from different species as the top hit that end up in the result list. An important technique that addresses this issue and significantly reduces data complexity is to cluster protein identifications by sequence similarity. Thereby a set of hundreds of protein identifiers can be transformed into a much smaller set of protein clusters. We have developed a tool for MDS Proteomics that facilitates clustering of thousands of proteins by sequence similarity (Fig. 10.6). The researcher can define the level of sequence similarity applied for the clustering and select additional parameters such as preferred species (e.g., human, mouse, rat), data quality, and statistics. Homologous sequences from different species are effectively grouped into protein clusters. The protein that represents this cluster is selected based on the species of choice, the statistical overview of how many times a respective protein was identified over different fractions or experimental repetitions, and a data quality score that represents an automated assignment of the experimental quality based on expert knowledge. Importantly, the automated assignments can be critically reviewed and changed. This clustering approach reduces the data complexity significantly and allows the researcher to focus on selected protein groups of interest.

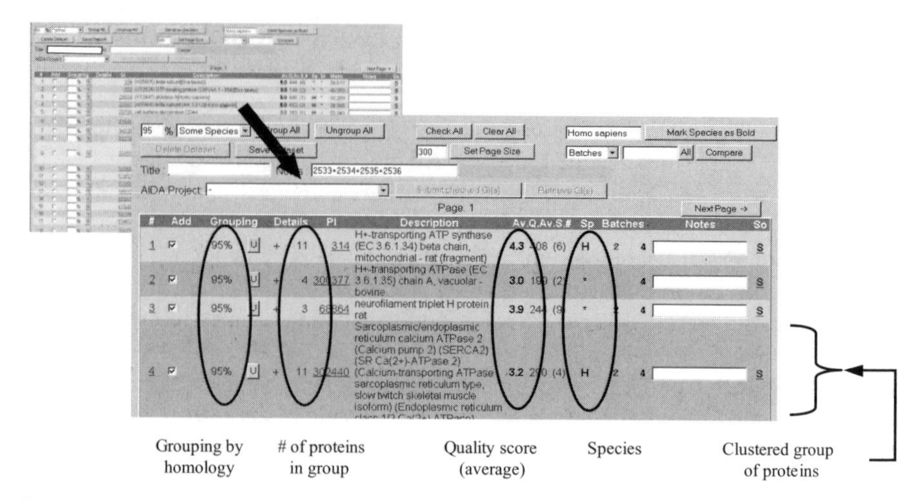

Figure 10.6. Screenshot of a software tool for the analysis of protein identifications from shotgun proteomics (developed for MDS Proteomics). The tool facilitates clustering and subsequent comparative analysis of thousands of protein identifications by various similarity criteria, thereby significantly reducing the complexity of the data set. Individual protein clusters can be examined in detail in the context of integrated public and proprietary data.

After reducing the complexity in this manner, the list of experimentally identified proteins can be further filtered to focus on specific subsets of proteins (or protein clusters). Ideally, all categories of information that are integrated in the data warehouse approach can be employed, such as annotation and computational predictions. Categories of annotation suitable for filtering include, for example, subcellular localization, conserved domains, and selected categories of gene ontology, pathway association, or isoform distinction. In addition, any type of computational prediction that is performed on all proteins in the database can be used in the filtering process. Typically, proteins containing transmembrane predictions or signal sequences can be grouped in this way. One could also use the results of subcellular localization predictions to bring those proteins scoring above a certain threshold to the top of the list. Combinations of filtering categories can also be applied, reminiscent of the sorting functionality of an Excel spreadsheet. For analyses that are very time-consuming, such as isoform distinction, it is extremely valuable to provide a first quick way of filtering the data, while leaving the final in-depth analysis to later stages.

The combination of the clustering and various filtering categories allows the user to efficiently extract a particular subset of proteins from the experimental results. The capability of the bioinformatics tools to zoom in on information for each individual protein facilitates a more detailed analysis of this subset. Additional bioinformatic analyses including comparative analysis, statistical analysis, isoform distinction, and mapping of experimentally identified proteins to other resources such as the literature can be carried out (Fig. 10.7).

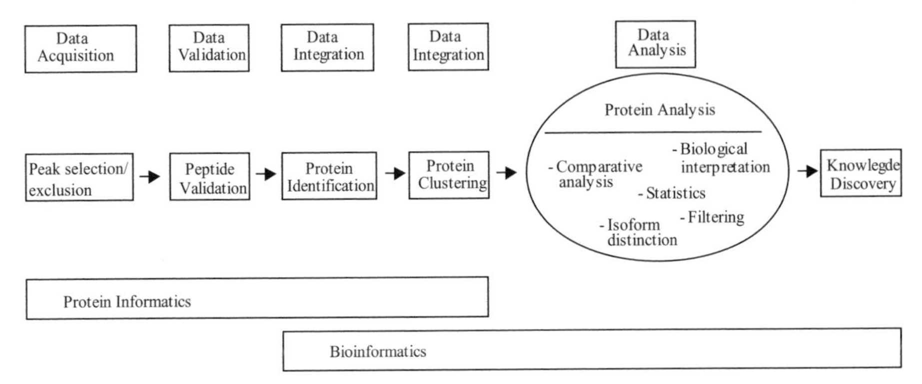

Figure 10.7. Overview of the steps in complex mixture analysis that involve protein informatics and bioinformatics applications. Categories of various process steps and examples of respective applications are shown in the upper and lower panel. These are mapped onto protein informatics or bioinformatics.

Comparative analysis of sets of proteins is a key component for the data analysis of complex mixtures and can include comparison of different fractions of the same experimental sample or from different treatments, or even from data sets merged from many experiments. The conceptually simple task of comparing results of lists of proteins becomes very difficult without relying on a standardized categorization. Using the data integration approach outlined in the previous section, we can seamlessly apply the protein categorization to the experimentally identified proteins, thereby efficiently grouping them on a sequence-based level. An efficient way of overcoming redundancy and complexity issues is to perform a comparative analysis based on protein clusters. Thereby, the resolution level can be initially set above the level of alleles and fragments, leaving this level of detail for subsequent analysis of relevant subsets. Comparative analysis can also be used to monitor the efficiency of different experimental prefractionation approaches in purifying a specific subset of proteins, for example, plasma membrane proteins.

The interpretation of the data with respect to isoform distinction is very time-consuming but of biological relevance. The difference between two alternative splice isoforms can be important for the function of the protein or correlate to tissue-specific expression. Knowledge of the exact nature of the gene product can thus be crucial for a biological interpretation. Carrying out an accurate distinction between protein isoforms is often a tedious manual process since a number of different data types/lines of evidence have to be reviewed. Relying on the unified data model from the data warehouse, this task can be automated to a varying degree. The protein identifiers of experimental result lists can be mapped to the detailed categorized isoform annotation of the data warehouse (e.g., specifying that the protein represents a C-terminal fragment of splice variant 1, which is one of three possible splice variants for this gene. One can therefore select at which level of resolution the results shall be compared, e.g. at the isoform level or at the gene level (disregarding more detailed information about different isofroms, variants and fragments). A second solution to distinguish protein isoforms involves the generation of a multiple sequence alignment and graphically displaying the experimental peptide information onto this alignment. Proteins in a cluster of interest can thereby be inspected visually, which greatly facilitates isoform distinction analysis.

This illustrates how bioinformatics applications can facilitate knowledge discovery by visualizing experimental evidence together with relevant preprocessed and integrated information in one combined view. The data warehouse strategy we have chosen greatly enhances our ability for an efficient visualization of these types of data. Other examples of such applications that facilitate efficient data mining, but that are not described above, include a tool that enables the scientist to get a fast overview of relevant literature. It maps literature records to protein records using protein names and their synonyms and aliases. It could be employed to compare the results of large-scale protein–protein interaction studies against the body of scientific literature, in order to identify interactions that have been described previously.

An important consequence of using an unified data model is that any type of information that can be mapped to the unambiguous protein identifiers can be integrated. Since gene and/or alternative splice variant associations are known for a large percentage of the protein index entries, it becomes possible to integrate gene expression data sets. This is one step toward integrating a large number of datasets for a systems biology analysis capability.

EXCHANGE OF PROTEOMICS DATA

The collection and comparison of data generated by different technologies and laboratories could significantly improve our understanding of the complex proteome. However, the lack of generally accepted data format standards prevents interchangeable use of proteomics data. In analogy to earlier initiatives aiming to standardize gene expression data formats, the Human Proteome Organization (HUPO) has formed an initiative to recommend standardized formats for proteomics data (*http://psidev.sourceforge.net/*). Data standards are initially aimed at mass spectrometry data, protein–protein interaction data, as well as a more general proteomics format. The data formats try to accommodate data from different experimental methods, while sharing common components such as analyzed system, sample description, and details of the relevant proteins. Such a format will support publications with detailed experimental results and will allow data exchange and the comparison of related proteomics experiments on a global level.

SUMMARY AND OUTLOOK

Industrial expression proteomics approaches aim to identify a significant portion of the proteome, notably potential drug targets and less abundant proteins. Gel-free parallel protein-based complex mixture analyis, or shotgun proteomics, offers the advantage over 2D gel-based approaches to more readily identify relevant target classes of membrane proteins. Shotgun proteomics approaches rely on contributions from various disciplines including biology, biochemistry, analytical, protein informatics, and bioinformatics. Appropriate fractionation steps are critical prerequisites for a reduction of the immense complexity of the proteome and an in-depth mass spectrometric analysis. Comprehensive databases are crucial to maximize the number of proteins that can be identified.

The critical bottleneck, however, remains data analysis and knowledge discovery from the massive amounts of data that are generated by industrial proteomics facilities. Lacking commercial software tools that adequately bridge the gap between experimental results and bioinformatic data analysis, we have used a data warehouse strategy capable of overcoming this bottleneck. Key components include the control of all protein sequences in a protein index, computational analysis of their sequence similarity relationships to each other to control redundancy and homology, and their detailed annotation through integration of relevant information from a large variety of external databases.

Experimental results consisiting of lists of mere protein names and identifiers can now be mapped and transformed into a rich information source. A reduction of the data complexity can be achieved by a combination of clustering proteins based on sequence similarity, as well as filtering by a large number of annotation categories. Tools that provide an overview of the entire data set, but at the same time are capable of zooming in and displaying the wealth of annotation for any protein, enable the scientist to efficiently prioritize the data and focus detailed analysis on a subset of proteins relevant to the respective biological question. The standardization of protein sequence data serves a dual purpose by both allowing efficient comparison of large experimental data sets and enabling the flexible and scalable integration of any type of data such as gene expression and large-scale protein–protein interaction. Such a strategy is an important step toward enabling a systems biology analysis capability and brings proteomics closer to capitalizing on its advantages over gene-centric technologies.

Future developments that will impact the widespread use of shotgun proteomics include more sensitive instrumentation, as well as more thorough automation of the steps from data acquisition up to the delivery of results in a rich biological context. This includes strategies and tools to remove low-quality spectra, to apply an automated validation using expert knowledge, and to improve the reliability of protein identifications. A crucial factor is the availability of software solutions that seamlessly integrate the experimental data with the wealth of relevant biological knowledge. This will allow shotgun proteomics to be applied by a large number of groups that lack an advanced bioinformatics and data integration capability.

By focusing on bioinformatics relevant to shotgun proteomics, we have not covered other areas of proteomics where bioinformatics plays a critical role, such as the application of advanced pattern recognition algorithms for biomarker discovery or analysis of protein microarray data.

Acknowledgments

The authors would like to thank colleagues at MDS Inc.—Denmark who contributed to the concepts and ideas outlined in this chapter. Furthermore, we would like to acknowledge Alexandre Podtelejnikov, Dan B. Kristensen, Jacek R. Wisniewski, and Jan Brønd for helpful discussions during the writing of this chapter.

REFERENCES

Aebersold, R. (2003). Constellations in a cellular universe. *Nature* **422**:115–116.

Anderson, N.L., and Anderson, N.G. (2002). The human plasma proteome: History, character, and diagnostic prospects. *Mol. Cell Proteomics* **1**:845–867.

Apweiler, R., Bairoch, A., Wu, C.H., Barker, W.C., Boeckmann, B., Ferro, S., Gasteiger, E., Huang, H., Lopez, R., Magrane, M., Martin, M.J., Natale, D.A., O'Donovan, C., Redaschi, N., and Yeh, L.S. (2004). UniProt: The Universal Protein knowledgebase. *Nucleic Acids Res.* **32**:D115–D119.

Bunkenborg, J., Pilch, B.J., Podtelejnikov, A.V., and Wiśniewski, J.R. (2004). Screening for N-glycosylated proteins by liquid chromatography mass spectrometry. *Proteomics* (in press). (2004), **4**(2):454–465.

Croft, L., Schandorff, S., Clark, F., Burrage, K., Arctander, P., and Mattick, J.S. (2000). ISIS, the intron information system, reveals the high frequency of alternative splicing in the human genome. *Nat. Genet.* **24**:340–341.

Ficarro, S.B., McCleland, M.L., Stukenberg, P.T., Burke, D.J., Ross, M.M., Shabanowitz, J., Hunt, D.F., and White, F.M. (2002). Phosphoproteome analysis by mass spectrometry and its application to *Saccharomyces cerevisiae*. *Nat. Biotechnol.* **20**:301–305.

FlyBase Consortium (2003). The FlyBase database of the *Drosophila* genome projects and community literature. *Nucleic Acids Res.* **31**:172–175.

Gevaert, K., Goethals, M., Martens, L., Van Damme, J., Staes, A., Thomas, G.R., and Vandekerckhove, J. (2003). Exploring proteomes and analyzing protein processing by mass spectrometric identification of sorted N-terminal peptides. *Nat. Biotechnol.* **21**:566–569.

Gevaert, K., Van Damme, J., Goethals, M., Thomas, G.R., Hoorelbeke, B., Demol, H., Martens, L., Puype, M., Staes, A., and Vandekerckhove, J. (2002). Chromatographic isolation of methionine-containing peptides for gel-free proteome analysis: Identification of more than 800 *Escherichia coli* proteins. *J. Mol. Cell Proteomics* **1**:896–903.

Gronborg, M., Kristiansen, T.Z., Stensballe, A., Andersen, J.S., Ohara, O., Mann, M., Jensen, O.N., and Pandey, A. (2002). A mass spectrometry-based proteomic approach for identification of serine/threonine-phosphorylated proteins by enrichment with phospho-specific antibodies: Identification of a novel protein, Frigg, as a protein kinase A substrate. *Mol. Cell Proteomics* **1**:517–527.

Gygi, S.P., Rist, B., Gerber, S.A., Turecek, F., Gelb, M.H., and Aebersold, R. (1999). Quantitative analysis of complex protein mixtures using isotope-coded affinity tags. *Nat. Biotechnol.* **17**:994–999.

Hanson, B.J., Schulenberg, B., Patton, W.F., and Capaldi, R.A. (2001). A novel subfractionation approach for mitochondrial proteins: A three-dimensional mitochondrial proteome map. *Electrophoresis* **22**:950–959.

Harris, T., Lee, R., Schwarz, E., Bradnam, K., Lawson, D., Chen, W., Blasier, D., Kenny, E., Cunningham, F., Kishore, R., Chan, J., Muller, H., Petcherski, A., Thorisson, G., Day, A., Bieri, T., Rogers, A., Chen, C., Spieth, J., Sternberg, P., Durbin, R., and Stein, L. (2003). WormBase: A cross-species database for comparative genomics. *Nucleic Acids Res.* **31**:133–137.

Hubbard, T., Barker, D., Birney, E., Cameron, G., Chen, Y., Clark, L., Cox, T., Cuff, J., Curwen, V., Down, T., Durbin, R., Eyras, E., Gilbert, J., Hammond, M., Huminiecki, L., Kasprzyk, A., Lehvaslaiho, H., Lijnzaad, P., Melsopp, C., Mongin, E., Pettett, R., Pocock, M., Potter, S., Rust, A., Schmidt, E., Searle, S., Slater, G., Smith, J., Spooner, W., Stabenau, A., Stalker, J., Stupka, E., Ureta-Vidal, A., Vastrik, I., and Clamp, M. (2002). The Ensembl genome database project. *Nucleic Acids Res.* **30**:38–41.

Jain, K.K. (2003). Proteomics—Technologies, Markets and Companies. Jain PharmaBiotech Report. Basel, Switzerland.

Kambhampati, D. (2003). *Protein Microarray Technology.* Wiley-VCH, Weinheim, Germany.

Kristensen, D.B., Podtelejnikov, A.V., Brønd, J.C., Nielsen, M.L.,Olsen, J.V., Wiœniewski, J.R., and Bennett, K.L. (2003). Multiple LC-MS exclusion list analyses: A tool to enhance protein identification from complex biological samples. Proceedings of the 51st ASMS Conference on Mass Spectrometry and Allied Topics, June 8–12, Montreal, Quebec, Canada (Abstract A031566).

Kristensen, D.B., Brond, J.C., Nielsen, P.A., Andersen, J.R., Sorensen, O.T., Jorgensen, V., Budin K., Matthiesen, J., Veno, P., Jespersen, H.M., Ahrens, C.H., Schandorff, S., Ruhoff, P.T., Wisniewski, J.R., Bennett, K.L., and Podtelejnikov, A.V. (2004). Experimental peptidde identification repository (EPIR): An integrated peptide-centric platform for validation and mining of tandem mass spectrometry data. *Mol. Cell Proteomics* **3**:1023–1038.

Leung, A.K., Andersen, J.S., Mann, M., and Lamond, A.I. (2003). Bioinformatic analysis of the nucleolus. *Biochem. J.* **376**:553–569.

Levanon, E.Y., and Sorek, R. (2003). The importance of alternative splicing in the drug discovery process. *TARGETS* **2**:109–114.

Link, A.J., Eng, J., Schieltz, D.M., Carmack, E., Mize, G.J., Morris, D.R., Garvik, B.M., and Yates, J.R. 3rd. (1999). Direct analysis of protein complexes using mass spectrometry. *Nat. Biotechnol.* **17**:676–682.

Martin, S.E., Shabanowitz, J., Hunt, D.F., and Marto, J.A. (2000). Subfemtomole MS and MS/MS peptide sequence analysis using nano-HPLC micro-ESI fourier transform ion cyclotron resonance mass spectrometry. *Anal. Chem.* **72**:4266–4274.

Mironov, A.A., Fickett, J.W., and Gelfand, M.S. (1999). Frequent alternative splicing of human genes. *Genome Res.* **9**:1288–1293.

Mootha, V.K., Bunkenborg, J., Olsen, J.V., Hjerrild, M., Wisniewski, J.R., Stahl, E., Bolouri, M.S., Ray, H.N., Sihag, S., Kamal, M., Patterson, N., Lander, E.S., and Mann, M. (2003). Integrated analysis of protein composition, tissue diversity, and gene regulation in mouse mitochondria. *Cell* **115**:629–640.

Nakai, K., and Horton, P. (1999). PSORT: A program for detecting sorting signals in proteins and predicting their subcellular localization. *Trends Biochem. Sci.* **24**:34–36.

Nesvizhskii, A.I., Keller, A., Kolker, E., and Aebersold, R. (2003). A statistical model for identifying proteins by tandem mass spectrometry *Anal. Chem.* **75**:4646–4658.

Nielsen, H., Engelbrecht, J., Brunak, S., and von Heijne, G. (1997). Identification of prokaryotic and eukaryotic signal peptides and prediction of their cleavage sites. *Protein Engr.* **10**:1–6.

O'Farrell, P.H. (1975). High resolution two-dimensional electrophoresis of proteins. *J. Biol. Chem.* **250**:4007–4021.

Olsen, J.V., Andersen, J.R., Nielsen, P.A., Nielsen, M.L., Figeys, D., Mann, M., and Wisniewski, J.R. (2004). Hystag—a novel proteomic quantification tool applied to differential display analysis of membrane proteins from distinct areas of mouse brain. *Mol. Cell Proteomics* (2004), **3**(1):82–92.

Ong, S.E., Blagoev, B., Kratchmarova, I., Kristensen, D.B., Steen, H., Pandey, A., and Mann, M. (2002). Stable isotope labeling by amino acids in cell culture, SILAC, as a simple and accurate approach to expression proteomics. *Mol. Cell Proteomics* **1**:376–386.

Pandey, A., Podtelejnikov, A.V., Blagoev, B., Bustelo, X.R., Mann, M., and Lodish, H.F. (2000). Analysis of receptor signaling pathways by mass spectrometry: Identification of vav-2 as a substrate of the epidermal and platelet-derived growth factor receptors. *Proc. Natl. Acad. Sci. USA* **97**:179–184.

Patterson, S.D. (2003). Data analysis—the Achilles heel of proteomics *Nat. Biotechnol.* **21**:221–222.

Perkins, D.N., Pappin, D.J., Creasy, D.M., and Cottrell, J.S. (1999). Probability-based protein identification by searching sequence databases using mass spectrometry data. *Electrophoresis* **20**:3551–3567.

Pruitt, K., and Maglott, D. (2001). RefSeq and LocusLink: NCBI gene-centered resources. *Nucleic Acids Res.* **29**:137–140.

Thanaraj, T.A., Stamm, S., Clark, F., Riethoven, J.J., Le Texier, V., and Muilu, J. (2004). ASD: The Alternative Splicing Database. *Nucleic Acids Res.* **32**:D64–D69.

Tyers, M., and Mann, M. (2003). From genomics to proteomics *Nature* **422**:193–197.

Wang, S., and Regnier, F.E. (2001). Proteomics based on selecting and quantifying cysteine containing peptides by covalent chromatography. *J. Chromatogr. A* **924**:345–357.

Washburn, M.P., Wolters, D., and Yates, J.R. 3rd (2001). Large-scale analysis of the yeast proteome by multidimensional protein identification technology. *Nat. Biotechnol.* **19**:242–247.

Wolters, D.A., Washburn, M.P., and Yates, J.R. 3rd (2001). An automated multidimensional protein identification technology for shotgun proteomics. *Anal. Chem.* **73**:5683–5690.

11

PROTEIN ARRAYS

David S. Wilson and Steffen Nock

Absalus, Inc.
Mountain View, CA

Industrial Proteomics: Applications for Biotechnology and Pharmaceuticals, edited by Daniel Figeys
ISBN 0-471-45714-0

INTRODUCTION

The discipline known as *molecular biology* has historically defined itself by focusing on the individual genes and their products within living organisms. The power of this science lies in its precision, being able to reduce phenotypic activities to tangible molecular species. Such a focused approach does, however, have its limitations in that complex, multigenic phenotypes often cannot be understood without simultaneous consideration of numerous gene products and their interactions with each other. The advent of deoxyribonucleic acid (DNA) microarrays, however, has altered the perspective of modern biologists, providing them with a systemwide view of biological processes coupled with molecular precision. DNA microarrays allow for the simultaneous measurement of messenger ribonucleic acid (mRNA) expression from thousands of genes in a single experiment (Schena et al., 1995). Such arrays have already been used to profile complex diseases such as leukemia and breast and prostate cancer (Dhanasekaran et al., 2001; Golub et al., 1999; Perou et al., 2000).

The functional products of nearly all genes, however, are proteins, which are therefore the targets of pharmaceutical interventions. mRNA expression profiling microarrays provide only a partial picture of the complex interplay of biomolecules in a cell or an organism. The correlation between the abundance of mRNA and the encoded proteins is poor (Anderson and Seilhamer, 1997; Gygi et al., 1999). Furthermore, mRNA expression data provide no information about the activity, posttranslational modification, or localization of proteins. Proteomics, defined as the analysis of the protein complement expressed by a genome (Pennington et al., 1997), is required to accurately complete the quantitative description of the precise state of a complex biological system.

In order to take an analogous systems-oriented approach at the proteomic level, protein microarrays have been developed. Miniaturization and multiplexation dramatically increase the rate of data generation, allowing the researcher to examine a wider, less biased segment of the proteome when dissecting biological events. There are a variety of types of protein arrays, each one being tailored to a particular type of application. Expression profiling arrays, consisting of immobilized protein capture molecules such as antibodies, provide information about protein abundance, while functional arrays help to decipher protein–protein, protein–DNA, and protein–small molecule interactions as well as enzymatic profiles. Protein arrays have the necessary breadth, throughput, robustness, flexibility, and reproducibility to become essential tools for the analysis of biological phenomena (Jenkins and Pennington, 2001).

ARRAY CONFIGURATIONS

Proteins, unlike DNA, are very fragile biomolecules whose function is entirely dependent on the integrity of their three-dimensional structure. Slight changes in the environment, such as pH shifts, dehydration, elevated temperature, exposure to certain material surfaces, and so forth, often lead to reduced or complete loss of activity. Therefore, it is very important to immobilize proteins using appropriate dispensing conditions as well as surface immobilization chemistries. The biophysical properties of proteins must be taken into account when building a miniaturized protein analysis platform. Furthermore the development and production of proteins is much more complicated and time-consuming than it is for DNA. DNA capture probes for arrays can be designed in silico and easily synthesized using either polymerase chain reaction (PCR) or chemical synthesis. Antibodies or other protein capture molecules, however, cannot be predicted using the amino acid sequence of the target protein. They rather must be developed using immunization or other approaches. Another hurdle for protein arrays lies in the analyzed sample itself. Proteins cannot be amplified and labeled using methods such as PCR. The burden for high-sensitivity detection in expression profiling arrays therefore shifts to the development of high-affinity, highly specific capture molecules. The next section of this review will outline these issues in reference to developing functional protein arrays.

SOLID VERSUS LIQUID ARRAYS

The historical roots of protein microarrays are the macroscopic immunoassays, generally performed in 96-well microtiter plates. In the late 1950s, work by Ekins in London (Ekins, 1960) and Berson and Yalow in New York (Yalow and Berson, 1960) developed the first sensitive immunoassay, based on radioisotope detection. In the late 1980s, Roger Ekins speculated that microspots of antibodies on a solid support could lead to multiplexed, extremely sensitive, and quantitative assays for protein abundance (Ekins et al., 1990). It was not until the late 1990s, however, that microarrays were developed, initially for mRNA expression profiling and, by the end of the decade, for proteins.

At present there are three major types of configuration for multiplexed protein analysis platforms: (i) the classical two-dimensional (2D) array format, comparable to DNA arrays, (ii) encoded particle-based systems, and (iii) multiplexed, solution-based homogeneous assays with reporter molecules. Some properties of the different systems are compared in Table 11.1. The 2D arrays and particle-based configurations rely on a solid support for protein immobilization, whereas the solution-based homogenous assay is performed entirely in solution. An example of the latter configuration is the e-Tag system from Aclara Biosciences, where fluorescent e-Tag reporters are coupled to biological probes such as antibodies. When the e-Tag reporter-probe conjugate becomes bound to its target molecule, the linkage between the reporter and the e-Tag is cleaved. The basis for this target-bound-specific cleavage is that a second target-binding molecule is also included in the assay and is conjugated to a factor that can cleave the e-Tag from the probe, but only when the cleaving agent is in close proximity to the probe–e-Tag conjugate. The target thus serves to bring the e-Tag probe conjugate and the cleaving factor together in a ternary ("sandwich") complex and facilitates the cleavage. In such an assay, the amount of cleavage is indicative of the concentration of the target species.

This platform can be multiplexed in the following manner. Each probe is attached to a unique e-Tag reporter molecule. After performing the assay, the cleaved probes are

TABLE 11.1. Comparison of Multiplexed Protein Analysis Platforms

	Two-Dimensional Array	Particle-Based Array	Solution-Based Homogenous Assay
Assay type	Direct capture, sandwich assay, functional assays	Direct capture, sandwich assay	Sandwich assay
Assay format	Nonequilibrium[a]	Equilibrium if washing step is omitted	Real equilibrium
Dispensing	Spotting	Batch incubation	n.a.[b]
Detection	Fluorescence, radioactivity, SPR, mass spectrometry,	Fluorescence	Fluorescence
Multiplexing	High	Medium	Low to medium

[a] Zeptosens offers a fluorescence planar waveguide system that does not require a washing step and therefore allows for measurements under equilibrium conditions (Pawlak et al. 2002).
[b] n.a., not applicable.

separated from each other using capillary electrophoresis. The migrating position of each reporter is known, which allows one to translate the fluorescent signal of each reporter into the degree to which each probe was bound by its target molecule. With proper calibration, one can draw conclusions about the abundance of the different target proteins in the sample. This system has the advantage that the assay is homogenous and performed under equilibrium conditions, meaning that no potentially disruptive washing steps are required at any point in the assay. By contrast, most 2D arrays and particle-based systems require the inclusion of washing steps after incubation with the sample, and this often disturbs the equilibrium and lowers the sensitivity of the assay. For the e-Tag system, two binding reagents have to be developed for each analyte. One carries the e-Tag, whereas the other carries the reactivity that allows the release of the tag once they are in close proximity. This system is fairly new and no peer-reviewed publications are available at this point.

Two-dimensional protein arrays rely on positional coding to identify a matrix of protein-binding agents arrayed on a chip surface. They have been used to profile numerous analytes in complex protein mixtures as well as to measure protein–protein interactions and enzymatic activities (see below). Haab et al. (2001), for example, used an array of 115 different capture antibodies to measure the cognate analytes in a mixture of fluorescently labeled protein antigens. Particle-based systems employ beads or nanoparticles as the solid support for the protein-binding agent. Each particle is derivatized with one particular capture reagent. Each particle is also labeled in some way so as to distinguish it from other particles that represent other specificities. The most common way to encode beads relies on the incorporation of different ratios of two or more fluorophores (Kettman et al., 1998). Alternatively, nanobars can be encoded with "bar codes"—stripes of different metals along the axis of the particle (Nicewarner-Pena, et al., 2001). Unlike 2D arrays, where the protein-binding agents must be individually spotted onto each array, particle-based system allow for batch production of particles carrying the same specificity. This makes the production process much easier, less expensive, faster, and reliable. In addition, whereas spotting microscopic amounts of liquid onto a flat surface can result in damaging evaporation, no such issues occur for the derivatization of particles in suspension.

Two-dimensional arrays allow for incorporation of a variety of detection systems. Confocal fluorescence scanners offer very high sensitivity and good dynamic range, but these systems rely on either sample labeling or the use of sandwich assays. The 2D arrays also allow for the use of truly label-independent detection methods such as surface plasmon resonance (SPR) and mass spectrometry (MS). MS can offer the additional advantage of providing detailed information as to the identity and modification state of a captured protein. Particle-based systems and homogenous arrays are generally limited to fluorescence as the readout. This limits their applicability in discovery research, when the goal is to determine the identity of proteins.

Particle-based and 2D arrays provide similar opportunities for measuring biochemical species and activities. For expression profiling studies, immobilized protein-binding agents can capture analytes directly from fluorescently labeled protein mixtures. To obtain higher sensitivity, a sandwich-type assay using a fluorescently labeled detection antibody can be used. Functional assays, however, are not generally transferable onto bead-based or homogenous liquid array systems. The different assay formats are illustrated in Figure 11.1.

One advantage of 2D arrays over liquid or particle-based systems is the degree of multiplexing. A 2D array can easily be built of hundreds or thousands of features, whereas particle-based systems are often limited by the number of individually encoded features. In all of these systems, a major limitation with respect to multiplexing lies in the specificity of the protein-capture agents. Array systems with many features are much more prone to nonspecific interactions than low-density arrays. It is therefore extremely important to choose the protein-binding agents very carefully and preferentially add some redundancy into the assay.

The key elements for the construction of a protein array are (a) the nature of the solid support and the surface coating, (b) the way in which the proteins are applied to the solid support, (c) the detection of the binding of the target molecules to the array, and (d) the generation and isolation of proteins for the array.

OPTIMIZING SURFACES FOR PROTEIN IMMOBILIZATION

Surfaces used for protein arrays can be roughly broken down into two categories—those based on physical adsorption and those based on covalent/quasi-covalent interactions. Physical adsorption is used to attach proteins to hydrophobic plastics such as polystyrene or highly charged surfaces such as polylysine or nitrocellulose. These interactions are noncovalent, involve multiple attachment points, and are difficult to control. Also, while some proteins remain largely active when immobilized in this way, others can be almost totally inactivated. Therefore, when arraying a large number of proteins

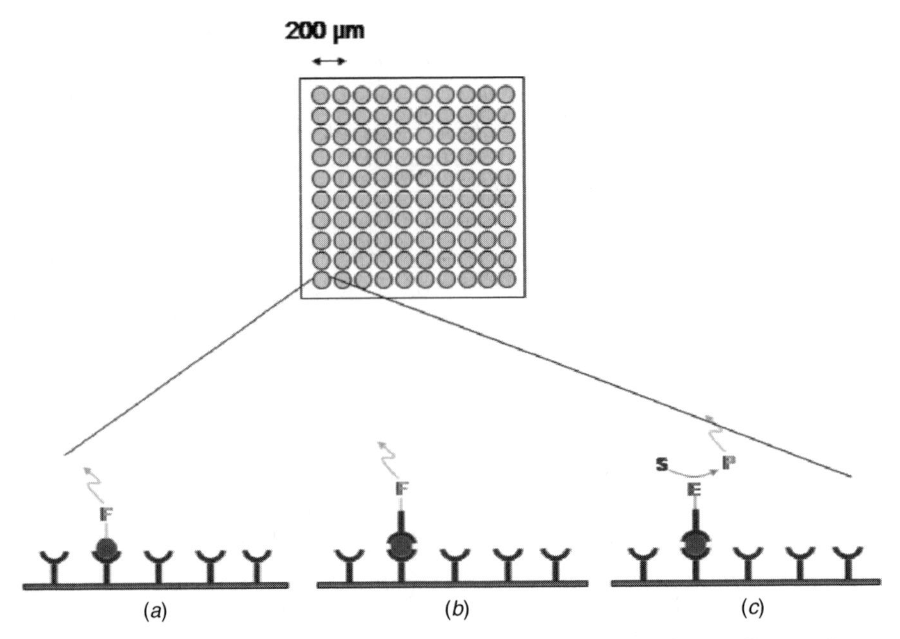

Figure 11.1. Schematic representation of the most commonly used assay formats for protein expression profiling. (*a*) direct labeling of the protein analyte using fluorescent dyes. (*b*) Sandwich-type assay with detection of the target protein by a fluorescently labeled detection antibody or (*c*) an enzyme-conjugated detection antibody, allowing for an ELISA-based detection with a fluorescent, chemiluminescent, or colorimetric readout (S = substrate, P = product).

by physical adsorption, it is likely that many of them will not retain activity. It is often observed that proteins denature on a surface due to too many contact points (Butler et al., 1992, 1993) or can bleed off from these surfaces during the course of the assay. The latter difficulty can be alleviated by performing a covalent cross-linking reaction subsequent to immobilizing the antibody noncovalently (Miller et al., 2003), although this may inactivate some of the attached proteins. One advantage of these surfaces, however, is the ease of use. No protein modification needs to be performed prior to immobilization.

A preferred method involves a small number of covalent or strong noncovalent (i.e., biotin-avidin) bonds, such that the number of potentially inactivating interactions with the surface is limited. Optimally, the attachment point is far away from the binding or active site of the protein, leaving its function largely unaltered. Examples include covalent attachment of proteins (MacBeath and Schreiber, 2000), immobilization of biotinylated proteins onto streptavidin-coated surfaces (Ruiz-Taylor et al., 2001), and immobilization of histidine-tagged proteins onto Ni^{2+}-chelating surfaces (Zhu et al., 2001). The immobilization method used can also affect the density of proteins packed onto the surface. The intensity of specific signal produced on a feature of an array is a function of the capture agent surface density and fractional activity. Peluso et al. (2003)

investigated how these two factors are affected by the orientation of the capture agents on the surface by comparing randomly versus specifically oriented capture agents based on both full-sized antibodies and Fab′ fragments. They showed that specific orientation of capture agents consistently increased the analyte-binding capacity of the surfaces, with up to 10-fold improvements over surfaces with randomly oriented capture agents.

Flat surfaces have an inherent limitation on the density of proteins that can be immobilized per unit area. To go beyond this, it is necessary to build from the surface a gel, mesh, or brush structure onto which the proteins are immobilized. One potential drawback may be poor penetration into and out of the structure, leading to slow equilibration times and/or higher background signal. This approach is reviewed in the reference by Mirzabekov and Kolchinsky (2002) and a commercial slide based on this concept is sold by Perkin-Elmer Life Sciences (Hydrogel Coated Slide). The hydrogel is based on polyacrylamide and the proteins are immobilized noncovalently, but the exact surface chemistry has not been disclosed. In the product literature, the manufacturers claim that the slides can immobilize antibodies at a density and activity capable of binding 13.5 pmol/cm^2 of antigen, which is five- to sixfold higher than obtainable on a flat surface with a layer of densely packed, oriented antibodies (Peluso et al., 2003). This novel hydrogel surface was utilized by Miller et al. (2003), as described below, and found to provide better signal-to-noise ratios than various surface chemistries (polylysine, nitrocellulose, aminosilane, aldehyde-functionalized monolayers) on flat glass slides. The hydrogel also appears to protect proteins from denaturation due to dehydration or other factors, as evidenced by the fact that the percentage of antibodies that were active on the hydrogel surface was about three times higher than on the flat polylysine-coated surface (Miller et al., 2003).

Derivatization chemistries for particles are also well established. Capture chemistries developed for 2D surfaces can be readily adapted to nanobars and beads (Zhou et al., 2001).

SPOTTING TECHNOLOGIES

The basic requirements for the arraying procedure are reproducibility, spatial addressability, retention of protein activity, and speed. Currently there are a variety of printing tools available. Most of these tools, however, were originally developed for DNA spotting and are not ideal for printing proteins. Printing methods can be divided into two types—contact and noncontact. In contact mode, a tiny needle or capillary picks up a small volume of protein solution and then makes contact with the derivatized surface, thus transferring a subnanoliter volume (Haab et al., 2001; Joos et al., 2000; Moody et al., 2001). The use of ink-jet print-heads, in contrast, allows for noncontact printing. Picoliter amounts of protein solution are applied to the surface without direct physical contact between the printing device and the array (Roda et al., 2000). A printing device based on piezo technology is currently marketed by Perkin-Elmer Life Sciences. Other printing methods include electrospray deposition on slightly conductive surfaces (Avseenko et al., 2002; Moerman et al., 2001; Morozov and Morozova, 1999),

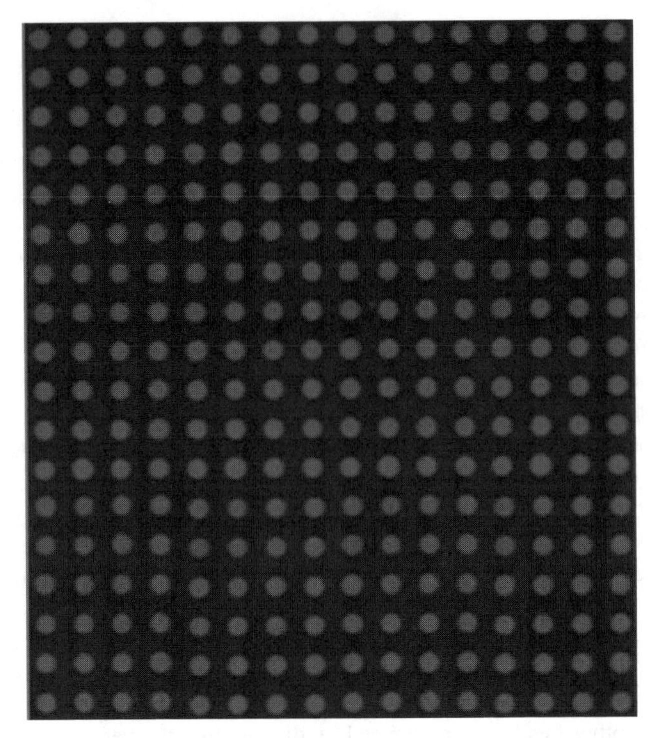

Figure 11.2. Protein array printed using TopSpot technology. A 24-position print-head was used to print a 4 × 3 antibody array using directly labeled Cy3 goat IgG (0.66 mg/mL) in 1× PBS onto ArrayLink Hyphob slides. Integral signal intensity CV <4%. (With kind permission from IMTEK, Chair for MEMS Applications, University of Freiburg, Germany and HSG-IMIT, Villingen-Schwenningen, Germany)

deposition by a hydrogel stamper inked with an aqueous protein solution (Martin et al., 1998), and direct application of protein solutions via microfluidic networks (Bernard et al., 2000; Rowe et al., 1999a). All of these methods, however, face the issue of protein dehydration during the long serial printing process. One solution to this problem was proposed by MacBeath and Schreiber (2000). They spotted in a humidified chamber and added glycerol to the protein solution in order to prevent evaporation. Whether the arraying methods described above are robust and versatile enough to produce protein arrays in a manufacturing environment remains to be seen. Another approach to overcome evaporation and to speed up the printing process is the so-called Top Spot Technology developed by IMTEC. It created a highly parallel dispensing tool to print liquids in the subnanoliter range. The print-heads carry up to 96 nozzles, allowing for the simultaneous deposition of 96 different proteins at a time with one piezo stroke. An example of a protein array created using this technology is given in Figure 11.2.

Particle-based systems do not face this issue because protein immobilization is done in a batch mode by incubating the activated particles with the different protein solutions (Carson and Vignali, 1999).

DETECTION TECHNOLOGIES

The final step in a chip experiment is to detect the bound proteins at each feature. Up to the present time, nearly all reports of protein arrays rely on detection by fluorescence or chemiluminescence. Different charge coupled device (CCD)-based cameras or laser scanners with confocal detection optics are currently available for use with planar arrays. Flow-cytometry-based instruments are available for bead-based systems (Fulton et al., 1997). In order to use fluorescence-based detection, fluorophores must be introduced into the sample. In the mRNA expression profiling field, the sample is labeled concomitantly with polymerase-based amplification, using fluorescent deoxyribonucleutide triphosphate (dNTP) analogs. Unfortunately, there is no equivalent reaction available for protein labeling. Proteins can, however, be fluorescently labeled using chemical reactions.

Haab et al. (2001) have taken a chemical labeling approach to introduce fluorescent dyes into complex protein mixtures. They borrowed a concept from the mRNA expression profiling technology of comparing two different samples by labeling them with different fluorophores, mixing them together, and then applying them to an array (Fig. 11.1*a*). If the two dyes have different spectral properties, they can be distinguished by a fluorescence scanner. This two-sample/two-color concept can be used to compare normal versus treated samples, normal versus pathological samples, or to quantify the amounts of different proteins in an experimental sample by comparing it to a mock sample with known amounts of each protein. Analogous to a DNA chip experiment, one protein mixture was labeled with the dye Cy3, the other with Cy5. After labeling, the two protein samples were mixed together and then applied to the array. After incubation and washing, the relative intensity of Cy3 and Cy5 fluorescence on each spot on the array was determined using a fluorescence reader. They observed that 60 percent of the antibodies could give a qualitative indication of the presence of their cognate antigen at the highest concentration used. Quantitative data could only be collected for 23 percent of the antibodies. The sensitivity was limited in part by nonspecific binding of labeled proteins to the array. This is one of the disadvantages of whole-sample labeling. There are a number of "sticky" proteins in most natural protein mixtures, especially those derived from blood, and these produce signal throughout the array. Also, the labeling procedure, by replacing charged lysine residues with bulky hydrophobic dyes, can mask antigenicity and can dramatically lower the solubility of proteins, thus resulting in nonspecific surface adsorption. Another issue with whole-sample labeling is that it is difficult to control the degree of labeling and this compromises reproducibility. Furthermore some proteins label more efficiently than others, leading to unequal sensitivities for different proteins in the sample.

An alternative to the direct labeling of analytes is the use of a sandwich assay. This assay format uses two antibodies against the same target protein, each one recognizing different epitopes (Figs. 11.1*b* and 11.1*c*). In a typical assay setup, a capture antibody is immobilized on the solid support. Once the target protein is captured, a second antibody (the *detection antibody*) that carries a fluorescent label or an enzyme that can produce a fluorescent, chemiluminescent, or colored product is applied to the array. Upon binding of the detection antibody to the cognate analyte on the array, the amount of captured target protein can be determined by the fluorescence intensity relative to a standard curve. The advantage of this sandwich setup lies in the fact that two binding events have to take place in order to produce a specific signal on the array. Furthermore the sensitivity of the system is higher because the detection antibody can carry multiple fluorophores or an enzyme that amplifies a signal. The disadvantage of this approach is that it requires two specific binding agents for each protein that is to be measured. Several examples have been described of protein arrays and bead-based systems using sandwich pairs to detect, for example, human cytokines (Carson and Vignali, 1999; Huang, 2001; Huang et al., 2001; Lin et al., 2002; Mendoza et al., 1999; Moody et al., 2001; Wiese et al., 2001).

Schweitzer et al. (2000) developed another way to increase the sensitivity of a sandwich assay. They used oligonucleotide-labeled detection antibodies in combination with immunoRCA (rolling circle amplification). Upon localization of the detection antibody on the array, the oligonucleotide on the antibody is extended by a DNA polymerase using a single-stranded circular DNA template. This polymerization step creates a long, single-stranded DNA sequence tethered to the specific feature of the array. Hybridization of fluorescently labeled oligonucleotides to this DNA leads to a high density of fluorophores on the feature. Using this approach, Schweitzer et al. (2002) were able to detect cytokine concentrations in the low picomolar regime. This corresponds to detection limits usually obtained using a classical 96-well plate enzyme-linked immunosorbent assay (ELISA) with enzymatic signal amplification. A typical example of an array developed using immunoRCA is shown in Figure 11.3.

Alternative strategies for the detection of protein-binding events on array surfaces rely on planar waveguide technology in conjunction with fluorescent labels, as described

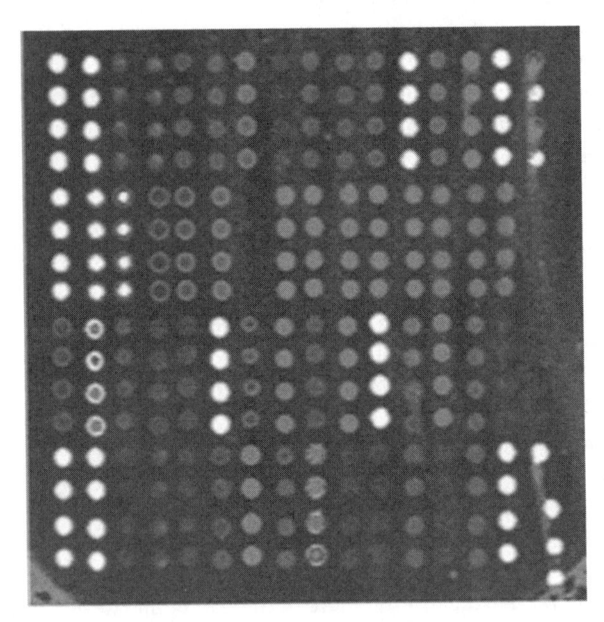

Figure 11.3. (See color insert.) Fluorescence image of a multiplexed cytokine sandwich assay using rolling circle amplification of the signal. Signal intensity is represented in pseudocolor. (With kind permission from Molecular Staging, Inc.)

by Rowe et al. (1999b, 1999c) and Duveneck et al. (2003). With this technology, only the surface-bound fluorescently labeled molecules are detected, so no washing step is required. This allows one to take measurements under equilibrium conditions.

Several groups are also working on interfacing protein arrays with mass spectrometry (MS). In contrast to fluorescence and other optical readouts, MS gives additional molecular weight information about the surface-bound protein. The most widely used system relies on matrix-assisted laser desorption ionization (MALDI) or surface enhanced laser desorption and ionization (SELDI-MS). The target may be captured either by surface-immobilized antibodies (Borrebaeck et al., 2001) or by different chromatographic matrices and analyzed directly by MS (Merchant and Weinberger, 2000). One limitation of this system is the relative low sensitivity in comparison to fluorescence-based methods and the need to treat proteolytically the surface-captured proteins prior to MALDI analysis. It is also difficult to obtain quantitative data based on a purely MS-based approach. To overcome these limitations, MS has been combined with SPR biosensors (Nelson et al., 2000).

CONTENT

The number of protein array platforms continues to grow at an astonishing rate, and the completion of the Human Genome Project has revealed hundreds of thousands of proteins that biologists would like to measure. The bottleneck at the present time in applying array technology to proteomics is the limited number of biological reagents that are available. The development of such reagents has not increased in proportion to the platforms and genomics information. In particular, the types of reagents most useful for protein arrays—sandwich antibodies that recognize native proteins—are in great demand. Although hundreds of companies offer thousands of different antibodies, there are only about 150 to 200 sandwich pairs available to human proteins.

The resources required to develop an antibody sandwich pair should not be underestimated. It involves (i) finding an appropriate source of a complementary DNA (cDNA) and verifying its sequence, (ii) cloning the cDNA into an expression vector, (iii) expressing the protein (generally at least 2 mg) in a functional form and purifying sufficient quantities for binding agent development, (iv) making antibodies or other binding agents, and (v) identifying and validating a sandwich pair of binding agents. A typical cost for outsourcing this entire project is in the range of $25,000 to $50,000 per antigen and may require up to a year to complete.

There have been several attempts to shortcut this procedure. For example, it may be possible to bypass the protein production by immunizing with DNA, which subsequently gets expressed inside the animal. This method may not be robust enough, however, as there is a large variation in expression levels of different coding sequences. Recently, the use of in vivo DNA electroporation has been described and may increase the robustness of this technology to the point where it can be relied upon to immunize animals (Zhang et al., 2002). In such a case, once the antibodies are made, they are validated using mammalian cell lines transfected with DNA encoding the antigen, and compared with nontransfected cells for reactivity with the antibodies in question. It is still too early to know whether this technology will be robust enough to replace conventional immunization methods.

Public and private collections of cDNAs include the Harvard Institute of Proteomics, the IMAGE Consortium, the Mammalian Gene Collection, the American Tissue Culture Collection (ATCC), as well as a number of private companies such as Open Biosystems and Invitrogen. Currently, there are no companies supplying similar collections of purified recombinant human proteins or antibodies.

Using synthetic peptide antigens generally produces lower affinity antibodies than using the full-length protein, but this approach may be preferable for generating antibodies against specific phosphoproteins, for example, that would otherwise be difficult to obtain in large quantities. One still needs to produce some full-length protein to validate the reagents in this case, however.

In vitro methods for creating specific protein-binding agents have also been described. These include phage display and ribosome display of antibody fragments (Hanes et al., 2000), mRNA display (Wilson et al., 2001; Xu et al., 2003), and systematic evolution of ligands by exponential enrichment (SELEX) of nucleic acid libraries (Brody and Gold, 2000; Robertson and Ellington, 2001). As in monoclonal antibody development, however, the majority of resources are used in generating the antigen and screening through all of the binding agents that are created to identify those with sufficient affinity and specificity. One must conduct a detailed work flow cost analysis on these methods before concluding that they are more efficient than "traditional" approaches. To this date, none of these novel methods has proven itself to be more economical, or capable of generating better binding agents, than traditional monoclonal antibody methods. For this reason, companies such as QualityProteomics are focusing on streamlining the monoclonal approach to increase the efficiency, especially of the postimmunization antibody-screening process. Through assay miniaturization and other process improvements, the development cost per antibody can be reduced by 5- to 10-fold. Under these circumstances, it becomes economically feasible to begin creating the biological content for analyzing hundreds to thousands of human proteins in any of the various array formats described in the previous section.

The Human Proteome Organization (HUPO) is a nonprofit, international organization dedicated to proteomics and is coordinating an *antibody initiative*, with the goal of collecting and distributing antibodies against thousands of proteins from humans and model organisms. This landmark project will depend on generous government and industry support due to the high cost of development for such reagents.

An alternative to making protein-based protein-binding agents is to isolate them from libraries of random nucleic acid sequences (Petach and Gold, 2002) by SELEX. High-affinity RNA or DNA molecules, called *aptamers*, can be culled out of random sequence libraries of up to 10^{15} members in size. In addition, it is possible to select from bromodeoxyuridine-containing libraries for photoaptamers—nucleic acids that photo-cross-link to their target proteins (Petach and Gold, 2002). It has been proposed that microarrays of such photo-cross-linking agents could be used to capture specific target proteins by covalent interactions, allowing for stringent post-cross-linking washing to remove nonspecifically bound proteins. This technology is being developed by Somalogic, Inc. Figure 11.4 shows a chip based on this technology.

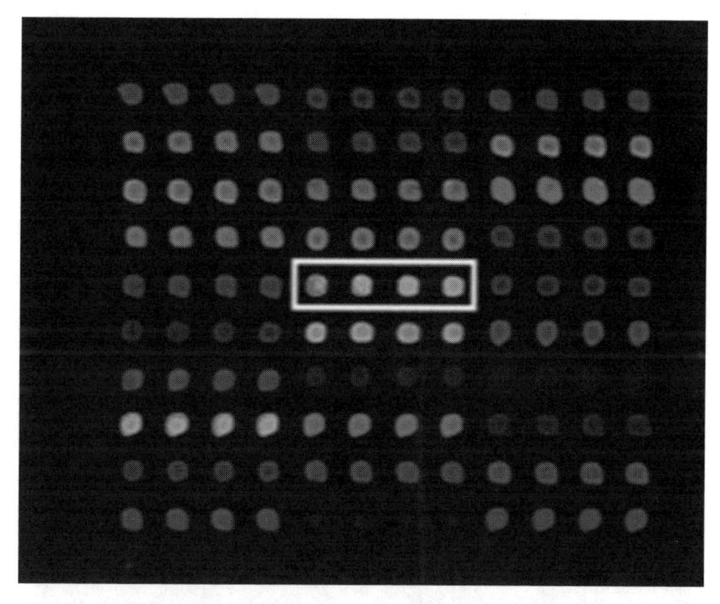

Figure 11.4. (See color insert.) Fluorescence image of a photoaptamer microarray. Human serum, diluted fourfold into assay buffer, was incubated on an array of 24 unique aptamers, spotted in quadruplicate, for 60 min. The array was washed, photo-cross-linked, and washed again to remove unbound proteins. The boxed spots are the von Willebrand factor protein captured on the cognate aptamer. (With kind permission from Somalogic, Inc.)

EXPRESSION PROFILING

Arrays of >100 Specificities

The term *expression profiling* implies that one is surveying a large number of genes in a relatively unbiased manner for correlations in expression levels with important biological phenomena, such as the onset of disease or the execution of a developmental event. With few exceptions, protein arrays have to date only been used to monitor relatively small numbers of proteins in parallel, mainly due to the lack of developed antibody content for these platforms. As mentioned above, Haab et al. (2001) arrayed 115 antibodies in a proof-of-principle experiment. There are also a small number of commercially available arrays with a large number of antibody specificities that focus on a particular subset of human proteins, including a signal transduction array (400 antibodies), an apoptosis array (150 antibodies), and a cell cycle array (60 antibodies) from Hypromatrix and a 512-antibody array from BD Clontech with antibodies against human proteins in a variety of classes (signal transduction, cell cycle regulation, gene transcription, and apoptosis). For these arrays, the experimentalist must use a random protein-labeling protocol such as the one described above. Figure 11.5 shows an expression-profiling experiment using an array of 512 different antibodies.

Figure 11.5. (See color insert.) Pseudocolor fluorescence image of a 512-member antibody array incubated with a Cy5-labeled extract of RANTES-treated macrophages and a Cy3-labeled extract of RANTES treated lymphocytes. (With kind permission form BD-Biosciences Clontech.)

Application of these high-density arrays includes an analysis of protein expression changes resulting from exposure to radiation of cancer cells. Sreekumar et al. (2001) used commercially available components to fabricate arrays with 146 different antibodies against proteins involved in stress response, cell cycle progression, and apoptosis. These off-the-shelf components were (i) glass slides coated either with polylysine (for physical adsorption of antibodies) or aldehyde-bearing compounds (for random coupling to lysine residues on antibodies), (ii) spotting robots designed for making DNA arrays, and (iii) antibodies against the proteins of interest. To compare irradiated to nonirradiated cells, they used the two-color labeling method outlined above. The authors observed five previously known protein up-regulation events, thus validating their array-based approach. They also observed up-regulation of five other important regulatory proteins not previously known to increase in abundance upon radiation exposure, including DFF40/CAD, DFF45/ICAD, and STAT1-alpha. Surprisingly, radiation decreased the abundance of the known cancer marker carcenoembryonic antigen. This report demonstrates the feasibility of using off-the-shelf components to create antibody microarrays that reveal important regulatory phenomena.

One unresolved question in the above analysis is what the status is for antibodies that did not provide a correlation with the radiation treatment. In some cases, the antibodies may not have been active in binding to the antigens, while in other cases the abundance of the antigens may have been below the detection limit. It is therefore difficult to know how many antigens are actually being assayed on a given array.

Miller et al. (2003) thoroughly and elegantly addressed these questions with respect to an array of 184 unique antibodies spotted onto the polyacrylamide-based hydrogel surface mentioned above. They sought to identify serum biomarkers associated with prostate cancer. Initially, to determine which antibodies were providing accurate data on the abundances of their cognate antigens, they looked for statistical correlations for each antibody between experiments in which the 2 fluorescent labels were swapped with respect to the 2 labeled samples. From the 184 specificities, only 8 of the antibodies showed a robust correlation, and these also showed good correlations with the relative abundances of the antigens as measured by macroscopic ELISA. All except one of these recognize antigens that are known to have high serum concentrations. For the vast majority of antibodies, therefore, the antigens were presumably below the detection limit. Despite the limitations, they identified 5 proteins whose serum levels were correlated with the presence of prostate cancer: one of the molecules, von Willebrand factor, had a higher abundance in prostate cancer patients, while the others had a lower abundance.

The low fraction (4 percent in the case above) of antibodies that provide reliable data on antigen abundance is a major drawback with using direct sample labeling. It may be more advantageous to use amplification systems to increase the signal strength for the double-labeling systems. For example, the two samples under comparison could be labeled with two different haptens, after which they would be mixed together and then applied to the antibody arrays, as described above for the two-color fluorescent-labeling experiments. In the latter case, however, the arrays could be incubated with antibodies to the two different haptens, where the two antibodies are labeled with spectrally distinct fluorophores. To maximize the signal strength, these detection antibodies could carry multiple fluorophores.

Signal amplification was used in combination with direct sample labeling by Knezevic et al. (2001), but without dual labeling. Each experimental tissue sample was randomly biotinylated, applied to an array of 368 antibodies against signaling proteins, and then developed using a biotin-binding reagent conjugated to an enzyme that creates a fluorescent precipitate at the position of the array to which it is localized. Signal amplification derived from the multiple turnover of the enzyme. Using this approach, they characterized protein expression levels from human tissue derived from squamous cell carcinoma of the oral cavity. Because such tissues are based on multiple cell types, they further refined their analysis by using laser capture microdissection to obtain protein from specific microscopic cellular populations. They performed expression analysis on either the epithelial cells (cancerous versus normal) or the stromal cells in contact with them. They thus identified 11 different proteins that changed expression levels. In some cases, the change was restricted to the cancer cells, in other cases to the stroma, and in some cases changes were observed in both cell populations. On going from a normal to a cancerous state, one of the proteins, Rsk, decreased in epithelial cells and simultaneously increased in stroma. Such a biomarker could have been overlooked if the tissue microdissection was not performed.

Focused Arrays

With the exception of cases mentioned above, most published reports of protein expression profiling arrays focus on a smaller number (5 to 75) of specificities. In such cases, it is possible to build highly sensitive, sandwich-based immunoassays, focused on a particular class of proteins with a high chance of being relevant to the biological process under investigation. As mentioned above, Schweitzer et al. (2000) constructed such an array to measure expression levels of 75 cytokines, and other groups have reported similar products, but with a smaller number of specificities (Huang, 2001; Huang et al., 2001; Lin et al., 2002; Mendoza et al., 1999; Moody et al., 2001; Pawlak et al., 2002; Tam et al., 2002; Wiese et al., 2001). Zyomyx now offers a commercial array for the quantitative measurement of 30 human cytokines based on the sandwich approach, using a chip that can analyze 6 different samples in an automated workstation capable of processing 12 chips. Pierce and Proteoplex offer arrays within the wells of microtiter plates for the high-throughput (HTP) analysis of samples for the abundances of smaller numbers of cytokines.

Of particular importance for the HTP measurement of up to about 25 biomarkers will be the bead-based systems such as those developed by Luminex or BD Biosciences. Quantitative antibody arrays for the measurement of 6 (Chen et al., 1999) or 15 (Carson and Vignali, 1999) human cytokines have been reported. Numerous bioreagent companies (such as those listed on the Luminex web site) are now commercializing such kits for the quantitative measurement of multiple biomarkers, including cytokines, in parallel. The samples are loaded into the wells of 96-well plates and read consecutively in a flow cytometry device. Because these kits are based on sandwich assays, they are very sensitive, easy to use, and quantitative. Several new kits are being released every year.

Using such focused cytokine detection arrays, it has been possible to uncover the complex changes in cytokine abundance as a result of exposure to proinflammatory factors such as lipopolysaccharide or tumor necrosis factor-α (Schweitzer et al., 2002). Such arrays have also delineated the differences between TH1 and TH2 helper cells in terms of cytokine production (Carson and Vignali, 1999), as well as uncovered an unexpected relationship between vitamin E levels and the cytokine monocyte chemoattractant protein-1 (Lin et al., 2002).

Integrating Expression Profiling Platforms into the Biomarker Discovery Pipeline

The degree of multiplexing may change over the course of a program in biomarker discovery. At the beginning, one would like to profile as many different proteins as possible to find correlations with a biological process or disease state. Arrays for measuring hundreds to thousands of different proteins may be most appropriate at this point. The data from such experiments, however, will likely be qualitative in nature, and because of the significant amount of labor and cost associated with processing each array, the sample size will be limited. To validate these leads, a higher throughput and a more quantitative platform may be more appropriate, such as lower density arrays with <20 specificities, based on sandwich assays. The bead-based systems, capable of handling samples from 96-well plates with high throughput, may be more suited for this phase. In some cases, further validation may utilize standard ELISA methods, which are a high throughput, highly sensitive, and inexpensive. In this way, as a project moves from a discovery to a validation and finally a clinical diagnostic assay, the number of specificities and the cost per sample will decrease, with a simultaneous increase in precision and throughput. For diagnostic applications, in most cases the eventual clinical assay will probably be based on single analyte measurements, by ELISA, for example. In the future, however, there may be cases where patterns of protein abundances, rather than the levels of individual markers, may be used in diagnostics. An example of this was provided by Petricoin et al. (2002), who used mass spectrometry to identify a pattern of mass/charge peak intensities from serum that correlate strongly with the presence of ovarian cancer. The predictive value of this pattern was much higher than that of the best single analyte biomarker, "cancer antigen 125."

Protein Activity Arrays and Antigen Arrays

By spotting antigens onto arrays and exposing them to human serum, it is possible to measure a person's immune response, thus providing information as to autoimmune disorders (Joos et al., 2000), allergy (Kim et al., 2002), or pathogen exposure. Robinson et al. (2002) arrayed 196 known autoantigens and incubated them with serum from healthy individuals and from patients suffering from six different autoimmune conditions. The patient antibody binding could easily be monitored with detection agents that recognize the human antibodies. The authors observed different antigen reactivity patterns in the various conditions, suggesting that such a multiplexed assay could be helpful in diagnosing these sometimes confused pathologies.

Proteome-scale screening of the activity of proteins is also being accomplished using microarrays. Zhu et al. (2001) made expression constructs for nearly all of the 5800 genes in yeast, expressed them, and arrayed them onto glass slides. They then interrogated this array for binding to regulatory molecules such as calmodulin and certain lipids. In this way, they identified many of the gene products with these binding characteristics. With any approach that examines such a large number of proteins, of course, it is impractical to determine what fraction of them is active and at sufficiently high densities on the surface to observe activity. Nevertheless, this type of array should be powerful for identifying numerous binding activities in the genomes of model organisms. Protometrix, Inc., is attempting to build similar arrays of human proteins for commercial release. Arrays focused on kinase specificity have also been reported by the same group (Zhu et al., 2000), allowing for the parallel analysis of the substrate specificity of all the known yeast kinases.

CONCLUSIONS AND FUTURE PROSPECTS

With the large number of protein microarray platforms available today, the burden has now shifted to the biological content to breath life into these devices. Platforms differ mainly in the degree of multiplexing and the sample throughput. In a biomarker discovery program, the researcher may switch between platforms with the evolving stage of a project. Fortunately, the biological reagents incorporated into these platforms will generally be transferable between them. For protein microarrays to be useful in proteomics research, this content will need to be developed and made available for incorporation into several different platforms. This is the main bottleneck in the protein microarray field today. Because the antibody production industry is highly fragmented, effort must be made to consolidate much of the existing and future content for incorporation into these platforms.

The drug discovery pipeline will also grow increasingly dependent on protein microarray technology for target identification and validation, and for investigating the specificity of lead compounds toward members of gene families of kinases, phosphatases, receptors, and so forth. Focused microarrays will play an increasingly important role of defining the mechanism of action of drugs and in managing undesirable side effects. Databases of drug effects on the abundances of signaling molecules, for instance, may be useful in designing drug combinations that maximize effectiveness while limiting adverse reactions. Arrays will also find use in the stratification of patient populations with respect to drug responsiveness and adverse reactions.

Within a few years, protein microarrays may be so ubiquitous in the basic research, drug development, and diagnostic industries that it would not occur to anyone to write a review on the technology.

REFERENCES

Anderson, L., and Seilhamer, J. (1997). A comparison of selected mRNA and protein abundances in human liver. *Electrophoresis* **18**:533–537.

Avseenko, N.V., Morozova, T.Y., Ataullakhanov, F.I., and Morozov, V.N. (2002). Immunoassay with multicomponent protein microarrays fabricated by electrospray deposition. *Anal. Chem.* **74**:927–933.

Bernard, A., Renault, J.P., Michel, B., Bosshard, H.R., and Delamarche, E. (2000). Microcontact printing of proteins. *Adv. Mat.* **12**:1067–1070.

Borrebaeck, C.A., Ekstrom, S., Hager, A.C., Nilsson, J., Laurell, T., and Marko-Varga, G. (2001). Protein chips based on recombinant antibody fragments: A highly sensitive approach as detected by mass spectrometry. *Biotechniques* **30**:1126–1132.

Brody, E.N., and Gold, L. (2000). Aptamers as therapeutic and diagnostic agents. *J. Biotechnol.* **74**:5–13.

Butler, J.E., Ni, L., Brown, W.R., Joshi, K.S., Chang, J., Rosenberg, B., and Voss, E.W. (1993). The immunochemistry of sandwich ELISAs—VI. Greater than 90% of monoclonal and 75% of polyclonal anti-fluorescyl capture antibodies (CAbs) are denatured by passive adsorption. *Mol. Immunol.* **30**:1165–1175.

Butler, J.E., Ni, L., Nessler, R., Joshi, K.S., Suter, M., Rosenberg, B., Chang, J., Brown, W.R., and Cantarero, L.A. (1992). The physical and functional behavior of capture antibodies adsorbed on polystyrene. *J. Immunol. Methods* **150**:77–90.

Carson, R.T., and Vignali, D.A. (1999). Simultaneous quantitation of 15 cytokines using a multiplexed flow cytometric assay. *J. Immunol. Methods* **227**:41–52.

Chen, R., Lowe, L., Wilson, J.D., Crowther, E., Tzeggai, K., Bishop, J.E., and Varro, R. (1999). Simultaneous quantification of six human cytokines in a single sample using microparticle-based flow cytometric technology. *Clin. Chem.* **45**:1693–1694.

Dhanasekaran, S.M., Barrette, T.R., Ghosh, D., Shah, R., Varambally, S., Kurachi, K., Pienta, K.J., Rubin, M.A., and Chinnaiyan, A.M. (2001). Delineation of prognostic biomarkers in prostate cancer. *Nature* **412**:822–826.

Duveneck, G.L., Bopp, M.A., Ehrat, M., Balet, L.P., Haiml, M., Keller, U., Marowsky, G., and Soria, S. (2003). Two-photon fluorescence excitation of macroscopic areas on planar waveguides. *Biosens. Bioelectron.* **18**:503–510.

Ekins, R.P. (1960). *Clin. Chim. Acta* **5**:453–462.

Ekins, R., Chu, F., and Biggart, E. (1990). Multispot, multianalyte, immunoassay. *Ann. Biol. Clin. (Paris)* **48**:655–666.

Fulton, R.J., McDade, R.L., Smith, P.L., Kienker, L.J., and Kettman, J.R. (1997). Advanced multiplexed analysis with the FlowMetrix system. *Clin. Chem.* **43**:1749–1756.

Golub, T.R., Slonim, D.K., Tamay, P., Huard, C., Gaasenbeek, M., Mesirov, J.P., Coller, H., Loh, M.L., Downing, J.R., Caligiuri, M.A., Bloomfield, C.D., and Lander, E.S. (1999). Molecular classification of cancer: Class discovery and class prediction by gene expression monitoring. *Science* **286**:531–537.

Gygi, S.P., Rochon, Y., Franza, B.R., and Aebersold, R. (1999). Correlation between protein and mRNA abundance in yeast. *Mol. Cell Biol.* **19**:1720–1730.

Haab, B.B., Dunham, M.J., and Brown, P.O. (2001). Protein microarrays for highly parallel detection and quantitation of specific proteins and antibodies in complex solutions. *Genome Biol.* **2**:4.1–4.13.

Hanes, J., Schaffitzel, C., Knappik, A., and Pluckthun, A. (2000). Picomolar affinity antibodies from a fully synthetic naive library selected and evolved by ribosome display. *Nature Biotechnol.* **18**:1287–1292.

Huang, R. (2001). Detection of multiple proteins in an antibody-based protein microarray system. *J. Immunol. Methods* **255**:1–13.

Huang, R.P., Huang, R., Fan, Y., and Lin, Y. (2001). Simultaneous detection of multiple cytokines from conditioned media and patient's sera by an antibody-based protein array system. *Anal. Biochem.* **294**:55–62.

Jenkins, R.E., and Pennington, S.R. (2001). Arrays for protein expression profiling: Towards a viable alternative to two-dimensional gel electrophoresis? *Electrophoresis* **1**:13–29.

Joos, T.O., Schrenk, M., Hopfl, P., Kroger, K., Chowdhury, U., Stoll, D., Schorner, D., Durr, M., Herick, K., Rupp, S., Sohn, K., and Hammerle, H. (2000). A microarray enzyme-linked immunosorbent assay for autoimmune diagnostics. *Electrophoresis* **21**:2641–2650.

Kettman, J.R., Davies, T., Chandler, D., Oliver, K.G., and Fulton, R.J. (1998). Classification and properties of 64 multiplexed microsphere sets. *Cytometry* **33**:234–243.

Kim, T.E., Park, S.W., Cho, N.Y., Choi, S.Y., Yong, T.S., Nahm, B.H., Lee, S., and Noh, G. (2002). Quantitative measurement of serum allergen-specific IgE on protein chip. *Exp. Mol. Med.* **34**:152–158.

Knezevic, V., Leethanakul, C., Bichsel, V.E., Worth, J.M., Prabhu, V.V., Gutkind, J.S., Liotta, L.A., Munson, P.J., Petricoin, III E.F., and Krizman, D.B. (2001). Proteomic profiling of the cancer microenvironment by antibody arrays. *Proteomics* **1**:1271–1278.

Lin, Y., Huang, R., Santanam, N., Liu, Y.G., Parthasarathy, S., and Huang, R.P. (2002). Profiling of human cytokines in healthy individuals with vitamin E supplementation by antibody array. *Cancer Lett.* **187**:17–24.

MacBeath, G., and Schreiber, S.L. (2000). Printing proteins as microarrays for high-throughput function determination. *Science* **289**:1760–1763.

Martin, B.D., Gaber, B.P., Patterson, C.H., and Turner, D.C. (1998). Direct protein microarray fabrication using a hydrogel "stamper." *Langmuir* **14**:3971–3975.

Mendoza, L.G., McQuary, P., Mongan, A., Gangadharan, R., Brignac, S., and Eggers, M. (1999). High-throughput microarray-based enzyme-linked immunosorbent assay (ELISA). *Biotechniques* **27**:778–788.

Merchant, M., and Weinberger, S.R. (2000). Recent advancements in surface-enhanced laser desorption/ionization-time of flight-mass spectrometry. *Electrophoresis* **21**:1164–1177.

Miller, J.C., Zhou, H., Kwekel, J., Cavallo, R., Burke, J., Butler, E.B., Teh, B.S., and Haab, B.B. (2003). Antibody microarray profiling of human prostate cancer sera: Antibody screening and identification of potential biomarkers. *Proteomics* **3**:56–63.

Mirzabekov, A., and Kolchinsky, A. (2002). Emerging array-based technologies in proteomics. *Curr. Opin. Chem.* **6**:70–75.

Moerman, R., Frank, J., Marijnissen, J.C., Schalkhammer, T.G., and van Dedem, G.W. (2001). Miniaturized electrospraying as a technique for the production of microarrays of reproducible micrometer-sized protein spots. *Anal. Chem.* **73**:2183–2189.

Moody, M.D., van Ardel, S.W., Orencole, S.F., and Burns, C. (2001). Array-based ELISAs for high-throughput analysis of human cytokines. *Biotechniques* **31**:186–194.

Morozov, V.N., and Morozova, T.Y. (1999). Electrospray deposition as a method to fabricate functionally active protein films. *Anal. Chem.* **71**:1415–1420.

Nelson, R.W., Nedelkov, D., and Tubbs, K.A. (2000). Biosensor chip mass spectrometry: A chip-based proteomics approach. *Electrophoresis* **21**:1155–1163.

Nicewarner-Pena, S.R., Freeman, R.G., Reiss, B.D., He, L., Pena, D.J., Walton, I.D., Cromer, R., Keating, C.D., and Natan, M.J. (2001). Submicrometer metallic barcodes. *Science* **294**:137–141.

Pawlak, M., Schick, E., Bopp, M.A., Schneider, M.J., Oroszlan, P., and Ehrat, M. (2002). Zeptosens' protein microarrays: A novel high performance microarray platform for low abundance protein analysis. *Proteomics* **2**:383–393.

Peluso, P., Wilson, D.S., Do, D., Tran, H., Venkatasubbaiah, M., Quincy, D., Heidecker, B., Poindexter, K., Tolani, N., Phelan, M., Witte, K., Jung, L.S., Wagner, P., and Nock, S. (2003). Optimizing antibody immobilization strategies for the construction of protein microarrays. *Anal. Biochem.* **312**:113–124.

Pennington, S.R., Wilkins, R.W., Hochstrasser, D.F., and Dunn, M.J. (1997). Proteome analysis: From protein characterization to biological function. *Trends Cell Biol.* 7:168–173.

Perou, C.M., Sorlie, T., Eisen, M.B., van de Rijn, M., Jeffrey, S.S., Rees, C.A., Pollack, J.R., Ross, D.T., Johnsen, H., Akslen, L.A., Fluge, O., Pergamenschikov, A., Williams, C., Zhu, S.X., Lonning, P.E., Borresen-Dale, A.L., Brown, P.O., and Botstein, D. (2000). Molecular portraits of human breast tumours. *Nature* **406**:747–752.

Petach, H., and Gold, L. (2002). Dimensionality is the issue: Use of photoaptamers in protein microarrays. *Curr. Opin. Biotechnol.* **13**:309–314.

Petricoin, E.F., Ardekani, A.M., Hitt, B.A., Levine, P.J., Fusaro, V.A., Steinberg, S.M., Mills, G.B., Simone, C., Fishman, D.A., Kohn, E.C., and Liotta, L.A. (2002). Use of proteomic patterns in serum to identify ovarian cancer. *Lancet* **359**:572–577.

Robertson, M.P., and Ellington, A.D. (2001). In vitro selection of nucleoprotein enzymes. *Nature Biotechnol.* **19**:650–655.

Robinson, W.H., DiGennaro, C., Hueber, W., Haab, B.B., Kamachi, M., Dean, E.J., Fournel, S., Fong, D., Genovese, M.C., de Vegvar, H.E., Skriner, K., Hirschberg, D.L., Morris, R.I., Muller, S., Pruijn, G.J., van Venrooij, W.J., Smolen, J.S., Brown, P.O., Steinman, L., and Utz, P.J. (2002). Autoantigen microarrays for multiplex characterization of autoantibody responses. *Nat. Med.* **8**:295–301.

Roda, A., Guardigli, M., Russo, C., Pasini, P., and Baraldini, M. (2000). Protein microdeposition using a conventional ink-jet printer. *Biotechniques* **28**:492–496.

Rowe, C.A., Scruggs, S.B., Feldstein, M.J., Golden, J.P., and Ligler, F.S. (1999a). An array immunosensor for simultaneous detection of clinical analytes. *Anal. Chem.* **71**:433–439.

Rowe, C.A., Scruggs, S.B., Feldstein, M.J., Golden, J.P., and Ligler, F.S. (1999b). An array immunosensor for simultaneous detection of clinical analytes. *Anal. Chem.* **71**:433–439.

Rowe, C.A., Tender, L.M., Feldstein, M.J., Golden, J.P., Scruggs, S.B., MacCraith, B.D., Cras, J.J., Ligler, F.S. (1999c). Array biosensor for simultaneous identification of bacterial, viral, and protein analytes. *Anal. Chem.* **71**:3846–3852.

Ruiz-Taylor, L.A., Martin, T.L., Zaugg, F.G., Witte, K., Indermuhle, P., Nock, S., and Wagner, P. (2001). Monolayers of derivatized poly(L-lysine)-grafted poly(ethylene glycol) on metal oxides as a class of biomolecular interfaces. *Proc. Natl. Acad. Sci. USA* **98**:852–857.

Schena, M., Shalon, D., Davis, R.W., and Brown, P.O. (1995). Quantitative monitoring of gene expression patterns with a complementary DNA microarray. *Science* **270**:467–470.

Schweitzer, B., Roberts, S., Grimwade, B., Shao, W., Wang, M., Fu, Q., Shu, Q., Laroche, I., Zhou, Z., Tchernev, V.T., Christiansen, J., Velleca, M., and Kingsmore, S.F. (2002). Multiplexed protein profiling on microarrays by rolling-circle amplification. *Nature Biotechnol.* **20**:359–365.

Schweitzer, B., Wiltshire, S., Lambert, J., O'Malley, S., Kukanskis, K., Zhu, Z., Kingsmore, S.F., Lizardi, P.M., and Ward, D.C. (2000). Immunoassays with rolling circle DNA amplification: A versatile platform for ultrasensitive antigen detection. *Proc. Natl. Acad. Sci. USA* **97**:10113–10119.

Sreekumar, A., Nyati, M.K., Varambally, S., Barrette, T.R., Ghosh, D., Lawrence, T.S., and Chinnaiyan, A.M. (2001). Profiling of cancer cells using protein microarrays: Discovery of novel radiation-regulated proteins. *Cancer Res.* **61**:7585–7593.

Tam, S.W., Wiese, R., Lee, S., Gilmore, J., and Kumble, K.D. (2002). Simultaneous analysis of eight human Th1/Th2 cytokines using microarrays. *J. Immunol. Methods* **261**:157–165.

Wiese, R., Belosludtsev, Y., Powdrill, T., Thompson, P., and Hogan, M. (2001). Simultaneous multianalyte ELISA performed on a microarray platform. *Clin. Chem.* **47**:1451–1457.

Wilson, D.S., Keefe, A.D., and Szostak, J.W. (2001). The use of mRNA display to select high-affinity protein-binding peptides. *Proc. Natl. Acad. Sci. USA* **98**:3750–3755.

Xu, L., Aha, P., Gu, K., Kuimelis, R.G., Kurz, M., Lam, T., Lim, A.C., Liu, H., Lohse, P.A., Sun, L., Weng, S., Wagner, R.W., and Lipovsek, D. (2003). Directed evolution of high-affinity antibody mimics using mRNA display. *Chem. Biol.* **10**:91–92.

Yalow, R.S., and Berson, S.A. (1960). *J. Clin. Invest.* **39**:1157–1162.

Zhang, L., Nolan, E., Kreitschitz, S., and Rabussay, D.P. (2002). Enhanced delivery of naked DNA to the skin by non-invasive in vivo electroporation. *Biochim. Biophys. Acta* **1572**:1–9.

Zhou, H., Roy, S., Schulman, H., and Natan, M.J. (2001). Solution and chip arrays in protein profiling. *Trends Biotechnol.* **19**[10(Suppl.)]:S34–S39.

Zhu, H., Bilgin, M., Bangham, R., Hall, D., Casamayor, A., Bertone, P., Lan, N., Jansen, R., Bidlingmaier, S., Houfek, T., Mitchell, T., Miller, P., Dean, R.A., Gerstein, M., and Snyder, M. (2001). Global analysis of protein activities using proteome chips. *Science* **293**:2101–2105.

Zhu, H., Klemic, J.F., Chang, S., Bertone, P., Casamayor, A., Klemic, K.G., Smith, D., Gerstein, M., Reed, M.A., and Snyder, M. (2000). Analysis of yeast protein kinases using protein chips. *Nat. Genet.* **26**:283–289.

WEB RESOURCES

Institution	**URL**
Aclara Biosciences	*www.aclara.com*
ATCC	*www.atcc.org*
BD Biosciences	*www.bdbiosciences.com*
BD Clontech	*www.clontech.com*
Harvard Institute of Proteomics	*www.hip.harvard.edu*
Human Proteome Organization	*www.hupo.org*
Hypromatrix	*www.hypromatrix.com*
IMAGE Consortium	*http://image.llnl.gov/*
IMTEC	*www.imtek.uni-freiburg.de*
Invitrogen	*www.invitrogen.com*
Luminex	*www.luminexcorp.com*
Mammalian Gene Collection	*http://mgc.nci.nih.gov/*
Molecular Staging	*www.molecularstaging.com*
Open Biosystems	*www.openbiosystems.com*
Perkin-Elmer Life Sciences	*www.perkinelmer.com/lifesciences*
Pierce	*www.piercenet.com*
Proteoplex	*www.proteoplex.com*
Protometrix	*www.protometrix.com*
QualityProteomics	*www.qualityproteomics.com*
Somalogic	*www.somalogic.com*
Zyomyx	*www.zyomyx.com*

INDEX

Industrial Proteomics: Applications for Biotechnology and Pharmaceuticals, edited by Daniel Figeys
ISBN 0-471-45714-0